Mechatronic Systems Techniques and Applications

Volume 3

Energy and Power Systems

Gordon and Breach International Series
in Engineering, Technology and Applied Science

Volumes 1–5

Edited by Cornelius T. Leondes

Books on **Mechatronic Systems**
Techniques and Applications

Previously published in this series were volumes 1–6 on **Medical Imaging Systems Techniques and Applications** and volumes 7–15 on **Structural Dynamic Systems Computational Techniques and Optimization**

Forthcoming in the *Gordon and Breach International Series in Engineering, Technology and Applied Science*

Biomechanical Systems Techniques and Applications

Computer-Aided and Integrated Manufacturing Systems (CAIMS) Techniques and Applications

Expert Systems Techniques and Applications

Computer Techniques in Medical and Biotechnology Systems

Data Base and Data Communication Networks Techniques and Applications

Computer-Aided Design, Engineering and Manufacturing (CADEM) Systems Techniques

This book is part of a series. The publisher will accept continuation orders which may be cancelled at any time and which provide for automatic billing and shipping of each title in the series upon publication. Please write for details.

Mechatronic Systems Techniques and Applications

Volume 3
Energy and Power Systems

Edited by

Cornelius T. Leondes

Professor Emeritus
University of California
at Los Angeles

Gordon and Breach Science Publishers

Australia • Canada • France • Germany • India •
Japan • Luxembourg • Malaysia • The Netherlands •
Russia • Singapore • Switzerland

Amsteldijk 166
1st Floor
1079 LH Amsterdam
The Netherlands

British Library Cataloguing in Publication Data

Mechatronic systems techniques and applications :
    energy and power systems ; v. 3 – (Gordon and Breach
    international series in engineering, technology and applied
    science — ISSN 1026-0277)
    1. Power resources – Automatic control 2. Mechatronics
    I. Leondes, Cornelius T.
    629.8'9

ISBN 90-5699-677-0

CONTENTS

SERIES DESCRIPTION AND MOTIVATION

Many aspects of explosively growing technology are difficult or essentially impossible for one author to treat in an adequately comprehensive manner. Spectacular technological growth is made stunningly manifest by any number of examples, but, just to note one here, the Intel 486 IBM-compatible PC was first introduced in late 1989. At that time the price of this PC was in the $10,000 range and it was thought to be much too powerful for widespread use. By early 1992, a little more than two years later, the price had dropped to $1,000 and it was felt that much more power was needed, leading directly to the Pentium IBM-compatible PC. A similar price reduction pattern has followed for the Pentium computer, which was then replaced by the Pentium II. With the introduction in 1999 of the Pentium III processor, the pattern of rapid decrease in price is now evident for Pentium II-based computers. In fact, the decline in prices is accelerating, with Pentium computers available for less than $500, Pentium II PCs for less than $1,000, and new 600 MHz Pentium III-based computers selling for substantially less than Pentium II systems did at their introduction. This "power hungry" pattern will very likely continue into the foreseeable future, with a 1,000 MHz Pentium III processor expected sometime in 2000. The CD-ROM has now evolved to the DVD (Digital Versatile Disk) with data storage capability a full magnitude greater (7 Gb vs. 650 Mb). A DVD-ROM can hold a database of all the phone numbers and addresses in the United States, which would normally require multiple CD-ROMs. And the DVD format has room to grow. In any event, these examples and their clear implications with respect to the many application-oriented issues in diverse fields of engineering, technology and applied science and their continuing advances make it obvious that this series will fill an essential role in numerous ways for individuals and organizations.

Areas of major significance will be defined and world-class co-authors identified as contributors for essential volumes in respective areas. These areas will be determined by criteria including:

1. Will volumes fill important textbook voids in respective areas?

2. In some cases, a "time void" for an important area will clearly suggest the need for a volume. For example, the important area of Expert Systems might have a textbook void of several years that "requires" an important new volume.

3. Are these technology areas that simply cannot sensibly be treated comprehensively by a single author or even several co-authors?

Examples of areas requiring important volumes will be carefully defined and structured and might include, as the case arises, volumes in:

1. Medical imaging systems

2. Structural dynamic systems

3. Mechatronic systems

4. Biomechanical systems

5. Computer-aided and integrated manufacturing systems (CAIMS)

6. Expert systems

7. Computer techniques in medical and biotechnology systems

8. Data base and data communication networks

9. Computer-aided design, engineering and manufacturing (CADEM) systems.

One of the most important aspects of this series will be that, despite rapid advances in technology, respective volumes will be defined and structured to constitute works of indefinite or "lasting" reference interest.

SERIES PREFACE

The first industrial revolution, with its roots in James Watt's steam engine and its various applications to modes of transportation, manufacturing and other areas, introduced to mankind novel ways of working and living, thus becoming one of the chief determinants of our present way of life.

The second industrial revolution, with its roots in modern computer technology and integrated electronics technology — particularly VLSI (Very Large Scale Integrated) electronics technology, has also resulted in advances of enormous significance in all areas of modern activity, with great economic impact as well.

Some of the areas of modern activity created by this revolution are: medical imaging, structural dynamic systems, mechatronics, biomechanics, computer-aided and integrated manufacturing systems, applications of expert and knowledge-based systems, and so on. Documentation of these areas well exceeds the capabilities of any one or even several individuals, and it is quite evident that single-volume treatments — whose intent would be to provide practitioners with useful reference sources — while useful, would generally be rather limited.

It is the intent of this series to provide comprehensive multi-volume treatments of areas of significant importance, both the above-mentioned and others. In all cases, contributors to these volumes will be individuals who have made notable contributions in their respective fields. Every attempt will be made to make each book self-contained, thus enhancing its usefulness to practitioners in a specific area or related areas. Each multi-volume treatment will constitute a well-integrated but distinctly titled set of volumes. In summary, it is the goal of the respective sets of volumes in this series to provide an essential service to the many individuals on the international scene who are deeply involved in contributing to significant advances in the second industrial revolution.

PREFACE

Mechatronic Systems Techniques and Applications

Energy and Power Systems

The field of mechatronics (mechanics/electronics) has evolved as a highly powerful and most cost effective means for product realization. This is due to developments in powerful computers including microprocessors, Application Specific Integrated Circuits (ASICs), computational techniques, and advances in the product design process. End products cover a wide spectrum of fields such as manufacturing, transportation, energy and power systems, and a great variety of electromechanical systems. A number of descriptions of the broad field of mechatronics have been put forward on the international scene. One such description is: "the synergistic and optimal design, development and support of a wide variety of diverse engineering or industrial products and processes utilizing a wide variety of technologies and computers and computer processes." Recently, the technical committee on mechatronics formed by the International Federation for the Theory of Machines and Mechanisms, in Prague, Czech Republic, adopted the following definition or description of the term: "Mechatronics is the synergistic combination of precision mechanical engineering, electronic control and systems thinking in the design products and manufacturing processes." Whichever description is adopted, the general process and great significance of mechatronics are apparent.

This is the third set of volumes in the *Gordon and Breach International Series in Engineering, Technology and Applied Science*, and it consists of 5 distinctly titled and well-integrated volumes on *Mechatronic Systems Techniques and Applications* that can nevertheless be utilized as individual books. In any event, the great breadth of this field certainly suggests the requirement for 5 volumes for an adequately comprehensive treatment.

The set of volumes on mechatronics treats:

1. Industrial Manufacturing

2. Transportation and Vehicular Systems

3. Energy and Power Systems

4. Electromechanical Systems

5. Diagnostic, Reliability and Control System Techniques.

The first chapter in this volume, by Mielczarski and Zajaczkowski, notes that power systems belong to the largest systems constructed by our civilization. Electrical energy supply is a complex task consisting of three major steps: generation, transmission and distribution. Limiting energy storage capability of power systems means that supply must meet constantly changing demands every second. Power generation is an example of a complex electromechanical system in which energy from fuels such as coal, oil or gas is converted into electrical energy in a system: boiler-steam turbine-synchronous generator. Since large thermal power generating units cannot rapidly react to increasing demand, the problem of balance between mechanical and electrical powers on a shaft of a unit steam turbine-generator is one of the most important problems for power systems engineers. This chapter is a comprehensive treatment of the issues and techniques in this major area of significant application of modern mechatronic system techniques. Illustrative examples are included.

In chapter 2, Swidenbank et al. describe the great advances now being made in the application of computers to control and data monitoring of power plants. Power generation has tended to lag behind other process industries in this respect, mainly due to the emphasis on reliability, with a lower priority being given to efficiency and commercial performance. Environmental and commercial constraints, however, are forcing generator utilities to improve existing plant control performance. Electricity generation is a complex process with a unique disadvantage in relation to most other energy sources: It cannot be stored readily, and it falls to the control engineer to ensure the balance between output and demand. Many regulations and constraints are imposed in the generation of electrical power, with hundreds of control loops within the plant to ensure safety and stability. Although the majority of these loops are of the logic type, others require continuous control for desired stability and safety. In a fossil fuel power plant, the system is usually broken into sub-components for the purpose of controller design. Often, steam raising, prime mover and power generation are treated separately with controllers designed around each subsystem, making a power generation unit a highly complex interacting system. The problem of controller design is also compounded by the fact that the system is highly nonlinear and dependent upon generated load and plant component conditions. These factors have led to the investigation of advanced control strategies for power plant equipment. This area of major economic significance and other important aspects utilizes a number of significant mechatronic system techniques, treated in depth within this chapter.

The next contribution, by McDonald et al., is a comprehensive treatment of the application of Artificial Neural Networks (ANNs) to the short-term load forecasting problem in power systems. Chapter 3 offers examples in which ANNs provide an effective forecasting system, and presents ANN methodology techniques for the analysis of electrical plant data for condition monitoring purposes. As such, the issues and techniques examined play an essential role in the mechatronic aspects of modern large-scale electric power systems.

Certainly one of the most universally pervasive energy systems is that of total systems for the purpose of heating, ventilation and air conditioning (HVAC). According to Zhou, Rao and Chuang (chapter 4), processes provide a comfortable environment, but consume a great deal of energy. Many efforts have been put into increasing energy conservation since the energy crisis of 1973. On the other hand, conservation efforts have led to tight building envelopes and low ventilation rates, which cause poor indoor air quality (IAQ), the so-called "Sick Building Syndrome." Conflict exists between energy saving and IAQ improvements. In this chapter, an intelligent system approach is proposed to support HVAC operations aimed at improving energy conservation and IAQ control.

In chapter 5, Yang and Linkens present an integrated control, instrumentation and optimization scheme for reheat furnace operations that utilizes mechatronic system techniques. This chapter describes the concept of integration advocated in mechatronics, as it has been extensively exploited in development of advanced reheat furnace operation control from instrumentation, data collection to control and optimization. As a result of this integration, different control functions involved in the reheat furnace process can be carried out on a wider information basis. This leads to more efficient coordination, hence better overall operation performance. General requirements for the integration scheme are given, along with detailed techniques of algorithms, strategies and implementation.

Edwards and Spurgeon (chapter 6) present techniques for nonlinear control and observer theory for the problem of temperature control in a gas-fired furnace. The furnace under consideration can be thought of as a gas-filled enclosure containing a heat sink (such as a load), bounded by insulating walls. Heat input is achieved via a burner located in one of the end walls and combustion products are evacuated through a flue positioned in the roof. Although this represents the simplest design possible — a single burner arrangement — such a plant could legitimately represent an industrial furnace for the firing of ceramics. This design has been chosen as a starting point for the study of controllers for high temperature heating plants, principally because a prototype single burner furnace is available for controller evaluation purposes. Because of the pervasiveness of such systems in industry, this contribution is an essential element of this volume that also addresses energy systems.

In the final chapter, Sepehri and Lawrence tell us the human interface for controlling heavy-duty hydraulic equipment, for example excavators, has not changed significantly over many years. Traditionally two joysticks, each with two degrees of freedom, have been used by operators to control joint rates of the machine arms independently. Motivated by potential benefits such as reduction in learning/adaptation time, productivity gains, reduction in fatigue and enhanced safety, techniques are presented to develop, implement and examine new control interfaces in teleoperated control of such heavy-duty hydraulic machines.

The authors demonstrate benefits and performance evaluation of recently developed mechatronic systems and techniques through direct application to real-world machines or through computer simulations. The necessary implementation to retrofit these machines into teleoperated systems is outlined. The goal is to reduce the required level of operator skill, while maintaining machine performance and safety, by controlling the machine in task coordinates. In a task coordinate system, machines implement direction and speed is specified by the deflection of a 3D joystick handle. The appropriate control signals to the actuators are determined by a computer control system.

This volume on mechatronic systems techniques and applications in energy and power systems reveals the effectiveness of techniques available and, with further development, the essential role they will play in the future. The authors are all to be highly commended for their splendid contributions; these papers will provide a significant and unique reference source for students, research workers, practitioners, computer engineers and others on the international scene for years to come.

1 MECHATRONIC SYSTEMS TECHNIQUES IN THE IMPLEMENTATION OF NON-LINEAR CONTROLLERS OF SYNCHRONOUS GENERATORS

WLADYSLAW MIELCZARSKI[1] and ANTONI M. ZAJACZKOWSKI[2]

[1]*Monash University, Department of Electrical and Computer Systems Engineering, Centre for Electrical Power Engineering Clayton, Victoria 3168, Australia*
[2]*Optimal Control Laboratory, Department of Fundamental Research in Electrical Engineering, Institute of Electrical Engineering, Warsaw, Poland*

1.1. INTRODUCTION

Power systems belong to the largest systems constructed by our civilization. Electrical energy supply is a complex task consisting of three major steps: generation, transmission and distribution. Limiting energy storage capability of power systems means that the supply has to meet constantly changing demand in every second. Power generation is an example of complex electromechanical systems in which energy from fuels as coal, oil or gas is converted into electrical energy in a system: boiler-steam turbine-synchronous generator. Large thermal power generating units cannot rapidly react on increasing demand so the problem of balance between mechanical and electrical powers on a shaft of a unit steam turbine – generator is one of the most important problems for power systems engineers.

Most power generating units are controlled by voltage controllers with Power System Stabilizers (PSS) as supporting controllers. Generator speed is controlled by a speed controller which adjusts a valve of a steam turbine.

1

Existing controllers are designed with the assumption that a synchronous generator can be modelled by a single linear equation which is usually represented by a transmittance of a simple system with inertia. To improve decreasing performance of the controlled systems PSSs were introduced in the sixties. Their task is to feed a voltage controller with additional stabilizing signals. Further development of power systems and interconnections resulted in a number of reported problems with stability of power systems. It has become clear that the control of power generating units requires a new type of controllers which can be utilize existing knowledge on nonlinear models. In the eighties feedback linearization as a design method for nonlinear plant was introduced allowing implementation of nonlinear controllers for synchronous generators.

This chapter is organized in the following way. The introduction to power system stability is presented in Sub-Chapter 2. Models of steam turbines, synchronous generators and excitation systems are introduced subsequently in Sub-Chapters 3, 4 and 5. Simulation results of an electromechanical system steam turbine – synchronous generator is discussed in Sub-Chapter 6. A problem of synchronous generator state observation is presented in Sub-Chapters 8 and 9 as well as simulation results which confirm superiority of the nonlinear controllers

2.1. POWER SYSTEM STABILITY

Most power generators are driven by steam turbines transferring energy from the steam into electrical energy flowing from generators to the power system and subsequently to energy users. One of the major problems in operating a set of machinery – generator – turbine is to stabilise speed of a generator rotor when it is subjected to disturbances. There are two main categories of disturbances. The first one results from deviation in frequency and voltage magnitude of a power system. The second category includes oscillations resulting from operation of a steam turbine. Maintaining synchronism between various parts of a power system becomes more difficult, in particular, for interconnected power systems.

A typical configuration of a single machine system is shown in Figure 1. It includes a generator (G), a transformer (Tr), used to increase an output voltage, a tie line (T_1) and a power system represented by the voltage V_s and angular frequency ω_s.

A generator shaft is subjected to two main torques: a mechanical torque from a steam turbine and an anti-reaction force resulting from a magnetic flux between the stator and the rotor of a generator. This anti-reaction force represents the flow of electrical energy from generator windings to power

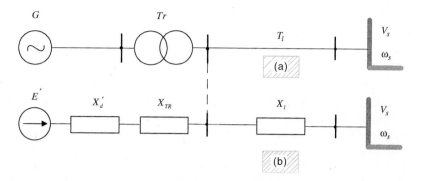

Figure 1. Single machine system, (a) configuration, (b) equivalent model

systems. Assuming that a rotor angular speed ω is close to a synchronous speed the balance of power (torques) on a generator shaft can be described as

$$\frac{d\omega}{dt} = d(\omega - \omega_s) - M_e + M_m, \qquad (2.1)$$

where

ω_s - synchronous angular speed,

d - damping coefficient,

M_e - electromagnetic torque corresponding to the electrical power P_e,

M_m - mechanical torque corresponding to the mechanical power P_m,

$D_t = d/dt$.

Electrical power transmitted between a generator and a power system throughout a tie line depends on generator and system voltages, reactance of the connection and angle between an internal voltage, (E') of a generator and a power system voltage V_s

$$P_{el} = \frac{E'V_s}{L}\sin\delta, \qquad (2.2)$$

where

E' - internal voltage of generator,

V_s - power system voltage,

$X = X'_d + X_{TR} + X_l$ - reactance of the connection,

δ - load angle (rotor angle), angle between the phasors $\overline{V}_s$ and $\overline{E}'_s$.

When a frequency of a power system is constant a load angle is changing with a speed of a rotor

$$D_t\delta = \omega_B(\omega - \omega_s), \qquad (2.3)$$

where

$$\omega_B = 2\pi f_B, \quad f_B - \text{base frequency},$$
$$\omega_s = 2\pi f_s, \quad f_s - \text{frequency of power system}.$$

Equation 2.2 and 2.3 describe dynamic electromechanical processes during production and transfer of electrical energy between a set of turbine-generator and a power system. Equation 2.2 expresses electrical power transfer throughout a tie line. Assuming an uncontrolled set of turbine-generator and typical parameters in per unit system as: $E' = 1.05$, $V_{s1} = 1.0$, $L = 0.4$ electrical energy flowing between a generator and a power system is represented by the curve $P_{el}(V_{s1})$ - Figure 2.

Disturbances in the power system, in particular short-circuits result in deviations of a power system voltage affecting power flow. If the power system voltage drops from the value V_{s1} to the value V_{s2} the relationship describing power flow is now represented by the new curve $P_{el}(V_{s2})$ - Figure 2. Before the disturbance, the point representing balance between mechanical and electrical powers was defined by the crossing point of two characteristics $P_{el}(V_{s1})$ and P_m. This point is denoted as A in Figure 2. When the power system voltage drops to the value V_{s2} electrical power that can be transmitted throughout a tie line decreases to the value P_B determined by the point B. The reduction in the electrical power transmitted with the constant mechanical power, as the system is not controlled, results in an access of mechanical power equal to the difference between $P_m - P_B$. This access, due to equation 2.1, causes the acceleration of the generator rotor and the angular speed ω of a rotor increases the load angle δ, due to the relationship 2.3.

When the load angle δ increases, the point determining balance between mechanical and electrical powers is moving along curve $P_e(V_{s2})$ to point C and afterwards to point D, where areas ABC and CDE become equal - Figure 2. These areas represent energy deviation during transient processes. In point D, the electrical power transmitted to the power system becomes larger than the mechanical power so, due to equation 2.1, the angular speed of the rotor starts decreasing and the working point is moving from point D to point C and towards point B. This process of power oscillations is gradually damped by magnetic forces from damping windings and mechanical friction so in the end a new balance appears in point C. Electrical power and load angle oscillations are shown in Figure 3 and Figure 4.

An uncontrolled set of turbine-generator is stable when during transient processes an area below the mechanical power line, representing access of mechanical energy on a generator shaft, is small enough that it can be compensated by an access of electrical energy when the load angle will increase above the new balanced value δ_C. The electrical power transmitted

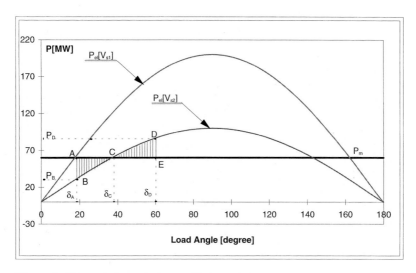

Figure 2. Energy balance during oscillations

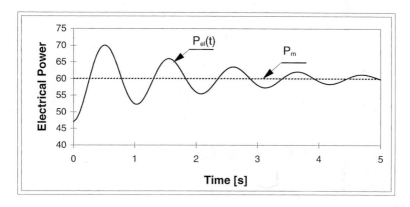

Figure 3. Deviation of electrical power

to the power system is a nonlinear function of the load angle and when this angle increases above 90 degree, the electrical power decreases. The further increase of the load angle means the mechanical power becomes larger than the electrical power and it results in further rotor acceleration and the lost of stability.

There is a need for controlling power generators and turbines to improve stability of generation and transmission. Stability of a single machine system

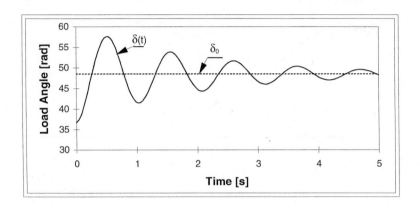

Figure 4. Deviation of load angle

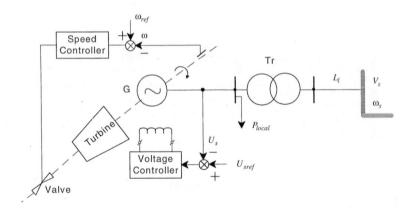

Figure 5. Structure of electromechanical controlled system

can be increased by the regulation of the internal generator voltage E' and the mechanical power produced by a steam turbine. Design of voltage and frequency (speed) controllers requires development of turbine and generator models and determination of their parameters. An application of optimal nonlinear controllers should be proceeded by the development of generator observers that can provide unmeasurable state variables. The controlled system has a structure as shown in Figure 5.

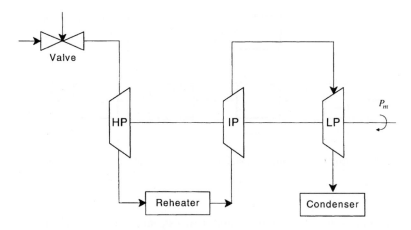

Figure 6. The model of a steam turbine

3.1. STEAM TURBINE MODELS

There are many types of steam turbines used in power stations. Models of turbines for stability and control studies are defined by IEEE Report [13]. One of the most common type of steam turbine is the turbine shown in Figure 6. It consists of three parts:

- HP – high pressure,
- IP – intermediate pressure,
- LP – low pressure.

Flow of steam from a boiler is controlled by a steam valve. After passing the high pressure part steam is reheated before it enters the IP part. Used steam after leaving a turbine is condensed. Energy carried by compressed steam throughout a turbine is gradually changed into mechanical energy on a common shaft of a unit turbine – generator.

An approximate linear model of the tandem compound, single reheat steam turbine used for stability and control studies is shown in Figure 7. The corresponding state and output equations are of the form

$$D_t P_{CO} = (-P_{CO} + P_{RH})/\tau_{CO}, \tag{3.1}$$

$$D_t P_{RH} = (-P_{RH} + P_{CH})/\tau_{RH}, \tag{3.2}$$

$$D_t P_{CH} = (-P_{CH} + \overline{P}_{GV})/\tau_{CH}, \tag{3.3}$$

$$P_m = F_{HP} P_{CH} + F_{IP} P_{RH} + F_{LP} P_{CO}, \tag{3.4}$$

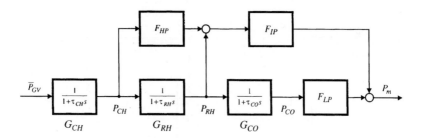

Figure 7. Approximate linear model of the tandem compound, single reheat steam turbine

where the following notation is used: τ_{CH} - steam chest time constant, τ_{RH} - steam reheater time constant, τ_{CO} - crossover time constant, F_{HP}, F_{IP}, F_{LP} - coefficients representing portions of energy produced by the high, intermediate and low pressure stages respectively, $\overline{P}_{GV}$ - steam valve output, P_m - mechanical power on the shaft of a generator [13].

Since time constants τ_{CH}, τ_{CO} are much smaller than τ_{RH} we can neglect inertial elements G_{CH}, G_{CO} and obtain the reduced-order model of a turbine:

$$D_t P_{RH} = r_7 P_{RH} + r_8 \overline{P}_{GV}, \qquad (3.5)$$

$$P_m = F_{HP} \overline{P}_{GV} + (F_{IP} + F_{LP}) P_{RH}, \qquad (3.6)$$

where

$$r_7 = -\frac{1}{\tau_{RH}}, \quad r_8 = \frac{1}{\tau_{RH}}, \qquad (3.7)$$

Flow of steam from a boiler to a turbine is controlled by a steam valve which in most cases is driven by a hydraulic system. An equivalent model of a steam valve with a primary speed controller is shown in Figure 8.

The notation used in Figure 8 is as follows: P_{GV} - valve position calculated in per unit system, P_{GV}^{Max}, P_{GV}^{Min} - maximum and minimum position of a valve, $\dot{P}_{GV}^{Max}$, $\dot{P}_{GV}^{Min}$ - maximum and minimum speed, τ_3 - valve time constant, τ_1, τ_2 - time constants of the speed controller, k - amplifying constant of speed controller, ω - angular speed of a rotor, ω_{ref} - reference angular speed. In practical solutions, speed controller time constants τ_1, and τ_2 are very small and they can be neglected in modelling processes. If we neglect time constants τ_1, τ_2, then we obtain the following state and output equations of the steam valve and speed controller

$$D_t P_{GV} = r_9 P_{GV} + r_{10}^\omega + \beta_2 u_{GV} + P_o/\tau_{GV}, \qquad (3.8)$$

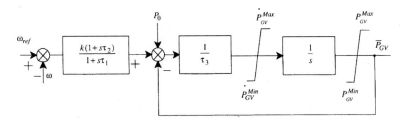

Figure 8. An equivalent model of a steam turbine valve and speed controller

$$\overline{P}_{GV} = \text{sat}(P_{GV}, P_{GV}^{Max}, P_{GV}^{Min}), \tag{3.9}$$

where

$$r_9 = -\frac{1}{\tau_3}, \quad r_{10} = -\frac{k_{GV}}{\tau_3}, \quad \beta_2 = \frac{k_{GV}}{\tau_3}, \tag{3.10}$$

and $u_{GV} = \omega_{ref}$ is assumed as the second control input of the turbogenerator system. Equations 3.5 and 3.8 are used for modelling of the steam turbine and the valve in the development of nonlinear controllers.

4.1. MODELS OF A SYNCHRONOUS GENERATOR

Typical synchronous generators driven by steam turbines have the synchronous angular speed of 1500 or 3000 rotations per minute (rpm) for 50Hz systems of 1800 or 3600 rpm for 60Hz systems. One directional magnetic flux of a rotor, created by an excitation winding, is coupled with three phase stator windings resulting in transmission of energy throughout an air gap between a rotor and a stator. As an excitation winding fixed at a rotor generates a magnetic flux, which has also a fixed position, shaft rotation results in varying mutual positions of an excitation winding and three phase windings of a stator. It means that mutual reactances of rotor and stator windings are functions of rotor position or in other words functions of time.

A problem of varying mutual reactance in modelling of synchronous generators can be overcome by an application of a special transformation which transforms the three phase model into the two phase model with two axes (d, q) changing time varying mutual reactances into constant parameters of the new model [4]. A relationship between currents in the three phase model and in the equivalent (d, q) model is in the form:

$$\mathbf{i}_{dq0} = \mathbf{P}(\gamma_r)\mathbf{i}_{abc}, \tag{4.1}$$

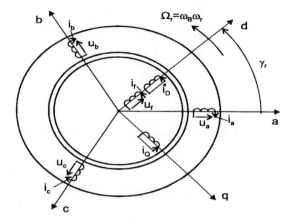

Figure 9. Equivalent circuits of a synchronous machine

The transformation matrix $\mathbf{P}(\gamma_r)$ is defined as

$$\mathbf{P}(\gamma_r) = \sqrt{2/3} \begin{bmatrix} \cos\gamma_r & \cos(\gamma_r - 2\pi/3) & \cos(\gamma_r + 2\pi/3) \\ \sin\gamma_r & \sin(\gamma_r - 2\pi/3) & \sin(\gamma_r + 2\pi/3) \\ 1/\sqrt{2} & 1/\sqrt{2} & 1/\sqrt{2} \end{bmatrix} \qquad (4.2)$$

where γ_r - represents a time varying angle between the axis of phase "a" and the new system axis "d" - Figure 9.

It can be noticed that the transformation matrix $\mathbf{P}(\gamma_r)$ is orthogonal i.e. $\mathbf{P}(\gamma_r)^{-1} = \mathbf{P}(\gamma_r)^T$, what means that this transformation is power invariant and we should expect to use the same power expression in either the (a, b, c) or the $(d, q, 0)$ coordinate systems. For a symmetrical three phase construction the zero component of an equivalent model can be neglected.

The equivalent (d, q) model presented in Figure 9 consists of five equivalent circuits:

- two stator circuits denoted as "d" and "q"
- two damping circuits denoted as "D" and "Q"
- excitation circuits denoted as "f".

Electromagnetic processes in these five circuits can be described mathematically as a set of differential equations

$$D_t \Psi_d = \omega_B(-R_s i_d - \omega \Psi_q - u_d), \qquad (4.3)$$

$$D_t \Psi_q = \omega_B(-R_s i_q + \omega \Psi_d - u_q), \qquad (4.4)$$

$$D_t \Psi_f = \omega_B(-R_f i_f + u_f), \qquad (4.5)$$

$$D_t \Psi_D = \omega_B(-R_D i_D), \qquad (4.6)$$

$$D_t \Psi_D = \omega_B(-R_Q i_Q), \qquad (4.7)$$

where

R_s, R_f, R_D, R_Q — resistances of equivalent circuits,

$\Psi_d, \Psi_q, \Psi_f, \Psi_D, \Psi_Q$ — magnetic fluxes of equivalent circuits,

i_d, i_q, i_f, i_D, i_Q — currents of equivalent circuits,

u_d, u_q — "d, q" components of stator voltage

There is a linear relation between magnetic fluxes and currents determined by two matrix equations

$$\begin{bmatrix} \Psi_d \\ \Psi_f \\ \Psi_D \end{bmatrix} = \begin{bmatrix} L_d & L_{md} & L_{md} \\ L_{md} & L_f & L_{md} \\ L_{md} & L_{md} & L_D \end{bmatrix} \begin{bmatrix} i_d \\ i_f \\ i_D \end{bmatrix}, \qquad (4.8)$$

$$\begin{bmatrix} \Psi_q \\ \Psi_Q \end{bmatrix} = \begin{bmatrix} L_q & L_{mq} \\ L_{mq} & L_Q \end{bmatrix} \begin{bmatrix} i_q \\ i_Q \end{bmatrix}. \qquad (4.9)$$

where $L_d, L_q, L_f, L_D, L_Q, L_{md}, L_{mq}$ - reactances of equivalent circuits [4].

There are many options in which state variables describing electromagnetic processes can be selected. State variables can be chosen as equivalent magnetic fluxes, equivalent currents or any combination of fluxes and currents. The most common approach is to use magnetic fluxes as the state variables [4]. Equivalent circuits representing the 'd, q' model of a synchronous generator and described mathematically by equations 4.3–4.9 are shown in Figure 10.

Electrical torque on a shaft of a synchronous generator can be expressed in the form:

$$M_e = \Psi_d i_q - \Psi_q i_d. \qquad (4.10)$$

Balance of mechanical and electrical torques on a shaft of a generator and transmission of electrical power throughout a tie line to a power system are described by the following equations

$$D_t \omega = D(\omega - \omega_s) - M_e + M_m, \qquad (4.11)$$

$$D_t \delta = \omega_B(\omega - \omega_s). \qquad (4.12)$$

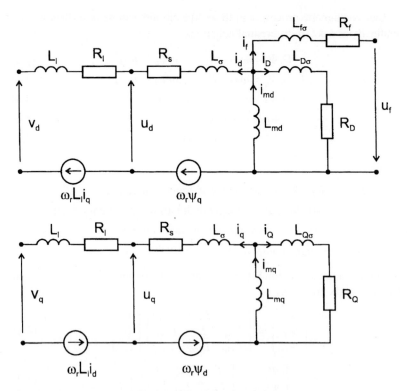

Figure 10. 'd, q' equivalent circuits of a synchronous generator

A set of differential equations comprising equations 4.3–4.7, 4.11 and 4.12 is known as the Park's model of synchronous generator or the seven dimensional model [4]. This model is the most comprehensive model used for stability and control study. It consists of seven nonlinear differential equations where nonlinearities are results of products of state variables.

In some cases, there is no need for detailed modelling of all electromagnetic processes so two reduced-order models have been derived. They result from the assumption that dynamic aspects of processes in some equivalent circuits can be negligible and these equivalents can be described with the use of algebraic equations. One of commonly used reduced-order models of a synchronous generator is the three dimensional model. It can be obtained neglecting an influence of damper windings and fast transients in stator windings [4,43]. The mathematically rigorous derivation of this model can be accomplished using the theory of singular perturbations [3], or the integral manifold approach [19].

The three dimensional model, called further the one-axis model, has the state coordinates δ, ω, e'_q and is described by the state equations of the form:

$$D_t\delta = \omega_B(\omega - \omega_s), \tag{4.13}$$

$$D_t\omega = D(\omega - \omega_s) + r_1 e'_q \sin\delta + r_2 \sin 2\delta + r_3 p_{RH} + r_4 \overline{P}_{GV}, \tag{4.14}$$

$$D_t e'_q = r_5 e'_q + r_6 \cos\delta + \beta_1 \overline{e}_{fD}, \tag{4.15}$$

where

$$r_1 = -\frac{1}{\tau_m}\frac{V_s}{L_l + L'_d}, \quad r_2 = -\frac{1}{\tau_m}\frac{1}{2}\frac{L'_d - L_q}{(L_l + L'_d)(L_l + L_q)}V_s^2,$$

$$r_3 = \frac{F_{IP} + F_{LP}}{\tau_m}, \quad r_4 = \frac{F_{HP}}{\tau_m},$$

$$r_5 = -\frac{1}{\tau'_{do}}\frac{L_l + L_d}{L_l + L'_d}, \quad r_6 = \frac{1}{\tau'_{do}}\frac{L_d - L'_d}{L_l + L'_d}V_s \quad \beta_1 = \frac{1}{\tau'_{do}},$$

and

$$\overline{e}_{fD} = \mathrm{sat}(e_{fD}, e_{fD}^{Max}, e_{fD}^{Min} \tag{4.17}$$

denotes the constrained output signal of an exciter.

It is also possible to derive another three dimensional model of a synchronous generator where the field current i_f is used as one of state variables.

$$D_t\delta = \omega_B(\omega - \omega_s), \tag{4.18}$$

$$D_t\omega = d_2(\omega - \omega_s)\cos^2\delta + p_1 i_f \sin\delta + p_2 \sin 2\delta + p_3 M_m, \tag{4.19}$$

$$D_t i_f = p_4(\omega - \omega_s)\sin\delta + p_5 i_f + p_6 u_f, \tag{4.20}$$

where coefficients d_2, $p_1 - p_6$ are defined as follows:

$$d_2 = -\frac{1}{R_Q}\frac{L_{mq}^2}{(L_l + L_q)^2}\frac{V_s^2}{\tau_m}, \quad p_1 = -\frac{L_{md}}{(L_l + L_d)}\frac{V_s}{\tau_m},$$

$$p_2 = -\frac{1}{2}\frac{L_d - L_q}{(L_l + L_d)(L_l + L_q)}\frac{V_s^2}{\tau_m}, \quad p_3 = \frac{1}{\tau_m},$$

$$p_4 = \omega_B\frac{L_{md}}{L_f(L_l + L_d) - L_{md}^2}V_s, \quad p_5 = -\omega_B\frac{L_l + L_d}{L_f(L_l + L_d) - L_{md}^2}R_f$$

$$p_6 = \omega_B\frac{L_l + L_d}{L_f(L_l + L_d) - L_{md}^2}. \tag{4.21}$$

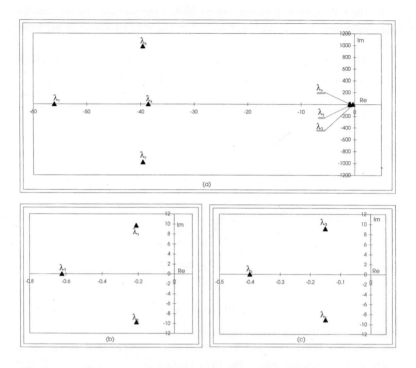

Figure 11. Location of eigenvalues, (a) - seven dimensional model M7_Park, (b) - three dimensional model M3_i_f, (c) - three dimensional model M3_e'_q

An eigenvalue analysis for seven and three dimensional models provides location of eigenvalues on the complex plane - Figure 11. This analysis has been carried out for a generator of 160 MVA [4]. Park's model is characterized by seven eigenvalues located on the left side of the complex plane. Positions of eigenvalues depend on the structure of the system and its parameters.

Eigenvalues $\lambda_7 = -56.1$ and $\lambda_4 = -38.6$ express transient processes in damper windings "D" and "Q". Eigenvalues $\lambda_6 = -39.7 + j985$ and $\lambda_5 = -39.7 + j985$ characterize transient processes in stator windings "d" and "q". Imaginary parts of these two conjugate complex numbers depend strongly on a tie line reactance. The value presented in Figure 11a has been obtained for the tie line reactance equal to 0.3 in pu. For the tie line reactance equal to the zero imaginary parts become equal to ω_s. The increase of the tie line reactance results in the increase of the values of the imaginary parts. Eigenvalues $\lambda_3 = -0.88 + j9.9$ and $\lambda_2 = -0.88 - j9.9$ characterize electromechanical transient processes. The imaginary part of these complex numbers is a function of a shaft inertia increasing for larger mechanical time constants τ_m. Eigenvalue $\lambda_1 = -0.22$ characterises transient processes resulting from excitation winding "f".

For linear models eigenvalues are constant but they are computed for nonlinear systems their positions are moving due to changes in points of linearization. Eigenvalues $\lambda_7, \lambda_6, \lambda_5, \lambda_4$ are located far left from the imaginary axis, so it is unlikely that their deviations are able to move them into the right side of the complex plane resulting in system instability. Eigenvalues $\lambda_3, \lambda_2, \lambda_1$ are dominant eigenvalues in the system considered and their deviations may result in instability of the system.

Equivalent three dimensional model, (M3-i_f), with a field current as a state variable has three eigenvalues: $\lambda_3 = -0.21 + j9.7$, $\lambda_2 = -0.21 - j9.7$, $\lambda_1 = -0.2$ located as shown in Figure 11b. Eigenvalues of the second equivalent model, (M3-e'_q), with e'_q as a state variable, model are shown in Figure 11c. Their values are: $\lambda_3 = -0.15 + j9.1$, $\lambda_2 = -0.15 - j9.1$, $\lambda_1 = -0.34$. Eigenvalue positions for these two reduced-order models are very similar, so behaviour of these models should be similar as well.

An eigenvalue analysis for both three dimentional models shows three dominant eigenvalues slightly shifted from their former positions in model M7 - Figure 11b and Figure 11c. It can be assumed that adequate control of the three dominant eigenvalues can assure stability of the system so the three dimensional model can be used for controller design. Such a controller should control only positions of dominant eigenvalues assuming that deviations of other eigenvalues cannot affect stability of the system.

As both reduced-order models can be used for design of observers and controllers, the M3-i_f model is more convenient for observer design as the field current which can be used as input measurement is the state variable, while the M3-e'_q model is more convenient for nonlinear controller design.

5.1. MODELS OF EXCITATION SYSTEMS

The main purpose of an excitation system is to provide direct current to the synchronous machine field windings aiming at the control of voltage and reactive power flow to enhance system stability. There are many excitation systems in use starting from the earliest based on direct current (DC) systems which have been gradually spearseded by alternating current (AC) systems and static (ST) excitation systems. Models of excitation systems for stability and control studies are provided by and IEEE Committee Reports [12,14].

One of the newest control system, type ST, is shown in Figure 12. The system presented utilizes transformers to convert voltage (and also current in

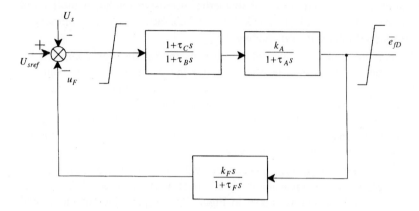

Figure 12. Type ST1 potential source-controlled rectifier exciter

compound systems) to an appropriate level. Rectifiers, either controlled and uncontrolled, provide the necessary direct current to the generator field. In many practical solutions most time constants of this excitation system have very small values comparing with time constant of a field winding so for the control study this excitation system can be represented as proportional element with signal limitation or an element with small inertia.

6.1. SIMULATION OF SYNCHRONOUS GENERATORS AND A STEAM TURBINE

Behaviour of three models of a synchronous generator has been simulated for transient processes caused by disturbances in a power system. A disturbance, like a short-circuit, is seen by a generator as a voltage drop behind a tie line. To investigate natural behaviour of the models the reference value signal ω_{ref} and an excitation voltage are assumed to be constant. It leads to the uncontrolled systems and allows to avoid an influence of controllers on generator behaviour.

Results of simulation are shown in Figure 13. Before the disturbance, simulated as a power system voltage drop, all three models were in steady state conditions. The voltage drop results in decrease the electrical power transmitted and subsequently in an increase of angular speed and load angle. After 0.25 [s] the disturbance is cleared and the power system voltage returns to its initial value - Figure 13a. Deviations of model state variables are decaying gradually - Figure 13b,c and d. It can be seen that the best damping

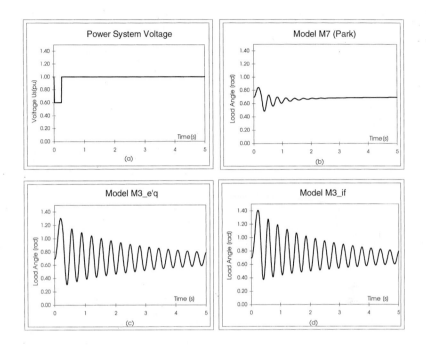

Figure 13. Simulation of transient processes for uncontrolled models

of oscillations has been obtained for the seven dimensional model, (M7_Park). It results from damping processes in damper and stator windings. Both three dimensional models do not include these windings so oscillations are less damped and transient processes take longer time to be stabilize.

It can be derived that a controller, which can stablize three dimensional models, should also, or even better, stabilize the seven dimensional model as its action will be supported by damper and stator windings. Such a controller should also stabilize a real generator as the seven dimensional model precisely describes the construction and parameters of a synchronous generator. The reduced order three dimensional models can be used for observer and controller design.

Simulation results a steam turbine and a turbine valve are shown in Figure 14. A transient process resulting in load angle increase causes reaction of a steam valve as angular speed of a generator, which is an input signal for primary speed controller, oscillates - Figure 14a. The reaction of a steam turbine is smaller as a large time constant of a reaheater equal $\tau_{RH} = 7[s]$ results in large inertia of the system. When the reheater time constant becomes

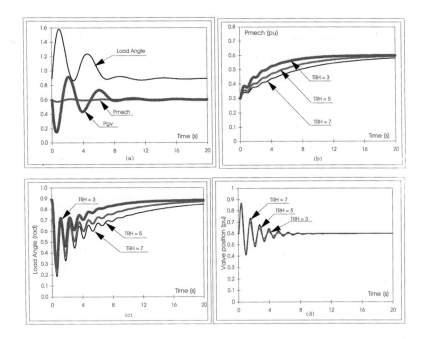

Figure 14. Simulation of steam turbine and valve

smaller the turbine reaction becomes faster in reaching a steady state - Figure 14b and Figure 14c. The time constant of a reheater does not effect the valve reaction as it is controlled by the difference between the reference angular speed and generator rotorspeed - Figure 14d.

The time constant of a reheater is much larger that time constants of electrical processes in a generator so the turbine regulation does affect initial stages of transient processes but its effect is visible in longer time, especially, in reaching a new steady state.

7.1. OBSERVERS OF SYNCHRONOUS GENERATORS

An application of optimal linear or nonlinear controllers requires the information on the entire state vector used for control. If some state variables are unmeasurable, observers have to be implemented for estimation of unknown state variables. A typical structure of an observer of a nonlinear system is shown in Figure 15. This type of observer can be defined as a nonlinear observer with a linear feedback loop [23,24].

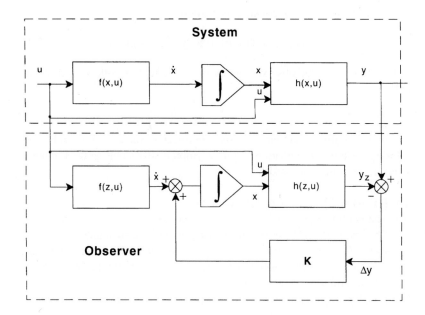

Figure 15. Structure of an observer for nonlinear system

Generally, an observer is a model of a system with additional signals which are functions of differences between measurable outputs of the system and the estimates produced by the observer. Differences between the signals measured and estimated are amplified in gain coefficient module **K** and fed back to the system model. An observer is properly designed if the error between state "**x**" and its estimate "**z**" is decaying to zero with the assumed time constant. The main aim of observer design is to prove that the estimation error will decay in the conditions assumed. Nonlinear system can be described by a set of differential equations

$$D_t \mathbf{x} = \mathbf{f}(\mathbf{x}, \mathbf{u}),$$
$$\mathbf{y} = \mathbf{h}(\mathbf{x}, \mathbf{u}) \tag{7.1}$$

where $\mathbf{x} = \mathbf{x}(t) \in \Re^n$ is the nonlinear system state vector, $\mathbf{y} = \mathbf{y}(t) \in \Re^m$ is the output vector. A nonlinear observer of the system above is in the following form

$$D_t \mathbf{z} = \mathbf{f}(\mathbf{z}, \mathbf{u}) + \mathbf{K}[\mathbf{h}(\mathbf{x}, \mathbf{u}) - \mathbf{h}(\mathbf{z}, \mathbf{u})], \tag{7.2}$$

where $\mathbf{z} = \mathbf{z}(t) \in \Re^n$ is the observer state vector and **K** is the gain matrix of dimension $m \times n$.

The error, $(\mathbf{e} = \mathbf{z} - \mathbf{x})$, equation is as follows:

$$D_t \mathbf{e} = \mathbf{f}(\mathbf{z}, \mathbf{u}) - \mathbf{f}(\mathbf{x}, \mathbf{u}) + \mathbf{K}[\mathbf{h}(\mathbf{x}, \mathbf{u}) - \mathbf{h}(\mathbf{z}, \mathbf{u})]. \qquad (7.3)$$

The stability of the observer can be proved using the following theorem [23].

THEOREM 7.1

For the nonlinear system 7.1, the nonlinear observer 7.2 is stable ie. solutions of the error equation 7.3 are asymptotically stable in the Lapunov sense, if the following conditions hold.

(i) $\mathbf{f}(\mathbf{x}, \mathbf{u})$ and $\mathbf{h}(\mathbf{x}, \mathbf{u}) \in C^\infty(t, \infty)$ in G.
(ii) The nonlinear system 7.1 is locally observable.
(iii) The real parts of the eigenvalues of $\mathbf{R}_s$ are chosen such that $\text{Re}\{\lambda_i(\mathbf{R}_S)\} = 1 < 0$ for $i = 1, 2, \ldots, n$, where $\mathbf{R}_s = \mathbf{A} - \mathbf{KC}$ and $\mathbf{A}, \mathbf{C}$ are matrices resulting from the linear approximation of functions $\mathbf{f}(\mathbf{x}, \mathbf{u})$ and $\mathbf{h}(\mathbf{x}, \mathbf{u})$, respectively.
(iv) $|\Delta x_i| < h_x$ and $|\Delta u_i| < h_u$ in G, where h_x, h_u are positive constants.

Most state variables used in models of a synchronous generators are unmeasureable or are difficult to measure. The typical variables measured are: the stator currents, output voltages, active and reactive powers and field current. It is possible to measure angular speed of a generator rotor but small oscillations resulting from shaft mechanical vibrations applified in a high gain controller may cause generator oscillations or even instability so angular rotor speed should be estimated. It is also difficult to measure the load angle as in a single machine system this angle is calculated as a phase shift between "q" axis of the rotor and the equivalent voltage phasor $\overline{V}_s$ which represents an entire power system. Using measurable signals it is possible to design a family of observers for assumed models of a synchronous generator and to prove their stability. The three dimensional model and the field current as the measured signal are used to illustrate estimation of state variables carried out by the observer. Simulation results are shown in Figures 16–18.

The first test assumed that a generator is in steady state conditions and the observer is turned on to start estimation processes. As the role of an observer is to calculate unknown variables so it is obvious that the intitial state of the observer state vector differs from the real state vector. The error appearing during the intial stage when the observer is determing the unknown variables and the time period required for observer convergence ie. when error between estimation and real values is equal to zero or small negligible values are measures of the observer performance. Figure 16 demonstrates

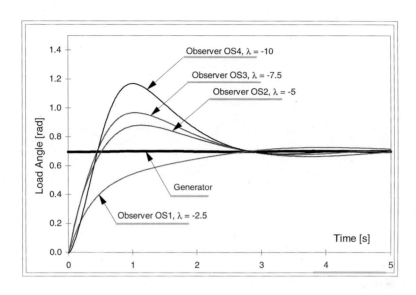

Figure 16. Observer transient estimation for a generator in the steady state

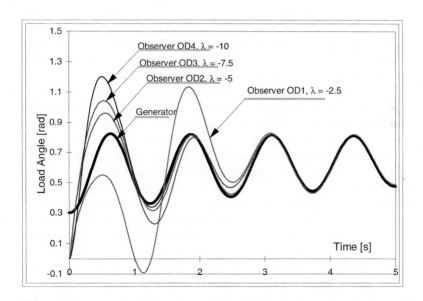

Figure 17. Observer transient estimation for a generator during the transient state

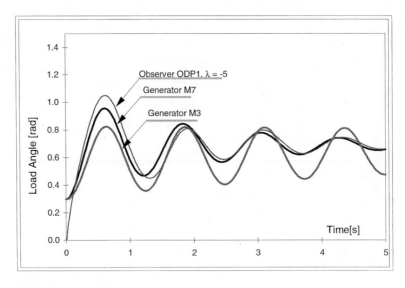

Figure 18. State estimation of seven dimensional model

observer transient processes for the assumed initial observer variables equal to zero for four cases of observer eigenvalues. All observer eigenvalues $\lambda_1 = \lambda_2 = \lambda_3 = \lambda$ have the same values. A full order observer designed for the three dimensional generator model has been used in this test. It estimates the entire state vector. To demonstrate the observer performance one of the state vector variables i.e the load angle has been selected.

The second test has been carried out to investigate observer behaviour during generator oscillations. It combines transients in the observer with transients in the system being observed. The observer has been designed for the three dimensinal model and such a model has been used to simulate a generator. The observer starts estimation from the origin as in the first test and its error is close to zero after 2–3 seconds (Figure 17).

Generally, the larger negative eigenvalues are assigned the faster estimation can be obtained but large negative eigenvalues lead to overregulation and may result in observer instability. An observer overregulation and time of convergence can be significantly reduced when the second measurable signal is fed to the observer. It leads to the application of vary fast observers and can be obtained by the special method of observer design [26].

The three dimensional model used in two previous tests does not represent precisely a real generator so the third test has been carried out. The full order observer designed for the reduced-order model estimates the state of

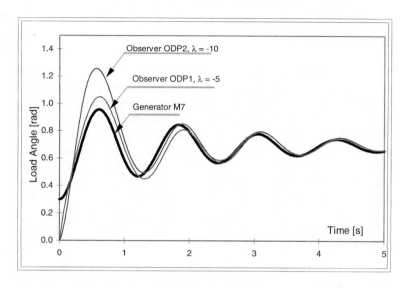

Figure 19. State estimation during generator oscillations

the seven dimensional model of a generator. This test investiagtes an effect of difference between the equivalents taken for design and the real construction.

Figure 18 shows how the observer designed for the three dimesional model M3_i_f estimates state variables when a signal fed comes from the seven dimensional model which simulates a real system. The observer estimates precisely state variables of the model M7. It shows that the observer designed for the reduced order equivalent can compute the state of the system from which measurements are provided.

The coincidence of transients in observer estimation and oscillations of a generator is shown in Figure 19. It reveals the phenomenon of over-regulation for large negative eigenvalues. In this case, both observers have been designed for the reduced order model and fed from the seven dimensional model M7.

Estimation of generator state variables results in some error between real and estimated states during transient processes appearing only when an observer is switched on the first time. When afterwards an observer reduces estimation errors to zero it provides the entire state vector without estimation error during steady state and transient processes. An analysis of estimation errors and their effects on the controller performance shows that an impact of estimation errors on control is negligible [31].

8.1. NONLINEAR CONTROLLERS FOR A SINGLE MACHINE SYSTEM

8.1.1. Introduction

From the beginning of power systems, a voltage controller of a generator was designed assuming linear representation of a synchronous generator. When fast high gain voltage controllers were implemented and when it was necessary to transmit electrical energy on long distances there appeared a need for an additional stabilization with the use of power system stabilizers (PSS). The constants used for PSS setting always depend on the electrical location of a unit in the system, the dynamic characteristics of the system and other units [4]. There is also an unsolved problem of PSS coordination in large systems or during inter-system power exchange.

The approach of controller design for linear representation of a synchronous generator and additional improvement of control with PSSs suffers from two main disadvantages:

- linear equivalent of a synchronous generator does not represent an entire dynamics of the system,
- voltage controllers and PSS do not use the entire state vector for the system control.

There is a strong need for the development of nonlinear controllers that can be designed taking into account nonlinearity of the controlled system and can employ the entire information on the controlled system by feeding a controller with the entire state vector.

The first attempts in this direction aimed at the application of linear optimal controllers (LOC). A number of works have shown that LOC provides better means than conventional controllers in coordinated stabilization of a generator [5,9,43]. However, implementation of an optimal controller generally requires information about the entire state of a control plant. This cannot be satisfied in the case of a synchronous generator due to impossibility of measuring damper winding fluxes and the angle δ between the q-axis and the power system voltage $\mathbf{v}_s$. Therefore the problem of synchronous generator state estimation arises.

Methodology of nonlinear observer design its application to the construction of nonlinear observers for four generator models has been presented [23,24]. Digital simulations confirmed the efficiency of the proposed methodology. Further progress in observer investigation allowed the construction of very fast linear and nonlinear observers that can estimate generator state variables in a dozen milliseconds [25,26].

Having efficient observers it is possible to apply LOC to synchronous generator stabilization but, it is well known [7,35], that an observer in a feedback loop may spoil robustness properties of LOC and reduce a region of stability. Considering this disadvantage, an attempt was made to design together a linear optimal controller and an observer for a synchronous generator so that the robustness of the resulting closed-loop system is recovered. This goal was accomplished by an additional output feedback loop, [27]. Although simulation results were quite satisfactory, we believed that the most proper approach to controller design for synchronous generators is an application of the modern theory of nonlinear systems [17,38]. Our conviction follows from the fact that design of linear optimal controllers is based on linear models resulting from linearization of nonlinear systems about some operating point $\mathbf{p}^o$. Thus a closed-loop system keeps assigned properties only in some neighbourhood of $\mathbf{p}^o$. In many cases, however, faults may cause that trajectories of generator state variables are moved far from the operating point, and due to the strong nonlinearity of the synchronous generator stability loss is possible [43].

The first attempts at an application of the modern nonlinear control theory to controller design for synchronous generators were made in the eighties [37,22]. The problem of noninteracting control of the load angle δ and the field flux linkage Ψ_f was considered [37]. However, it was not taken into account that in practice the field voltage $u_f (e_{fD})$ and the steam valve output P_{GV} are constrained signals, additionally the assumed small value for the equivalent time constant of a turbine, (0.3[s]), makes the problem far from practice. In the next attempt, the five-dimensional model of a generator was considered [22], and it was assumed that two inputs can be controlled: the field excitation voltage and the mechanical power. The last assumption is unrealistic from the technical point of view, because it is impossible to construct a mechanical power source that allows to control a mechanical power directly applied on a generator shaft. If one wants to use mechanical power P_m (or torque M_m) as an input of a generator, one should supplement a generator model by models of a steam or water turbine or a diesel engine [43], but in this case the multi-dimensional model of a turbine system has to be additionally considered and this complicates design. Thanks to the mentioned assumptions, a controller synthesis is easier but results are rather of a theoretical nature than of practical significance.

A conventional three-dimensional reduced-order model of a generator was considered [28,29]. Only one input signal, i.e. the field voltage u_f, is employed and it was assumed that the mechanical power on a generator shaft is constant during control. These assumptions are satisfied for large turbo-generators. Afterwards it was proved using feedback linearization that this model can be linearized by a nonlinear feedback and the most simple and convenient (for practical applications) form of this feedback was presented. It was also shown, that the same control law can be derived using Korobov's method [20]. The efficiency of the proposed nonlinear controller was evaluated by simulations under a number of conditions appearing in practice. In order to bring the problem closer to practical applications, nonlinear controllers with observers were examined. Limitation of the field voltage was also considered during simulations. Although, the problem of the nonlinear controller stability with an observer was not discussed, simulations showed very good transient properties of a synchronous generator with the nonlinear controller and an observer working under fault and post-fault conditions [30–32,45].

It is worth mentioning, that similar results to that of Mielczarski and Zajaczkowski were obtained independently by Ilic and Mak [15]. They considered the conventional third-order model of a generator described by the state variables δ, ω, e'_q and using engineering intuition they found new state coordinates $\tilde{x}_1 = \delta$, $\tilde{x}_2 = \Omega$, $\tilde{x}_3 = D_t\Omega$ and proved that there exists a linearizing feedback controller for the considered model of a generator.

In some works other methods of nonlinear controller design were applied to the problem of power system control. To mention just the most representative, we point to the work of Subbarao and Iyer [41], who applied the variable structure control for design of a multivariable controller for a turbogenerator. Since they employed the same model and assumed the same design specifications, their results have similar disadvantages as that of Singh [37]. The interesting direct linearization technique is proposed by Wang, et al. [42], and Gao L et al. [8], but they also did not provide simulation results for Park's model of a generator.

The main idea of an application of the feedback linearization method [17,18,36,10,11,36,38,40] for nonlinear controller design is to find a control feedback and new state coordinates such that a resulting closed-loop system described in the new state coordinates is locally equivalent to a linear system in the Brunovsky canonical form (see Figure 20).

Since the equivalent system becomes linear, it is relatively easy to design an external controller to implement required control strategy. Although, the idea of feedback linearization seems to be very simple, the design of a nonlinear controller requires in the first step checking necessary and sufficient conditions for feedback linearizatiom and, if this is the case, derivation of a smooth nonlinear one-to-one state transformation and a linearizing feedback. In general, since the new state coordinates must fulfill an overdetermined system of partial differential equations, this second step is hard or sometimes impossible to complete, but in the case of the reduced-order model of a single machine system we can obtain well defined and physically meaningful solution to the feedback linearization problem - see the next Subsections.

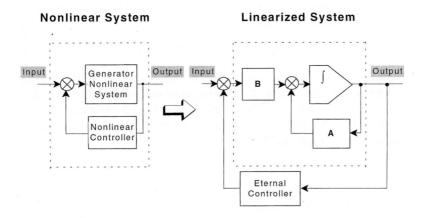

Figure 20. Transformation of a nonlinear system into a linear equivalent

8.1.2. The Reduced-order Model of a Single Machine System

In this Subsection we consider a controlled plant, called a single machine system, composed of a steam turbine with a steam valve and governor and a synchronous generator connected with a power system by a tie line as shown in Figure 5. We assume that the components of the single machine system are modelled by the state equations 4.13–4.15 of the one-axis model of a synchronous generator, the state equation 3.5 of the reduced-order model of the steam turbine and the state equation 3.8 of the steam valve with the speed controller. Additionally, we assume that the state variable P_{GV} and the control signal e_{fD} are not constrained i.e. the following relations hold:

$$\overline{P}_{GV} = P_{GV}, \quad \overline{e}_{fD} = e_{fD}. \tag{8.1}$$

Let

$$\xi = [\xi_1 \dots \xi_5]^T = [\delta \omega e'_q P_{RH} P_{GV}]^T \tag{8.2}$$

denote the state vector of the single machine system. It is easy to combine the state equations of the single machine system into one state equation in the vector form:

$$D_t \xi = \varphi(\xi) + \mathbf{b}_1 e_{fD} + \mathbf{b}_2 u_{GV}, \tag{8.3}$$

where $\varphi : \mathfrak{R}^5 \to \mathfrak{R}^5$ is defined by the right-hand sides of the state equations 3.5, 3.8, 4.13–4.15 and constant vectors $\mathbf{b}_1, \mathbf{b}_2 \in \mathfrak{R}^5$ are given as

$$\mathbf{b}_1 = [0 0 \beta_1 0 0]^T, \tag{8.4}$$

$$\mathbf{b}_2 = [0 0 0 0 \beta_2]^T. \tag{8.5}$$

The point $\mathbf{p}^o = (\xi^o, e_{fD}^o, u_{GV}^o) \in \mathfrak{R}^5 \times \mathfrak{R} \times \mathfrak{R}$ is called an operating point of the single machine system if

$$\varphi(\xi^o) + \mathbf{b}_1 e_{fD}^o + \mathbf{b}_2 u_{GV}^o = \mathbf{0}. \tag{8.6}$$

Note, that the operating point of the single machine system can be computed on the basis of the given loading conditions of the generator [4,43].

For further considerations it is convenient to have the origin $(\mathbf{0}, 0, 0)$ of the $\mathfrak{R}^5 \times \mathfrak{R}^1 \times \mathfrak{R}^1$ space as an operating point of the single machine system. Linear change of state and control variables

$$\xi - \xi^o = \mathbf{x}, \quad e_{fD} - e_{fD}^o = u_1, \quad u_{GV} - u_{GV}^o = u_2, \tag{8.7a}$$

$$\xi = \mathbf{x} + \xi^o, \quad e_{fD} = e_{fD}^o + u_1, \quad u_{GV} = u_{GV}^o + u_2, \tag{8.7b}$$

results in the following state equation of the single machine system:

$$D_t \mathbf{x} = \mathbf{a}(\mathbf{x}) + \mathbf{b}_1 u_1 + \mathbf{b}_2 u_2 \tag{8.8}$$

where the smooth vector field

$$\mathbf{a}(\mathbf{x}) = \varphi(\xi^o + \mathbf{x}) + \mathbf{b}_1 e_{fD}^o + \mathbf{b}_2 u_{GV}^o \tag{8.9}$$

is given by the component functions

$$a_1(\mathbf{x}) = \omega_B x_2, \tag{8.10}$$

$$a_2(\mathbf{x}) = D x_2 + r_1(e_q'^o + x_3)\sin(\delta^o + x_1) + r_2\sin 2(\delta^o + x_1)$$
$$+ r_3 x_4 + r_4 x_5 + P_m^o/\tau_m, \tag{8.11}$$

$$a_3(\mathbf{x}) = r_5 x_3 + r_6\cos(\delta^o + x_1) + r_5 e_q'^o + \beta_1 e_{fD}^o, \tag{8.12}$$

$$a_4(\mathbf{x}) = r_7 x_4 + r_8 x_5, \tag{8.13}$$

$$a_5(\mathbf{x}) = r_9 x_5 + r_{10} x_2 \tag{8.14}$$

defined in some open neighbourhood X^o of the origin.

The model of the single machine system described by the state equation 8.8 is further used for synthesis of nonlinear controllers.

In order to synthesise a suitable controller for the single machine system one of several design methods can be used [7,38]. The choice of the method depends among other things on design specifications and on the structure of the system. Since we are interested in effective stabilization of the single machine system in the given operating point $(\xi^o, e_{fD}^o, u_{GV}^o)$ we describe briefly the feedback linearization method [10, 11, 17, 18, 36, 38, 39] employed further to design nonlinear controllers for the single machine system.

8.1.3. Feedback Linearization of a Class of Nonlinear Systems

Since the dynamics of the single machine system is described by the nonlinear state equations with two control inputs appearing linearly, we consider here a class of nonlinear systems modelled by the state equation of the form:

$$D_t \mathbf{x} = \mathbf{a}(\mathbf{x}) + \sum_{i=1}^{m} \mathbf{b}_i(\mathbf{x}) u_i, \tag{8.15}$$

where $\mathbf{x} \in X \subset \mathfrak{R}^n$ is the state vector, X is an open set containing the origin, $u_i \in \mathfrak{R}$ are control variables, $\mathbf{a}$ and $\mathbf{b}_i$, $i = 1, \ldots, m$ are smooth vector fields on X, $\mathbf{a}(\mathbf{0}) = \mathbf{0}$ and $D_t = d/dt$.

Further some standard differential geometric concepts are to be defined [1, 10, 17].

Let $\mathbf{a}$ be a smooth (C^∞) vector field on an open set $X \subset \mathfrak{R}^n$ and let $h : X \to \mathfrak{R}$ be a smooth function with gradient $\nabla h(\mathbf{x})$, $\mathbf{x} \in X$. The Lie derivative $L_a h$ of h in the direction $\mathbf{a}$ is defined as

$$L_\mathbf{a} h(\mathbf{x}) = \langle \nabla h(\mathbf{x}) \mathbf{a}(\mathbf{x}) \rangle = \sum_{i=1}^{m} D_i h(\mathbf{x}) a_i(\mathbf{x}), \tag{8.16}$$

where $\langle \cdot, \cdot \rangle$ denotes duality between one-forms and vector fields.

Let $\mathbf{a}$ and $\mathbf{b}$ denote smooth vector fields on X. The Lie bracket $ad_{\mathbf{ab}}^1 = [\mathbf{a}; \mathbf{b}]$ of these vector fields is a smooth vector field defined as

$$ad_\mathbf{a}^1 \mathbf{b}(\mathbf{x}) = [\mathbf{a}; \mathbf{b}](\mathbf{x}) = D\mathbf{b}(\mathbf{x})\mathbf{a}(\mathbf{x}) - D\mathbf{a}(\mathbf{x})\mathbf{b}(\mathbf{x}), \tag{8.17}$$

where $D\mathbf{a}(\mathbf{x})$, $D\mathbf{b}(\mathbf{x})$ denote $n \times n$ Jacobian matrices.

Successive Lie brackets are defined by

$$ad_\mathbf{a}^k \mathbf{b}(\mathbf{x}) = [\mathbf{a}; ad_\mathbf{a}^{k-1} \mathbf{b}](\mathbf{x}), \quad k = 1, 2, \ldots .$$
$$ad_\mathbf{a}^o \mathbf{b}(\mathbf{x}) = \mathbf{b}(\mathbf{x}). \tag{8.18}$$

A set of smooth vector fields $\{\mathbf{a}_1, \mathbf{a}_2, \ldots, \mathbf{a}_r\}$ on X is called involutive if there exist smooth functions $\gamma_{ijk} : X \to \mathfrak{R}$ such that

$$[\mathbf{a}_i; \mathbf{a}_j](\mathbf{x}) = \sum_{k=1}^{r} \gamma_{ijk}(\mathbf{x}) \mathbf{a}_k(\mathbf{x}), \quad i \neq j, \quad 1 \leq i, j \leq r. \tag{8.19}$$

Now we can formulate the following *feedback linearization problem* [11].

For the system 8.15 a smooth diffeomorphism

$$\mathbf{T} = (T_1, \dots, T_n, T_{n+1}, \dots, T_{n+m}) : X \times U \to \tilde{X} \times W$$

is searched such that

(I) $X \subset X, \tilde{X} \subset \mathfrak{R}^n, U \subset \mathfrak{R}^m, W \subset \mathfrak{R}^m$ are open sets containing the origins

(II) $\mathbf{T(0, 0)} = (\mathbf{0, 0})$

(III) $(T_1, \dots, T_n)$: $X \to \tilde{X}$ is one-to-one and the Jacobian matrix $(D_j T_i(\mathbf{x}))_{i,j=1,\dots,n}$ is non-singular for all $\mathbf{x}$ in X

(IV) $(T_{n+1}, \dots, T_{n+m})$: $X \times U \to W$ has a non-singular Jacobian matrix with respect to $\mathbf{u} = (u_1, \dots, u_m)$ for a fixed $\mathbf{x} \in X$

(V) $\tilde{x}_i = T_i(\mathbf{x}), i = 1, \dots, n$ and $w_i = T_{n+i}(\mathbf{x}, \mathbf{u}), i = 1, \dots, m$ are state and control variables of the linear system

$$D_t \tilde{\mathbf{x}} = \tilde{\mathbf{A}} \tilde{\mathbf{x}} + \tilde{\mathbf{B}} \mathbf{w} \tag{8.20}$$

in Brunovsky canonical form with Kronecker indices $k_1 \geq k_2 \geq \dots \geq k_m$.

According to (IV), we can express the feedback control in the form:

$$\mathbf{u(x, w)} = \mathbf{f(x)} + \mathbf{G(x)w}, \tag{8.21}$$

where $\mathbf{f} : X \to \mathfrak{R}^m, \mathbf{G} : X \to \mathfrak{R}^{m \times m}$ are smooth mappings and $\mathbf{G(x)}^{-1}$ exists for each $\mathbf{x} \in X$.

If the diffeomorphism $\mathbf{T}$ exists, then we say that the closed-loop system 8.15, 8.21 described in new coordinates $(\tilde{x}_1, \dots, \tilde{x}_n, w_1, \dots, w_m)$, is $\mathbf{T}$-equivalent to the linear system 8.20.

The following theorem specifies necessary and sufficient conditions for feedback linearization of nonlinear systems (we omit arguments for the sake of brevity).

THEOREM 8.1 *(Hunt, Su, Meyer, 1983b, [11])*

The feedback linearization problem is solvable if and only if in some open set $X \subset X \subset \mathfrak{R}^n$ containing the origin the vector fields $\mathbf{a}, \mathbf{b}_i, i = 1, \dots, m$ satisfy the following conditions:

FL 1. The set of vector fields

$$C = \{\mathbf{b}_1, [\mathbf{a; b}_1], \dots, ad_{\mathbf{a}}^{k_1-1}\mathbf{b}_1\mathbf{b}_2, [\mathbf{a; b}_2], \dots,$$
$$ad_{\mathbf{a}}^{k_2-1}\mathbf{b}_2, \dots, \mathbf{b}_m, [\mathbf{a; b}_m], \dots, ad_{\mathbf{a}}^{k_m-1}\mathbf{b}_m\} \tag{8.22}$$

spans $\Re^n \forall \mathbf{x} \in X$.

FL 2. The set of vector fields

$$C_j^I = \{\mathbf{b}_1, [\mathbf{a}; \mathbf{b}_1], \ldots, ad_{\mathbf{a}}^{k_j-2}\mathbf{b}_1\mathbf{b}_2, [\mathbf{a}; \mathbf{b}_2], \ldots,$$
$$ad_{\mathbf{a}}^{k_j-2}\mathbf{b}_2, \ldots, \mathbf{b}_m, [\mathbf{a}; \mathbf{b}_m], \ldots, ad_{\mathbf{a}}^{k_j-2}\mathbf{b}_m\} \tag{8.23}$$

is involutive in X for $j = 1, \ldots, m$

FL 3. The span of each C_j^I is equal to the span of $C_j^I \cap C$.

In the constructive proof of this theorem it was shown [11], that the component functions T_i, $i = 1, \ldots, n$ must satisfy the following set of partial differential equations:

$$\langle \nabla T_1, ad_{\mathbf{a}}^j \mathbf{b}_i \rangle = 0, \quad j = 0, \ldots, k_1 - 2, \quad i = 1, \ldots, m$$
$$\langle \nabla T_{\sigma+1}, ad_{\mathbf{a}}^j \mathbf{b}_i \rangle = 0, \quad j = 0, \ldots, k_2 - 2, \quad i = 1, \ldots, m \tag{8.24}$$
$$\langle \nabla T_{\sigma_{m-1}+1}, ad_{\mathbf{a}}^j \mathbf{b}_i \rangle = 0, \quad j = 0, \ldots, k_m - 2, \quad i = 1, \ldots, m$$

$$\langle \nabla T_l, \mathbf{a} \rangle = L_{\mathbf{a}} T_l = T_{l+1}, \tag{8.25}$$

for

$$l = 1, 2, \ldots, \sigma_1 - 1, \sigma_1 + 1, \ldots, \sigma_2 - 1,$$
$$\sigma_2 + 1, \ldots, \sigma_{m-1} - 1, \sigma_{m-1} + 1, \ldots, n - 1,$$

where

$$\sigma_1 = k_1, \ \sigma_2 = k_1 + k_2, \ldots, \sigma_m = \sum_{j=1}^m k_j = n, \tag{8.26}$$

and that the remaining components of $\mathbf{T}$ $T_{n+i} + w_i, i = 1, \ldots, m$ must satisfy the following equations:

$$\langle \nabla T_{\sigma_1}, \mathbf{a} \rangle \pm \sum_{i=1}^m \langle \nabla T_1, ad_{\mathbf{a}}^{k_1-1} \mathbf{b}_i \rangle u_i = T_{n+1},$$
$$\langle \nabla T_{\sigma_2}, \mathbf{a} \rangle \pm \sum_{i=1}^m \langle \nabla T_{\sigma_1+1}, ad_{\mathbf{a}}^{k_2-1} \mathbf{b}_i \rangle u_i = T_{n+2}, \tag{8.27}$$
$$\langle \nabla T_n, \mathbf{a} \rangle \pm \sum_{i=1}^m \langle \nabla T_{\sigma_{m-1}+1}, ad_{\mathbf{a}}^{k_m-1} \mathbf{b}_i \rangle u_i = T_{n+m},$$

where '$+$' is for k_i odd and '$-$' is for k_i even. From the system 8.27 we can find a linearizing feedback $(\mathbf{x}, \mathbf{w}) \to \mathbf{u}(\mathbf{x}, \mathbf{w})$.

8.1.4. Design of a Nonlinear Field Voltage Controller

In this Subsection we apply the feedback linearization method to the design of a field voltage controller $\mathbf{x} \to u_1(\mathbf{x})$ for the single machine system. We assume that the mechanical torque (power) on the shaft of a generator is constant i.e. the following equation hold:

$$M_m = M_m^o. \tag{8.28}$$

From this assumption follows that we can neglect the dynamics of the turbine and valve and we can consider the single input controlled plant, described by the state equation of the form:

$$D_t \mathbf{x} = \mathbf{a}(\mathbf{x}) + \mathbf{b} u_1, \tag{8.29}$$

where

$$\mathbf{x} = [x_1 x_2 x_3]^T = [\delta - \delta^o \omega - \omega^\bullet e_q' - e_q^{o'e}]^T, \tag{8.30}$$

is the state vector, $\mathbf{a} : X^o \to \Re^3$ is the smooth vector field with component functions

$$a_1(\mathbf{x}) = \omega_B x_2, \tag{8.31}$$

$$a_2(\mathbf{x}) = Dx_2 + r_1(e_q'^o + x_3)\sin(\delta^o + x_1) + r_2 \sin 2(\delta^o + x_1)$$
$$+ M_m^o/\tau_m, \tag{8.32}$$

$$a_3(\mathbf{x}) = r_5 x_3 + r_6 \cos(\delta^o + x_1) + r_5 e_q'^o + \beta_1 e_{fD}^o, \tag{8.33}$$

defined in some open set $X^o \subset \Re^3$ containing the origin and constant vector $\mathbf{b} \in \Re^3$ is given as

$$\mathbf{b} = [0 0 \beta_1]^T \tag{8.34}$$

State equation 8.29 describes the dynamics of the one-axis model of a synchronous generator in a neighbourhood X^o of the steady-state equilibrium point

$$(\xi^o, e_{fD}^o) = (\delta^o, \omega^o, e_q'^\bullet, e_{fD}^o) \in \Re^3 \times \Re.$$

This state equation belongs to the class of nonlinear systems with one control input $u \in \Re$ appearing linearly and described by the following state equation:

$$D_1 \mathbf{x} = \mathbf{a}(\mathbf{x}) + \mathbf{b}(\mathbf{x})u \tag{8.35}$$

where $\mathbf{x} = [x_1 \dots x_n]^T \in \Re^n$ denotes the state vector and $\mathbf{a}, \mathbf{b}$ are smooth vector fields.

For this kind of nonlinear systems the necessary and sufficient conditions for feedback linearization simplify to the following two conditions [10,17,38,39]:

FL 1. The set of vector fields

$$C = \{\mathbf{b}, [\mathbf{a}; \mathbf{b}], \ldots, ad_{\mathbf{a}}^{n-1}\mathbf{b}\} \qquad (8.36)$$

spans $\Re^n \forall \mathbf{x} \in X$, $X \subset X^o$ some open set containing the origin.

FL 2. The set of vector fields

$$C^1 = \{\mathbf{b}, [\mathbf{a}; \mathbf{b}], \ldots, ad_{\mathbf{a}}^{n-2}\mathbf{b}\} \qquad (8.37)$$

is involutive in X.

In the constructive proof of the single input version of the *Theorem 8.1* it was shown [39], that the component functions $T_i, i = 1, \ldots, n, n+1$ of the diffeomorphism $\mathbf{T}$ must satisfy the following system of partial differential equations:

$$\langle \nabla T_i, \mathbf{b} \rangle = 0, \quad i = 1, \ldots, n-1, \qquad (8.38)$$
$$\langle \nabla T_i, \mathbf{a} \rangle = L_{\mathbf{a}} T_i = T_{i+1}, \quad i = 1, \ldots, n-1,$$
$$\langle \nabla T_n, \mathbf{a} + \mathbf{b}u \rangle = L_{\mathbf{a}+\mathbf{b}u} T_n = T_{n+1}. \qquad (8.39)$$

It was also shown [39], that T_1 can be found from the over-determined system of partial differential equations

$$\langle \nabla T_1, ad_{\mathbf{a}}^k \mathbf{b} \rangle = 0, \quad k = 0, 1, \ldots, n-2, \qquad (8.40)$$
$$\langle \nabla T_1, ad_{\mathbf{a}}^{n-1} \mathbf{b} \rangle \neq 0. \qquad (8.41)$$

It is easy to notice that if T_1 is found then the remaining component functions of $\mathbf{T}$ can be calculated from equations 8.38 and 8.39.

In the case of system 8.29 we have $n = 3$ and after simple calculations we obtain

FL 1.

$$\text{rank}[\mathbf{b}|[\mathbf{a}; \mathbf{b}](\mathbf{x})|ad_{\mathbf{a}}^{3-1}\mathbf{b}(\mathbf{x})] = 3, \quad \forall \mathbf{x} \in X,$$

where

$$X = \{\mathbf{x} \in \Re^3 | \sin(\delta^o + x_1) \neq 0\}; \qquad (8.42)$$

FL 2.

Since $[\mathbf{b}; [\mathbf{a}; \mathbf{b}]](\mathbf{x}) = \mathbf{0} \forall \mathbf{x} \in X$ the set of vector fields C^1
$= \{\mathbf{b}, [\mathbf{a}; \mathbf{b}]\}$ is involutive in X.

Using the theory of partial differential equations Hunt, Su and Meyer (1983a), [10], developed a systematic procedure for the solution of the system 8.40, 8.41. This procedure is applicable to nonlinear systems described by the state equation 8.35, but in many cases obtaining a linearizing feedback

$$u(\mathbf{x}, w) = f(\mathbf{x}) + g(\mathbf{x})w \tag{8.43}$$

with smooth functions $f, g : X \to \Re$ and g invertible on X may be hard or even impossible.

However, for a class of systems, called pure-feedback systems, there is no need to apply the procedure.

Recall, that a single input nonlinear system 8.35 is called a pure-feedback system if vector fields $\mathbf{a}$, $\mathbf{b}$ are of the form:

$$\mathbf{a(x)} = \begin{bmatrix} a_1(x_1, x_2) \\ a_2(x_1, x_2, x_3) \\ \vdots \\ a_{n-1}(x_1, x_2, \dots, x_n) \\ a_n(x_1, x_2, \dots, x_n) \end{bmatrix}, \quad \mathbf{b(x)} = \begin{bmatrix} 0 \\ 0 \\ \vdots \\ 0 \\ b_n(x_1, x_2, \dots, x_n) \end{bmatrix}. \tag{8.44}$$

Korobov (1973), [20], was the first who showed that, for these systems the first component function of $\mathbf{T}$ is simply given as $T_1(\mathbf{x}) = x_1$, and that the remaining components of $\mathbf{T}$ can be obtained from equations 8.38 and 8.39.

It is quite straightforward to observe that the one-axis model of a synchronous generator described by the state equation 8.29 is an example of a pure-feedback system and that in this case the diffeorphism $\mathbf{T}$ is given by:

$$\tilde{x}_1 = T_1(\mathbf{x}) = x_1 = \delta - \delta^o, \tag{8.45}$$

$$\tilde{x}_2 = T_2(\mathbf{x}) = a_1(\mathbf{x}), \tag{8.46}$$

$$\tilde{x}_3 = T_3(\mathbf{x}) = \omega_B a_2(\mathbf{x}), \tag{8.47}$$

$$w = T_4(\mathbf{x}, u_1) = \langle \nabla T_3, \mathbf{a(x)} + \mathbf{b}u_1 \rangle. \tag{8.48}$$

From the last equation we can calculate the linearizing state feedback

$$u_1(\mathbf{x}, w) = \left(w - \omega_B \sum_{j=1}^{3} D_j a_2(\mathbf{x}) a_j(\mathbf{x}) \right) / \beta_1 \omega_B D_3 a_2(\mathbf{x}), \tag{8.49}$$

where

$$D_1 a_2(\mathbf{x}) = r_1(e_q'^{\bullet} + x_3) \cos(\delta^o + x_1) + r_2 \cos 2(\delta^o + x_1),$$
$$D_2 a_2(\mathbf{x}) = D, \tag{8.50}$$
$$D_3 a_2(\mathbf{x}) = r_1 \sin(\delta^o + x_1).$$

It is easy to see, that the linearizing control is well-defined if the partial derivative $D_3 a_2(\mathbf{x})$ is bouned away from zero i.e. the following conditions hold:

$$V_s > 0 \quad \text{and} \quad \sin \delta \neq 0. \tag{8.51}$$

Transformation*8.45–8.48 is the simplest possible one, and was already used by Mielczarski and Zajaczkowski [28–30,32], for the classical and so called improved reduced-order model of a synchronous generator. The new state variables $\tilde{x}_1, \tilde{x}_2, \tilde{x}_3$ can be interpreted as the generator angle deviation from the desired equilibrium value δ^o, the rotor speed deviation from synchronous speed ω_s and the rotor acceleration respectively. Note, that if the rotor angle is expressed in electrical radians and time is expressed in seconds all new state variables are expressed in SI units. Furthermore, from equation 8.42 it follows that the set X in which the linearizing feedback 8.49 exists can be described as follows:

$$X = \bigcup_{k \in \mathbf{Z}} X_k, \quad X_k = \left\{ \mathbf{x} \in \Re^3 | k\pi < \delta < (k+1)\pi \right\} \tag{8.52}$$

where $\mathbf{Z}$ denotes the set of integers.

In accordance with the formulation of the feedback linearization problem of Subsection 8.2. the one-axis model 8.29 of a synchronous generator with the nonlinear state feedback 8.49 is T-equivalent to the linear system in controller canonical form [2]:

$$D_t \tilde{\mathbf{x}} = \tilde{\mathbf{A}} \tilde{\mathbf{x}} + \tilde{\mathbf{b}} w = \begin{bmatrix} 0 & 1 & 0 \\ 0 & 0 & 1 \\ 0 & 0 & 0 \end{bmatrix} \tilde{\mathbf{x}} + \begin{bmatrix} 0 \\ 0 \\ 1 \end{bmatrix} w. \tag{8.53}$$

In order to guarantee the desired dynamic characteristics of the closed-loop system, some external controller $\tilde{\mathbf{x}} \rightarrow w(\tilde{\mathbf{x}})$ should be used. The simple choice is a linear controller

$$w(\tilde{\mathbf{x}}) = \alpha^T \tilde{\mathbf{x}}, \tag{8.54}$$

where $\alpha^T = [\alpha_1 \alpha_2 \alpha_3]^T$ is the vector of constant feedback gains computed according to the Ackermann's formula for pole placement of single-input linear systems.

According to variables 8.7b the control signal e_{fD} of a synchronous generator can be expressed in the form

$$e_{fD}(\mathbf{x}) = e_{fD}^o + u_1(\mathbf{x}). \tag{8.55}$$

---

*The same transformation was independently found by Ilic and Mak (1989)

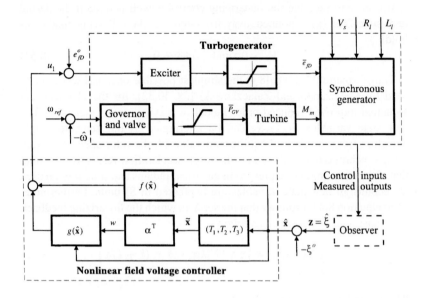

Figure 21. Turbogenerator with the nonlinear field voltage controller

The schematic diagram of the corresponding closed-loop system is shown in Figure 21. There are two feedback loops: the inner nonlinear loop responsible for feedback linearization and the outer linear loop that assigns poles s_1, s_2, s_3 of the linear system 8.53 to the desired locations in the left half of the complex plane. It is easy to notice, that there is considerable freedom in the choice of the form of the outer feedback loop. This choice depends on requirements imposed by the designer on dynamic characteristics of the closed-loop system and any controller that fulfills these requirements can be applied.

Stability of the linear closed-loop system $D_t \tilde{\mathbf{x}} = (\tilde{\mathbf{A}} + \tilde{\mathbf{b}}\alpha^T)\tilde{\mathbf{x}}$ and existence of the diffeomorphism (T_1, T_2, T_3) imply stability of the one-axis model of a synchronous generator with control feedback 8.55, providing that an initial state $\xi(0) = \xi^o + \mathbf{x}(0)$ of this model is in one of the disjoint sets X_k, $k \in \mathbf{Z}$. Thus in simulations it was assumed that

$$\xi(0) = \xi^o \in X_0 = \left\{ \xi \in \Re^3 | 0 < \delta < \pi \right\}. \qquad (8.56)$$

At this point we should admit, that the same nonlinear field voltage controller can be obtained using the input-output linearization method

[17,38]. However, we must remember that the main difficulty with application of this method lies in the problem of defining a suitable regulated output

$$y = c(\mathbf{x}), \tag{8.57}$$

for the system 8.35, where $c : X^o \to \Re$ is a smooth output function.

Recall, that the nonlinear system 8.35 with the output given by equation 8.57 has a relative degree $\rho \in \mathbf{N}$, $\mathbf{N}$ – the set of positive integers, if the following two conditions hold

IOL 1. $\mathrm{L_b}\mathrm{L_a^k}c(\mathbf{x}) = 0$ for all $\mathbf{x}$ in some neighbourhood of $\mathbf{x}(0) = \mathbf{x}^o$ and all $k < \rho - 1$,

IOL 2. $\mathrm{L_b}\mathrm{L_a^{\rho-1}}c(\mathbf{x}) \neq 0$.

If the nonlinear system has the relative degree $\rho \leq n$, then the feedback control

$$u(\mathbf{x}) = (w - \mathrm{L_a^\rho}c(\mathbf{x}))/\mathrm{L_b}\mathrm{L_a^{\rho-1}}c(\mathbf{x}) \tag{8.58}$$

yields the closed-loop system with the input-output dynamics described by a linear differential equation

$$\mathrm{D}_t^\rho y = w, \tag{8.59}$$

and is called an input-output linearizing control.

It can be shown [17], that if one has chosen the output function such that, it posses the relative degree equal to the dimension n of the system, then in the new state coordinates

$$\tilde{x}_1 = \mathrm{L_a^0}c(\mathbf{x}) = c(\mathbf{x}), \quad \tilde{x}_2 = \mathrm{L_a^1}c(\mathbf{x}), \dots \tilde{x}_n = \mathrm{L_a^{n-1}}c(\mathbf{x}) \tag{8.60}$$

the nonlinear system with the state feedback 8.58 can be described by the linear system $\mathrm{D}_t\tilde{\mathbf{x}} = \tilde{\mathbf{A}}\tilde{\mathbf{x}} + \tilde{\mathbf{b}}w$ in the controller canonical form.

If the relative degree is smaller than the dimension of the state space, then feedback 8.58 can still be applied but in this case stability of the $(n-\rho)$-dimensional zero dynamics of the closed-loop system should be considered. In the case of the one-axis model of a synchronous generator we can define as the regulated output y the control error of the rotor angle. This choice results in the output equation of the form:

$$y = c(\mathbf{x}) = x_1, \tag{8.60}$$

and using the input-output linearization method* we obtain the nonlinear feedback controller given by equations 8.49 and 8.55.

---

*Compare with our results concerning application of nonlinear field voltage controllers to a three machine power system (Zajaczkowski, *et al.*, 1995; Mielczarski, *et al.*, 1995)

It is worth mentioning here, that for this output function the one-axis model is the only one that includes transient processes in generator windings and has the relative degree equal to the dimension of the state space. If one considers higher order models with additional state equations describing transients in other generator windings and computes the relative degree associated with the regulated output 8.60 then one obtains the same result $\rho = 3$ for these higher order models including the most accurate Park's model. For example, the method of input-output linearization was applied to design a nonlinear field voltage controller for the two-axis model of a synchronous generator [30]. In this model there are two electromagnetic state variables e'_q, e'_d representing transient processes in the field and damper winding respectively (in derivation of the two-axis model only one damper winding in the quadrature axis is assumed).

It was also shown that, for this model the input-output linearizing control 8.58 yields the feedback system with the stable one-dimensional zero dynamics associated with the state variable e'_d.

The input-output linearization method can also be applied to design a state feedback for nonlinear system 8.15 with several regulated outputs y_i, $i = 1, \ldots , m$ defined by the output equations of the form:

$$y_i = c_i(\mathbf{x})$$

where c_i, $i = 1, \ldots , m$ are smooth real-valued functions defined in some open neighbourhood of the initial state $\mathbf{x}^o$.

Since we do not know *a priori* what regulated outputs should be defined for the one-machine system 8.8, we stay with the feedback linearization method in the next Subsection to design a stabilizing multivariable controller for this system.

8.1.5. Multivariable Controller for a Single Machine System

In this Subsection we consider a single machine system modelled by the state equations 8.8, 8.9–8.14 and apply the necessary and sufficient conditions for feedback linearization to obtain a linearizing feedback for this controlled plant. When dealing with multi-input systems described by the state equation 8.15, one is faced with the problem of choosing Kronecker indices k_j, $j = 1, \ldots , m$, which, in general, does not have a unique solution. In their well-known paper Hunt, Su and Meyer (1983b) [10], described a systematic method for choosing Kronecker indices for the system 8.15. This method is useful in the case of systems for which we have no additional information how to select Kronecker indices, but in the case of the single machine system the structure of this system and the results of the previous Subsection allow to choose the following Kronecker indices:

$$k_1 = 3, \quad k_2 = 2. \tag{8.61}$$

For these Kronecker indices we obtain the so called controllabity matrix $\mathbf{M}_C \in \Re^{5\times5}$ with column vectors

$$\mathbf{b}_1, [\mathbf{a}; \mathbf{b}_1], ad_{\mathbf{a}}^2\mathbf{b}_1, \mathbf{b}_2, [\mathbf{a}; \mathbf{b}_2] \in \Re^5, \tag{8.62}$$

where, as can be easily checked

$$[\mathbf{a}; \mathbf{b}_1](\mathbf{x}) = [0 - \beta_1 r_1 \sin(\delta^o + x_1) - \beta_1 r_5\, 0\, 0]^T, \tag{8.63}$$

$$\begin{aligned} ad_{\mathbf{a}}^2\mathbf{b}_1(\mathbf{x}) = [\beta_1 r_1 \omega_B \sin(\delta^o + x_1)\beta_1 r_1((D + r_5)\sin(\delta^o + x_1) \\ - \cos(\delta^o + x_1)a_1(\mathbf{x}))\beta_1 r_5^2\, 0\, \beta_1 r_1 r_{10}\sin(\delta^o + x_1)]^T, \end{aligned} \tag{8.64}$$

$$[\mathbf{a}; \mathbf{b}_2](\mathbf{x}) = [0 - \beta_2 r_4\, 0\, -\beta_2 r_8\, -\beta_2 r_9]^T, \tag{8.65}$$

The deterrminant of $\mathbf{M}_C$ can be expressed as

$$\det \mathbf{M}_C(\mathbf{x}) = \frac{\omega_B k_{GV}^2}{(L_l + L'_d)^2 \tau_{do}'^3 \tau_m^2 \tau_{RH} \tau_{GV}^2} V_s^2 \sin^2(\delta^o + x_1). \tag{8.66}$$

Providing that $V_s > 0$, $\det \mathbf{M}_C(\mathbf{x}) \neq 0$ in the union of disjoint sets X_k, $k \in \mathbf{Z}$ described by the relations $k\pi < \delta < (k + 1)\pi$, which are the same as respective relations obtained in the case of the one-axis model of the synchronous generator. In other words, vector fields $\mathbf{b}_1, [\mathbf{a}; \mathbf{b}_1], ad_{\mathbf{a}}^2\mathbf{b}_1, \mathbf{b}_2,$ $[\mathbf{a}; \mathbf{b}_2]$ span the space $\Re^5$ for $\mathbf{x} \in X_k, k \in \mathbf{Z}$. This means that, the condition FL 1 for feedback linearization of the single machine system is satisfied.

An interested reader may also check that the remaining two conditions are also satisfied in $X = \bigcup X_k, k \in \mathbf{Z}$ i.e.

FL 2. The sets of vector fields

$$C_1^I = \{\mathbf{b}_1, [\mathbf{a}; \mathbf{b}_1], \mathbf{b}_2, [\mathbf{a}; \mathbf{b}_2]\}, \quad C_2^I = \{\mathbf{b}_1, \mathbf{b}_2\} \tag{8.67}$$

are involutive in X.

FL 3. $\mathrm{span}C_1^I = \mathrm{span}C \cap C_1^I$ and $\mathrm{span}C_2^I = \mathrm{span}C \cap C_2^I$. \hfill (8.68)

For the Kronecker indices given by 8.61 we obtain the numbers

$$\sigma_1 = k_1 = 3, \quad \sigma_2 = k_1 + k_2 = 5, \tag{8.69}$$

and corresponding two systems of partial differential equations for the component functions T_1 and T_4

$$\langle \nabla T_1(\mathbf{x}), \mathbf{b}_1 \rangle = 0, \qquad \langle \nabla T_1(\mathbf{x}), \mathbf{b}_2 \rangle = 0,$$
$$\langle \nabla T_1(\mathbf{x}), [\mathbf{a}; \mathbf{b}_1](\mathbf{x}) \rangle = 0, \quad \langle \nabla T_1(\mathbf{x}), [\mathbf{a}; \mathbf{b}_2](\mathbf{x}) \rangle = 0, \qquad (8.70)$$
$$\langle \nabla T_4(\mathbf{x}), \mathbf{b}_1 \rangle = 0, \qquad \langle \nabla T_4(\mathbf{x}), \mathbf{b}_2 \rangle = 0. \qquad (8.71)$$

It is easy to notice, that

$$\tilde{x}_1 = T_1(\mathbf{x}) = x_1 = \delta - \delta^o \qquad (8.72)$$

satisfies the system 8.70, and that this component function is identical with that obtained for the case of the one-axis model of a synchronous generator (see the former Subsection). Using equation 8.25, we immediately have

$$\tilde{x}_2 = T_2(\mathbf{x}) = a_1(\mathbf{x}), \qquad (8.73)$$
$$\tilde{x}_3 = T_3(\mathbf{x}) = \omega_B a_2(\mathbf{x}). \qquad (8.74)$$

From the system 8.71 follow the conditions for partial derivatives $D_3 T_4$, $D_5 T_4$

$$\beta_1 D_3 T_4(\mathbf{x}) = 0, \quad \beta_2 D_5 T_4(\mathbf{x}) = 0.$$

These conditions allow to propose the component function T_4 in the form:

$$\tilde{x}_4 = T_4(\mathbf{x}) = k_\delta x_1 + k_{RH} x_4 = k_\delta(\delta - \delta^o) + k_{RH}(P_{RH} - P^o_{RH}), \quad (8.75)$$

where $k_\delta \geq 0$ and $k_{RH} > 0$ denote some weighing coefficients adjusted by the designer to obtain a desired behaviour of the closed-loop system.

It is obvious that, this function satisfies the partial differential equations 8.71 and that the last new state coordinate $\tilde{x}_5$ is defined by the following formula:

$$\tilde{x}_5 = T_5(\mathbf{x}) = k_\delta a_1(\mathbf{x}) + k_{RH} a_4(\mathbf{x}). \qquad (8.76)$$

Components of a linearizing feedback can be computed from the system 8.27, which in the case of the single machine system takes the form (we omit arguments for the sake of brevity):

$$\langle \nabla T_1, ad_\mathbf{a}^2 \mathbf{b}_1 \rangle u_1 + \langle \nabla T_1, ad_\mathbf{a}^2 \mathbf{b}_2 \rangle u_2 = w_1 - \langle \nabla T_3, \mathbf{a} \rangle, \qquad (8.77)$$

$$-\langle \nabla T_4, [\mathbf{a}; \mathbf{b}_1] \rangle u_1 - \langle \nabla T_4, [\mathbf{a}; \mathbf{b}_2] \rangle u_2 = w_2 - \langle \nabla T_5, \mathbf{a} \rangle, \qquad (8.78)$$

where

$$ad_\mathbf{a}^2 \mathbf{b}_2(\mathbf{x}) = [\beta_2 \omega_B r_4 \beta_2 (r_4(D + r_9) + r_3 r_8) \, 0 \, \beta_2 r_8(r_7 + r_9) \beta_2(r_4 r_{10} + r_9^2)]^T,$$

$$\langle \nabla T_1(\mathbf{x}), ad_\mathbf{a}^2 \mathbf{b}_1(\mathbf{x}) \rangle = \beta_1 \omega_B D_3 a_2(\mathbf{x}), \quad \langle \nabla T_1(\mathbf{x}), ad_\mathbf{a}^2 \mathbf{b}_2(\mathbf{x}) \rangle = \beta_1 \omega_B r_4,$$

$$\langle \nabla T_4(\mathbf{x}), [\mathbf{a}; \mathbf{b}_1](\mathbf{x}) \rangle = 0, \quad \langle \nabla T_4(\mathbf{x}), [\mathbf{a}; \mathbf{b}_2](\mathbf{x}) \rangle = -\beta_2 r_8 k_{RH},$$

$$\langle \nabla T_3(\mathbf{x}), \mathbf{a}(\mathbf{x}) \rangle = \omega_B \langle \nabla a_2(\mathbf{x}), \mathbf{a}(\mathbf{x}) \rangle = \sum_{j=1}^{5} D_j a_2(\mathbf{x}) a_j(\mathbf{x}),$$

$$\langle \nabla T_5(\mathbf{x}), \mathbf{a}(\mathbf{x}) \rangle = k_\delta \omega_B a_2(\mathbf{x}) + k_{RH}(r_7 a_4(\mathbf{x}) + r_8 a_5(\mathbf{x})),$$

$$D_1 a_2(\mathbf{x}) = r_1(e_q'^o + x_3)\cos(\delta^o + x_1) + 2r_2\cos 2(\delta^\bullet + x_1),$$

$$D_2 a_2(\mathbf{x}) = D, D_3 a_2(\mathbf{x}) = r_1\sin(\delta^o + x_1), D_4 a_2(\mathbf{x}) = r_3, D_5 a_2(\mathbf{x}) = r_4.$$

Solving equation 8.78 for unknown u_2 we obtain

$$u_2(\mathbf{x}, \mathbf{w}) = \frac{w_2 - k_\delta \omega_B a_2(\mathbf{x}) - k_{RH}(r_7 a_4(\mathbf{x}) + r_8 a_5(\mathbf{x}))}{\beta_2 r_8 k_{RH}}, \tag{8.79}$$

and taking this result into account, the first component of the linearizing feedback can be computed from equation 8.77 and expressed in the form:

$$u_1(\mathbf{x}, \mathbf{w}) = \frac{w_1 - \omega_B \langle \nabla a_2(\mathbf{x}), \mathbf{a}(\mathbf{x}) \rangle - \beta_2 \omega_B r_4 u_2(\mathbf{x}, \mathbf{w})}{\beta_1 \omega_B D_3 a_2(\mathbf{x})}, \tag{8.80}$$

Application of the feedback 8.79, 8.80 to the reduced-order model of the single machine system results in the **T**-equivalent linear system described the following state equation:

$$D_t \tilde{\mathbf{x}} = \tilde{\mathbf{A}} \tilde{\mathbf{x}} + \tilde{\mathbf{B}} \mathbf{w}, \tag{8.81}$$

where $\tilde{\mathbf{x}} = [\tilde{x}_1 \ldots \tilde{x}_5]^T$ is the state vector and

$$\tilde{\mathbf{A}} = \begin{bmatrix} 0 & 1 & 0 & 0 & 0 \\ 0 & 0 & 1 & 0 & 0 \\ 0 & 0 & 0 & 0 & 0 \\ 0 & 0 & 0 & 0 & 1 \\ 0 & 0 & 0 & 0 & 0 \end{bmatrix}, \quad \tilde{\mathbf{B}} = \begin{bmatrix} 0 & 0 \\ 0 & 0 \\ 1 & 0 \\ 0 & 0 \\ 0 & 1 \end{bmatrix}. \tag{8.82}$$

In order to achieve stability and desired dynamic behaviour of this system we apply the simple linear state feedback controller of the form:

$$\begin{bmatrix} w_1(\tilde{\mathbf{x}}) \\ w_2(\tilde{\mathbf{x}}) \end{bmatrix} = \begin{bmatrix} \alpha_1^T \tilde{\mathbf{x}} \\ \alpha_2^T \tilde{\mathbf{x}} \end{bmatrix}, \tag{8.83}$$

where

$$\alpha_1^T = [\alpha_{11} \alpha_{12} \alpha_{13} 0 0], \tag{8.84a}$$

$$\alpha_2^T = [0 0 0 \alpha_{24} \alpha_{25}] \tag{8.84b}$$

are vectors of constant feedback gains such that the closed-loop system $D_t \tilde{\mathbf{x}} = \tilde{\mathbf{A}} \tilde{\mathbf{x}} + \tilde{\mathbf{B}} \mathbf{w}(\tilde{\mathbf{x}})$ has its poles in the given positions in the complex plane.

As in the case of the one-axis model of a synchronous generator we should provide physical interpretation for the new state coordinates $\tilde{x}_i, i = 1, \ldots, 5$. Since the first three coordinates are the same as in the case of the one-axis model, they posses the same physical meaning. From the formula 8.75 with $k_{RH} = 1$ it follows that, if we define the reference signal for the dominant state variable P_{RH} of the turbine as

$$P_{RHref} = P_{RH}^o - k_\delta(\delta - \delta^o),\qquad(8.85)$$

then $\tilde{x}_4 = P_{RH} - P_{RHref}$ can be interpreted [47], as the control error of this state variable, and from this last formula we can conclude that, although, from the mathematical point of view, $k_\delta = 0$ is an acceptable choice for this weighing coefficient, it is not a reasonable choice if one wants to improve stabilization of the single machine system using the derived multivariable nonlinear controller instead of the field voltage controller of the former Subsection (see Subsection 8.6 concerning simulation results).

Similarly, after some calculations we can interpret $\tilde{x}_5$ as the signal proportional to the control error of valve position

$$\tilde{x}_5 = (P_{GV} - P_{GVref})/\tau_{RH},\qquad(8.86)$$

with the reference signal P_{GVref} expressed in the form:

$$P_{GVref} = P_{GV}^o + \tilde{x}_4 - k_\delta((\delta - \delta^o) + \tau_{RH}\omega_B(\omega - \omega_s)).\qquad(8.87)$$

The block diagram of the single machine system with the multivariable nonlinear controller is shown in Figure 22. As in the former case of the field voltage controller, two feedback loops can be distinguished: the inner nonlinear one for feedback linearization and the outer linear one which assigns poles $s_{11}, s_{12}, s_{13}, s_{24}, s_{25}$ of the system 8.81, 8.82 to the desired positions in the complex plane.

8.1.6. Simulation of a Single Machine System

To illustrate the dynamic behaviour of a single machine system equipped with the nonlinear controllers a number of simulation studies were performed. In all studies two systems were simulated: the first one, called reduced-order model (RO), with the single machine system modelled by the state equations 4.13, 4.15, 3.5, 3.8) and the second one, called full-order (FO) model, with a single machine system composed of the Park's model of a synchronous generator (see Section 4), a steam turbine model as shown in Figure 7, a steam valve and a speed controller described by state equation 3.8 and an exciter was modelled by the first order inertial element with the gain k_e

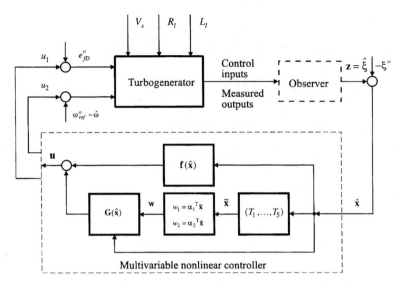

Figure 22. Turbogenerator with the multivariable nonlinear controller

and time constant τ_e. In both simulated systems the constraints 3.9, 4.17 on admissible values of the field voltage e_{fD} and steam valve position P_{GV} were included.

It was assumed that, the generator terminals were connected to a power system via a transformer and a tie line modelled by the equivalent resistance $R_l = R_{line}$ and inductance $L_l = L_{TR} + L_{line}$ - see Figure 23. In per unit system commonly used in modelling power system components, inductance L is equal reactance X so in many case these two variables are used interchangeably.

Simulations were performed for a single machine system characterized by the following nominal data [4]:

Synchronous generator

$S_N = 160$ [MVA],	$f_{sN} = 60$ [Hz],	$\omega_B = 377$ [rad/s],
$R_s = 0.00167$ [pu],	$L_d = 1.7$ [pu],	$L_q = 1.64$ [pu],
	$L_{md} = 1.55$ [pu],	$L_{mq} = 1.49$ [pu],
$R_f = 0.000742$ [pu],	$L_f = 1.65$ [pu],	$L_d' = 0.2439$ [pu],
$R_D = 0.0131$ [pu],	$L_D = 1.605$ [pu],	
$R_Q = 0.054$ [pu]	$L_Q = 1.526$ [pu],	
$\tau_m = 4.74$ [s],	$\tau_{do}' = 5.899$ [s],	

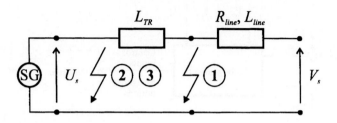

Figure 23. Locations of simulated faults

Exciter

$k_e = 1.0$ [pu],	$\tau_e = 0.02$ [s],
$e_{fD}^{Max} = 4.5$ [pu],	$e_{fD}^{Min} = -4.5$ [pu],

Transformer and tie line

$R_{TR} = 0.0$ [pu],	$L_{TR} = 0.1$ [pu],
$R_{line} = 0.02$ [pu],	$L_{line} = 0.3$ [pu],

Steam turbine

$\tau_{CH} = 0.2$ [s]	$\tau_{RH} = 8.0$ [s],	$\tau_{CO} = 0.4$ [s],
$F_{HP} = 0.3$ [pu]	$F_{IP} = 0.4$ [pu],	$F_{LP} = 0.3$ [pu],

Steam valve and governor

$k_{GV} = k = 20.0$ [pu],	$\tau_{GV} = \tau_3 = 0.1$ [s],
$P_{GV}^{Max} = 1.05$ [pu],	$P_{GV}^{Min} = 0.0$ [pu].

The operating point $\mathbf{p}^o$ of the system was determined by the following terminal conditions of the generator:

$$U_s^o = 1.0[pu], \quad P_e^o = 1.0[pu], \quad \cos\varphi^o = 0.85.$$

Using the elementary circuit theory, all remaining steady-state values of the system state and output variables can be computed.

CASE 1. The transient process resulting from a fault denoted as "1" in Figure 23 was simulated in three stages:

- the steady-state operation from 0.0 [s] to 0.1 [s]
- the fault ($V_s = 0.0$, $R_l = 0.0$, $L_l = 0.1$) at the transformer terminals from 0.1 [s] to 0.2 [s].
- post-fault stabilization ($V_s = V_s^o$, $R_l = 0.02$, $L_l = 0.4$) of the system from 0.2 [s] to 3.2 [s].

The following controlled systems were used:

1. the system with only speed controller (governor) denoted as No_C+GV
2. the system with a nonlinear field voltage controller and without the governor - NFVC
3. the system with a nonlinear field voltage controller and governor denoted as NFVC+GV
4. the system with a nonlinear multivariable controller denoted as NMC

Computations were executed with the constant gains of external controllers corresponding to the following pole locations:

$$\text{NFVC} - s_1 = s_2 = s_3 = -10,$$

$$\text{NMC} - s_{11} = s_{12} = s_{13} = -10, \; s_{21} = s_{22} = -20,$$

and the values of weighing coefficients of T_4 set to $k_\delta = 0.01$, $k_{RH} = 1.0$.

From the formulae 8.49. 8.79, 8.80 it follows, that the linearizing feedback controls depend on the coefficients $r_1 - r_3, r_5, r_6$ which are functions of parameters V_s, L_l representing disturbances occurring in the vicinity of the generator. Since these disturbances are not known, it was assumed that, these coefficient have their nominal values corresponding to the values V_s^o, L_l^o, which are assumed to be known. In other words, these coefficients were not adjusted, when a disturbance (fault) occurred.

The graphs of the rotor angle (load angle) δ [rad/s], rotor speed ω [pu], state variable e_q' [pu], terminal voltage U_s [pu], active power $P_e = u_d i_d + u_q i_q$ [pu], reactive power $Q_e = i_q u_d - i_q u_d$ [pu], electromagnetic torque $M_e = \Psi_d i_q - \Psi_q i_d$ [pu], exciter output $\bar{e}_{fD}$ [pu], mechanical torque M_m [pu], reheater output P_{RH} [pu], valve output $\overline{P}_{GV}$ [pu] and governor input $u_{GV} = \omega_{ref}$ [pu] obtained for RO-model of a single machine system are shown in Figures 24–26. The respective graphs for FO-model are shown in Figures 27–29.

Analysing the graphs presented we can notice that, in the case of the RO-model, the system with no controllers lost stability. The attachment of a speed controller suffices to preserve stability, but the system variables are poorly damped. The systems with the nonlinear controllers have much better dynamic behaviour and, as one can expect, the nonlinear multivariable controller is superior as compared with the field voltage controller.

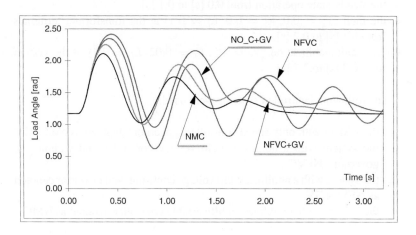

Figure 24. Load angle deviaton during fault and post-fault transient processes for the reduced model of a synchornous generator

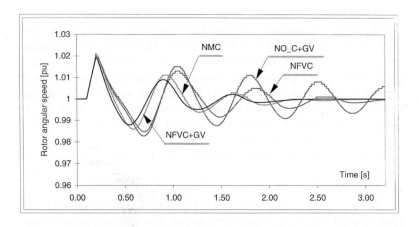

Figure 25. Rotor angular speed deviation during fault and post-fault transient processes for the reduced model of synchornous generator

It is worth mentioning here that, in the multivariable case, a control effort is shared by the two control channels and thus we observe the reduced control activity of the exciter. For the FO-model similar remarks can be formulated, but the stator and damper circuits enhance damping and produce high frequency oscillations during and after the fault.

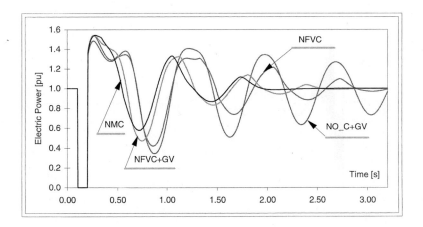

Figure 26. Output electric power deviation during fault and post-fault transient processes for the reduced model of synchornous generator

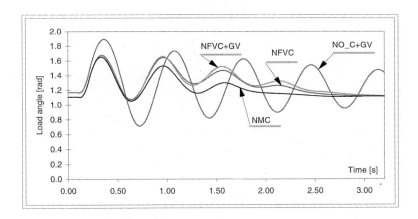

Figure 27. Load angle deviaton during fault and post-fault transient processes for the Park model of a synchornous generator

To improve the dynamics of the closed-loop systems with nonlinear controllers, one can propose shifting external controller poles more to the left. Our experience from simulations for different pole locations shows however, that such a proposal is not a very reasonable. This follows from existence of constraints on the exciter and steam valve outputs. If these constraints

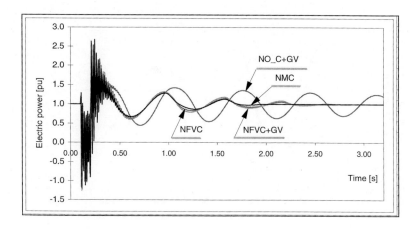

Figure 28. Output electric power deviation during fault and post-fault transient processes for the Park model of synchornous generator

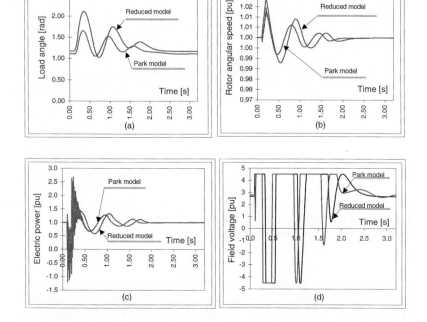

Figure 29. Fault and post-fault transient processes for two models for synchronous generators controlled by MNC.

are present, as is in the case in real turbogenerators, shifting poles to the left results in switching control inputs between upper and lower admissible values, and this generates undesirable oscillations of the turbogenerator state variables or even leads to stability loss. Hence, the pole location assumed is a compromise between modest control activity and satisfactory dynamic behaviour.

The nonlinear multivariable controller presented is the advanced version [47], of the multivariable controller formerly published [31]. The next two simulation studies were devoted to the investigation of the influence of k_δ and P_{GV}^{Max} on dynamic behaviour of the single machine system with such a controller.

In the first study the duration of the fault was the same as in the study described above, but in location **2**, Figure 23, i.e. the fault occurred directly at the generator terminals ($L_l = L_{TR} + L_{line} = 0$ during the fault). The influence of the weighing coefficient k_δ on dynamic performance of the rotor angle, terminal voltage, active power and valve output for the RO-model and FO-model is shown in Figures 30–31 and 32–33 respectively.

It can be noticed that, the system with $k_\delta = 0$ almost lost stability. This situation supplements our remarks concerning the physical interpretation of the component function T_4. It is easy to see, that zero value for this weighing coefficient causes no positive action performed via the mechanical control channel i.e. the external controller forces the mechanical torque (mechanical power) to remain unchanged, and therefore in this case we obtain the system with worse dynamic behaviour than a system where a nonlinear field voltage controller and a governor are controlled separately. For increasing values of k_δ we can observe that, system response tend to improved performance, but this tendency is limited by the influence of the aforementioned constraints and, in the case of FO-model, by inertia of a steam chest and a crossover which were neglected in the RO-model used for controller design.

From the presented graphs it follows that, for larger values of P_{GV}^{Max} we obtain better responses of the closed-loop system. This feature is particularly well demonstrated in the case of the RO-model. In the FO-model case, an additional influence of crossover and steam chest dynamics diminish the positive effects of larger values for P_{GV}^{Max}.

However, all results show that, the nonlinear multivariable controller guarantees the best dynamic performance of a single machine system, as compared to other simulated controllers.

9.1. CONTROLLING MULTIMACHINE SYSTEMS

A single machine system is controlled to stabilise transmission of electrical energy between a generator and the power system. The best performance

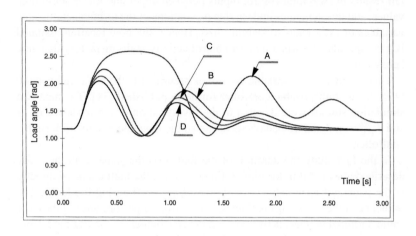

Figure 30. Fault and post-fault load angle deviation for different values of k_δ for the reduced model of a synchronous generator,
$A - k_\delta = 0.0, B - k_\delta = 0.005, C - k_\delta = 0.01, D - k_\delta = 0.015$

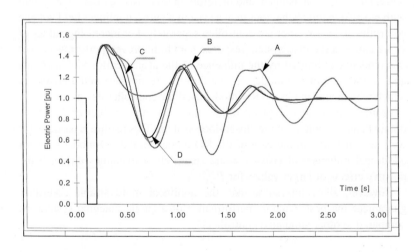

Figure 31. Fault and post-fault output electric power deviation for different values of k_δ for the reduced model of a synchronous generator,
$A - k_\delta = 0.0, B - k_\delta = 0.005, C - k_\delta = 0.01, D - k_\delta = 0.015$

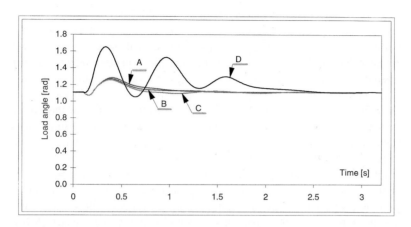

Figure 32. Fault and post-fault load angle deviation for different values of k_δ for the Park model of a synchronous generator, $A - k_\delta = 0.0$, $B - k_\delta = 0.005$, $C - k_\delta = 0.01$, $D - k_\delta = 0.015$

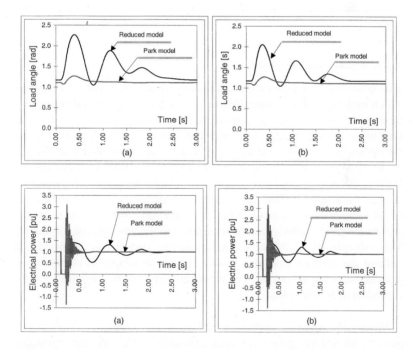

Figure 33. Fault and post-Fault transient processes for two models of a synchronous generator controlled by (a) - NFCV and (b) - NMC, $k_\delta = 0.015$

index of such a control is load angle of a generator, so naturally the control law of a single machine system is based on stabilization of load angle. In multimachine systems load angle of a generator is calculated as a difference between the axis "q" of this generator and the generator selected as the reference machine - slack generator. Design of a controller to stabilize the load angle to the reference generator and other generators in the system requires the knowledge of the state vector of all machines in the power system considered. Such a global control will require the global estimation of state variables and reliable transmission of data on line what can be possible technically but difficult to justify economically.

One of major obstacles in implementation of nonlinear optimal controllers is the difficulties in obtaining information on the vector state of other generators and determination of reference signals for controllers. An attempt has been made to apply the observation decopuled state space [6,16], originally introduced in [44], as a means of moving toward a decentralized design. See [21], where the nonlinear stream valving control is designed for multimachine systems. It leads to the use of decoupled reference rotor angles obtained by a nonlinear transformation which maps the conventional state space of the power system into a new observation decoupled state space [6].

9.1.1. Local Control Strategy

The more simple and direct approach is to derive a local control strategy in which each generator is controlled due to reference signals determining by the system control centre to achieve the optimal voltage levels and power flow in steady state conditions. If a disturbance is cleared without changing the optimal power flow conditions, generators are controlled with the use of the same reference signals before and after a fault. When a disturbance leads to the new optimal power flow, caused for example by disconnection of one of transmission lines, generators are initially controlled with the use of the same reference signals as before a fault and when the new reference signals are received the system is re-adjusted to the new optimal conditions. This sub-optimal control of the multimachine system can be introduced without the need for extensive investment into development of estimation and communication systems. The main philosophy of this approach can be expressed as follows [46]:

Having the desired steady-state values of the terminal voltage, active power and reactive power for each generator, design the corresponding non-linear field voltage controllers and then perform system stability studies with the non-linear controllers attached to different numbers of generators.

9.1.2. Steady-state Operation of a Synchronous Generator

Steady-state values of the terminal voltage and active and reactive power are used to compute desired steady-state value of the angle δ and to design non-linear field voltage controllers for each generator in the power system considered. A simple algorithm for computation of steady-state values of variables characterising the steady-state operation point $\mathbf{p}^o$ of a synchronous generator for the given terminal voltage U_s^o, active power P_e^o and reactive power Q_e^o is presented. In steady-state, the following system of equations is satisfied:

$$P_e^o = u_d^o i_d^o + u_q^o i_q^o, \tag{9.1}$$

$$Q_e^o = u_q^o i_d^o + u_d^o i_q^o, \tag{9.2}$$

$$u_d^o = - L_q i_q^o, \tag{9.3}$$

$$u_q^o = e_q'^o + L_d' i_d^o, \tag{9.4}$$

$$U_s^{o^2} = u_d^{o^2} + u_q^{o^2}. \tag{9.5}$$

Solving this system for unknowns u_d^o, u_q^o, i_d^o, i_q^o, $e_q'^o$ one obtains:

$$u_d^o = \frac{-U_s^o P_e^o}{\sqrt{P_e^{o^2} + \left(Q_e^o + U_s^{o^2}/L_q\right)^2}}, \tag{9.6}$$

$$u_q^o = \sqrt{U_s^{o^2} - u_d^{o^2}}, \tag{9.7}$$

$$i_q^o = - u_q^o/L_q \tag{9.8}$$

$$i_d^o = (P_e^o - u_q^o i_q^o)/u_d^o, \tag{9.9}$$

$$e_q'^o = u_q^o - L_d' i_d^o. \tag{9.10}$$

The remaining steady-state values of generator state and control variables are given as:

$$\omega^o = \omega_s, \tag{9.11}$$

$$\delta^o = \text{arctg}(-u_d^o/u_q^o), \tag{9.12}$$

$$e_{fD}^o = e_q'^o - (L_d - L_q') i_d^o. \tag{9.13}$$

9.1.3. Nonlinear Controller Design

According to this philosophy of multimachine system control we design nonlinear field voltage controllers for selected machines in the system. In the case of a multimachine system the reduced-order model of a generator does not include parameters of a tie line, and with this assumption we obtain the following formulae for the nonlinear field voltage controller:

$$e_{fD}(\mathbf{x}) = e_{fD}^o + u_1(\mathbf{x}, w) = \left(w - \omega_B \sum_{j=1}^{3} D_j a_2(\mathbf{x}) \right) / \beta_1 \omega_B D_3 a_2(\mathbf{x}),$$
$$\tag{9.14}$$

$$
\begin{aligned}
D_1 a_2(\mathbf{x}) &= r_1(e_q'^o + x_3)\cos(\delta^o + x_1) + r_2\cos 2(\delta^o + x_1), \\
D_2 a_2(\mathbf{x}) &= D, \\
D_3 a_2(\mathbf{x}) &= r_1\sin(\delta^o + x_1),
\end{aligned}
\tag{9.15}
$$

$$x_1 = \delta - \delta^o, \quad x_2 = \omega - \omega_s, \quad x_3 = e_q' - e_q'^o, \tag{9.16}$$

$$a_1(\mathbf{x}) = \omega_B x_2, \tag{9.17}$$
$$a_2(\mathbf{x}) = D_{x_2} + r_1(e_q'^o + x_3)\sin(\delta^o + x_1) + r_2\sin 2(\delta^o + x_1) + M_m^o/\tau_m, \tag{9.18}$$
$$a_3(\mathbf{x}) = r_5 x_3 + r_6\cos(\delta^o + x_1) + r_5 e_q'^o + \beta_1 e_{fD}^o, \tag{9.19}$$

$$r_1 = -\frac{U_s^o}{\tau_m L_d'}, \quad r_2 = -\frac{1}{\tau_m}\frac{1}{2}\frac{(L_d' - L_q)U_s^{o^2}}{L_d' L_q}, \tag{9.20}$$

$$r_5 = -\frac{1}{\tau_{do}'}\frac{L_d}{L_d'}, \quad r_6 = \frac{1}{\tau_{do}'}\frac{L_d - L_d'}{L_d'}U_s^o. \tag{9.21}$$

As in the case of the single machine system, a simple linear external controller

$$w(\tilde{\mathbf{x}}) = \alpha^T \tilde{\mathbf{x}}, \tag{9.22}$$

where

$$\tilde{x}_1 = x_1 = \delta - \delta^o, \quad \tilde{x}_2 = a_1(\mathbf{x}), \quad \tilde{x}_3 = \omega_B a_2(\mathbf{x}), \tag{9.23}$$

and $\alpha^T = [\alpha_1 \alpha_2 \alpha_3]^T$ is the vector of constant feedback gains responsible for pole placement of the equivalent linear system $D_t\tilde{\mathbf{x}} = \tilde{\mathbf{A}}\tilde{\mathbf{x}} + \tilde{\mathbf{b}}w$ in controller canonical form to the desired positions s_1, s_2, s_3 in the complex plane.

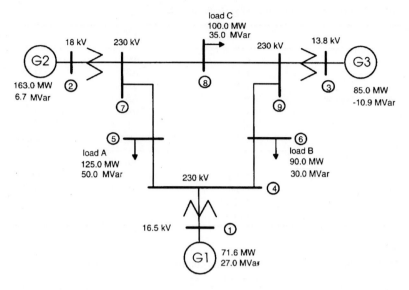

Figure 34. One line diagram of the three-machine power system

9.1.4. Three Machine System Data

The schematic diagram of the three machine, nine bus system, [4], chosen for stability studies is shown in Figure 34 and the nominal data for the generators are listed in Table 1. Parameters of excitation systems and PSS are shown in Table 2. Steady state conditions are presented in Table 3.

Table 1. Parameters of generators.

Generator	No.1	No.2	No.3
Rated [MVA]	247.5	192.0	128.0
Terminal voltage [kV]	16.5	18.0	13.0
Power factor	1.0	0.85	0.85
Type	hydro	steam	steam
Frequency [Hz]	60	60	60
L_d [pu]	0.1460	0.8958	1.1325
L_q [pu]	0.0969	0.8645	1.2578
L'_d [pu]	0.0608	0.1198	0.1813
L'_q [pu]	0.0969	0.1969	0.25
L_σ [pu]	0.0336	0.0521	0.0742
τ'_{do} [s]	8.96	6.0	5.89
τ'_{qo} [s]	0.0	0.535	0.6
τ_m [s]	47.28	12.8	6.02
D	0.0	0.0	0.0

Table 2. Parameters of excitation systems and PSS's.

	Generator	1	2
	k_a	480	480
	τ_a [s]	0.35	0.35
	k_f	0.65	0.65
Excitation Systems	τ_f [s]	4.5	4.5
	τ_r [s]	0.02	0.02
	e_{fD}^{Max}	3.5	3.5
	e_{fD}^{Min}	−3.5	−3.5
	k	50	30
	τ_1 [s]	0.1815	0.1482
PSS's	τ_2 [s]	0.0825	0.0494
	τ_w [s]	5	5
	τ_{lp} [s]	0.08	0.08

Table 3. Steady-state values of state and control variables of generators.

Generator	1	2	3
Angle δ [rad]	0.0625	0.904	0.864
Speed ω [pu]	1.0	1.0	1.0
EMF e_q' [pu]	1.056	0.788	0.768
Field voltage e_{fD} [pu]	1.082	1.79	1.403
Turbine torque M_m [pu]	0.716	1.63	0.85

It has been assumed that the values of the field voltage belong to the closed interval:

$$[e_{fD}^{Min}, e_{fD}^{Max}] = [-3.5, 3.5]$$

During the simulation studies, the generator No.1 was used as the reference machine as it can be considered as the representation of the remainder of the power system. Hence, this generator was not equipped with a controller.

There are some important factors which can have significant influence on system stability with non-linear field voltage controllers:

- Duration and location of the fault;
- Number of generators equipped with NFVC's;
- Location of poles of the external controllers.

To obtain conclusions concerning these factors, simulation studies have been carried out using the CYMSTAB^{TM} software.

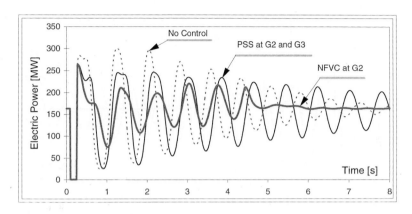

Figure 35. 10 cycle fault at bus 7. Electrical power output of G2

The first simulation study was performed for the case of a 10 cycle = 166 [ms] fault applied at the bus No. 7, which is very close to the terminals of generator No. 2. The simulation was carried out for the following different control structures of the system:

1. System without controllers - constant field voltages
2. System with the power system stabilizer (PSS) at generators No. 2 and No. 3
3. System with the NFVC attached to the generator No. 2

With the poles of the external controller assigned at $s_1, s_2, s_3 = -10$, this corresponds to the following values of feedback gains: $\alpha_1 = -1000, \alpha_2 = -300, \alpha_3 = -30$. The results of this simulation study are presented in Figure 35 showing the electrical power output of machine No. 2.

The response of the system without control clearly shows a very lightly damped oscillation. The addition of the PSS's on machines No. 2 and No. 3 improves the damping. However, superior damping is achieved by the addition of the NFVC on machine No. 2.

The second simulation study was performed for the case of the 10 cycle = 166 [ms] fault applied at the same bus as in the previous case, but with an additional NFVC on the generator No. 3.

Therefore, the structures of the respective systems were as follows:

1. System without controllers - constant field voltages;

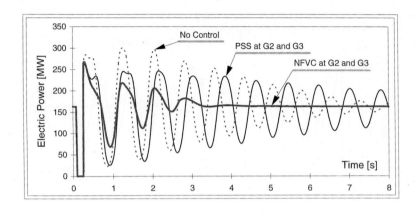

Figure 36. 10 cycle fault at bus 7. Electrical power output of G2

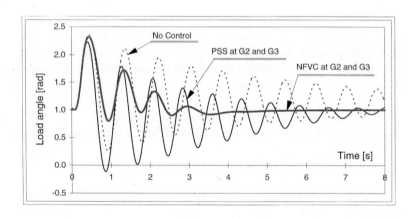

Figure 37. 10 cycle fault at bus 7. Load angle of G2

2. System with the power system stabilizers (PSS) at generator No. 2 and No. 3;

3. System with the NFVC's attached to the generator No. 2 and No. 3.

The poles of external controllers were the same for both generators: $s_1, s_2, s_3 = -10$, which corresponds to the following values of feedback gains: $\alpha_1 = -1000, \alpha_2 = -300, \alpha_3 = -30$. The results of this simulation study are presented in Figure 36 and Figure 37.

In all cases it has been observed that the systems with NFVC's possess the best dynamic performance, especially as compared with those with PSS's. The NFVC performed well even when the fault location was not at the machine with NFVC added [33].

The second simulation study undertaken for the New England 39 bus power system has shown better performance of nonlinear controllers in comparison to the PSS supported controller. It has also shown the robustness of the proposed controller. It is robust to structure and parameter deviations of the controlled system. In the study undertaken, the nonlinear controller designed for the three dimensional model of a generator was able to stablize machines simulated as four-dimensional models $[\delta, \omega, E'_q, E'_d]$ effectively. The nonlinear controller can also achieve rapid stabilization of the power system, independently of fault location [34].

10.1. CONCLUSION

An application of feedback linearization to power system control leads to a family of modern nonlinear controllers that can improve performance of synchronous generator. The proposed nonlinear controllers have simple structures which can be easily implemented with the use of digital or analog techniques. The new method proposed does not require additional communication networks or measurements. Nonlinear controllers can be implemented using existing input and output signals without the need for control system reconstruction. For mulitimachine systems, an external controller, which will be able to compensate differences between current signals on optimal power flow and the future signals worked out for a new network configuration, is required to enhance the performance of the controlled system.

ACKNOWLEDGEMENT

The research leading to control of multimachine systems has been supported by Australian Electricity Supply Industry Research Board, (AESIRB) and Energy Research and Development Corporation, (ERDEC) and has been carried out in collaboration with Dr M F Conlon, Project Supervisor and Dr O Roziman, Research Assistant. The authors wish to thank Dr Conlon and Roizman for their involvement in this work.

REFERENCES

1. Abraham, R., Marsden, J.E. and Ratiu, T., 1983, *Manifolds, tensor analysis and applications*, Addison-Wesley, Reading, Mass.

2. Ackermann, J., 1985, *Sampled-data control systems*, Springer, Berlin.
3. Ahmed-Zaid, S., Sauer, P.W., Pai, M.A. and Sarioglu, M.K., 1982, Reduced order modeling of synchronous machines using singular perturbation. *IEEE Trans. Circuits and Systems*, **CAS-29**, 782–786.
4. Anderson, P.M. and Fouad, A.A., 1994, Power system control and stability. *IEEE Press*, New York.
5. Anderson, J.H. and Raina, V.M., 1974, Power system excitation and governor design using optimal control theory. *Int. J. Control*, **19**, 289–308.
6. Chapman, J.W., *et al.*, 1993, Stabilizing a multimachine power system via decentralized feedback linearizing excitation control. *IEEE Trans. Power Systems*, **8**, 830–839.
7. Doyle, J.C. and Stein, G., 1979, Robustness with observers. *IEEE Trans. Automat. Contr*, **AC-24**, 607–611.
8. Gao, L., Chen, L., Fan, Y. and Ma, H., 1992, A nonlinear control design for power systems. *Automatica*, **28**, 975–979.
9. Harley, R.G., Lahoud, M.A. and Secker, A., 1986, Optimal and multivariable control of a turbogenerator. *Electric Power Systems Research*, **10**, 35–46.
10. Hunt, L., Su, R. and Meyer, G., 1983a, Global transformations of non-linear systems. *IEEE Trans. Automat. Contr.*, **AC-28**, 24–30.
11. Hunt, L., Su, R. and Meyer, G., 1983b, Design of multi-input nonlinear systems, in Brockett, R. and Millman, R. (eds), *Differential Geometric Control Theory*, Birkhauser, Boston.
12. IEEE Committee, 1968, Computer representation of excitation systems. *IEEE Trans. Power Appar. Syst.*, **PAS-87**, 1460–1468.
13. IEEE Committee, 1973, Dynamic models for steam and hydro turbines in power system studies. *IEEE Trans. Power Appar. Syst.*, **PAS-92**, 1904–1915.
14. IEEE Committee, 1981, Excitation system models for power system stability studies. *IEEE Trans. Power Appar. Syst.*, **PAS-100**, 494–509.
15. Ilic, M. and Mak, F.K., 1989, A new class of fast nonlinear voltage controllers and their impact on improved transmission capacity. In *Proc. of the American Control Conf.*, Pittsburgh, PA, 1246–1251.
16. Ilic, M.D. and Chapman, J.W., 1992, "Damping Mulitmachine Power System Oscillations with Feedback Linearizing Excitation Control", American Control Conference, Chicago, 1734–1735.
17. Isidori, A., 1989, *Nonlinear Control Systems*. Springer, Berlin, second edition.
18. Jakubczyk, B. and Respondek, W., 1980, On linearization of control systems. *Bull. Acad. Polon. Sci. Ser. Sci. Math. Astronom. Phys.*, **28**, 517–522.
19. Kokotovic, P.V. and Sauer, P.W., 1989, Integral manifold as a tool for reduced-order modeling of nonlinear systems: a synchronous machine case study. *IEEE Trans. Circuits and Systems*, **36**, 403–410.
20. Korobov, W., 1974, Controllability, stability of some nonlinear systems. *Differencialnyje uravnienija*, **IX**, 614–619 (in Russian).
21. Lu, Q. and Sun, Y.Z., 1989, Nonlinear stabilizing control of multimachine systems. *IEEE Trans. Power Systems*, **4**, 236–241.
22. Marino, R., 1984, An example of nonlinear regulator. *IEEE Trans. Automat. Contr.*, **AC-29**, 276–279.
23. Mielczarski, W., 1987a, Observing the state of a synchronous generator, Part 1. *Int. J. Control*, **45**, 987–1000.
24. Mielczarski, W., 1987b, Observing the state of a synchronous generator, Part 2. *Int. J Control*, **45**, 1001–1021.
25. Mielczarski, W., 1988a, Very fast linear and non-linear observers, Part 1. *Int. J. Control*, **48**, 1819–1831.
26. Mielczarski, W., 1988b, Very fast linear and non-linear observers, Part 2. *Int. J. Control*, **48**, 1833–1842.
27. Mielczarski, W. and Zajaczkowski, A.M., 1987, Robust optimal control of the synchronous generator. In *Preprints of 10-th World IFAC Congress*, Munich, FRG, **2**, 72–78.

28. Mielczarski, W. and Zajaczkowski, A.M., 1989, Nonlinear controller for a synchronous generator. *IFAC Nonlinear System Design Symposium 89*, Capri, Italy.
29. Mielczarski, W. and Zajaczkowski, A.M., 1990, Nonlinear stabilization of a synchronous generator. In *Preprints of 11-th World IFAC Congress*, Tallin, Soviet Union, **6**, 118–122.
30. Mielczarski, W. and Zajaczkowski, A.M., 1991, Design of a field voltage controller for a synchronous generator using feedback linearization. *Optimal Control A&M*, **12**, 73–88.
31. Mielczarski, W. and Zajaczkowski, A.M., 1994a, Multivariable nonlinear controller of a synchronous generator. *Optimal Control A&M*, **15**, 49–65.
32. Mielczarski, W. and Zajaczkowski, A.M., 1994b, Nonlinear field voltage control of a synchronous generator using feedback linearization. *Automatica*, **30**, 1625–1630.
33. Mielczarski, W., Zajaczkowski, A.M., Conlon, M.F. and Roizman, O., "Nonlinear Controllers for Three Machine Power Systems", *Proceeding of CONTROL'95*, Melbourne, 23–25 October 1995, 581–533.
34. Mielczarski, W., Zajaczkowski, A.M., Conlon, M.F. and Roizman, O., "Stabilizing a Multimachine System via Nonlinear Voltage Controller with the Use of a Loacal Control Strategy", to be published in *Optimal Control A&M*.
35. Okada, T., Kihara, M. and Furihata, H., 1985, Robust control system with observer. *Int. J. Control*, **41**, 1207–1219.
36. Respondek, W., 1985, Geometric methods in linearization of control systems. In *Math. Control Theory Banach Center Publ.*, **14**, 453–467.
37. Singh, S.N., 1980, Nonlinear state-variable-feedback excitation and governor-control design using decoupling theory. *IEEE Proceedings*, **127**, Pt.D, 131–141.
38. Slotine, J-J.E. and Li, W., 1991, *Applied nonlinear control*, Prentice-Hall, Englewood Cliffs, NJ.
39. Su, R., 1982, On linear equivalents of nonlinear systems. *Systems & Control Letters*, **2**, 48–52.
40. Su, R. and Hunt, L.R., 1986, A canonical expansion for nonlinear systems. *IEEE Trans. Automat. Contr.*, **AC-31**, 670–673.
41. Subbarao, G.V. and Iyer, A., 1993, Nonlinear excitation and governor control using variable structures. *Int. J. Control*, **57**, 1325–1342.
42. Wang, Y., Hill, D., Middleton, R.H. and Gao, L., 1993, Transient stability enhancement and voltage regulation of power systems. *IEEE Trans. Power Systems*, **PWRS-8**, 620–627.
43. Yu, Y., 1983, *Electric power system dynamics*, Academic Press, New York.
44. Zaborszky, J. and Whang, K.W., 1981, "Local feedback stabilization of large interconnected power systems in emergencies", *Automatica*, **17**, 673–686.
45. Zajaczkowski, A.M. and Mielczarski, W., 1993, Decoupling control of synchronous generators. In *Proc. First European Conference on Power Systems Transients*, Lisbon, Portugal, 92–99.
46. Zajaczkowski, A.M., Mielczarski, W., Conlon, M.F. and Roizman, O., 1995, Application of non-linear field voltage controllers to a three-machine power system. *Research report no 1995-R-06*, Department of Electrical and Computer Systems Engineering, Monash University, Victoria, Australia.
47. Zajaczkowski, A.M., 1996. Multivariable nonlinear control of a synchronous generator. In *Proc. 19-th Seminar on Fundamentals of Electrotechnics and Circuit Theory, SPETO'96*, 15–18, 05, Ustron, Poland.

2 MECHATRONIC TECHNIQUES IN BOILER TURBINE GENERATOR CONTROL

E. SWIDENBANK, M.D. BROWN, D. FLYNN, G.W. IRWIN and B.W. HOGG

Department of Electrical and Electronic Engineering, Ashby Building, The Queen's University of Belfast, BT9 5AH, UK

2.1. INTRODUCTION

Reduced cost of computer hardware, and the availability of a software literate work force have led to significant increases in the use of digital control in many manufacturing and process industries. As confidence is gained in the reliability of embedded computer systems, engineers and managers are beginning to realise the potential for better performance through advanced control. Conventional control technology cannot cope well with changing plant conditions and dynamics. PID controllers must therefore be constantly re-tuned to provide good control at a nominal setpoint. Novel control strategies can deal with plant non-linearity, and ageing of plant components, automatically. This is carried out by a learning process, whereby the controller detects changes in plant operating conditions, and updates the control appropriately. Self-tuning control consists of two tasks; selection of feedback gains, and modelling the plant. For computational simplicity, initial studies have been made using linear modelling techniques. The use of non-linear models is now being considered, together with artificial intelligence techniques for modelling and control.

Great advances are now being made in the application of computers to control and data monitoring of power plant. Power generation has tended to lag other process industries in this respect, mainly due to the emphasis on reliability, with a lower priority being given to efficiency and commercial performance. Environmental and commercial constraints, however, are forcing generator utilities to improve existing plant control performance.

The generation of electricity is a complex process, which has a unique disadvantage in relation to most other energy sources. Electricity cannot be readily stored, and it falls to the control engineer to ensure the balance between output and demand. Many regulations and constraints are imposed in the generation of electrical power, and hundreds of control loops are used within the plant to ensure safety and stability. Although the majority of these loops are of the logic type, others require continuous control for desired stability and safety. In fossil fuel power plant, the system is usually broken into sub-components for the purpose of controller design. Often, steam raising, prime mover, and power generation are treated separately, with controllers designed around each subsystem. In practice, a power generation unit is a highly complex interacting system. The problem of controller design is also compounded by the fact that the system is highly non-linear, and is dependant upon generated load, as well as plant component conditions.

These factors have led to the investigation of advanced control strategies for power plant equipment. The research team within The Department of Electrical and Electronic Engineering at The Queen's University of Belfast, have been working in power plant modelling and control for many years. Links have been established with generating utilities, as well as manufacturers of power generation equipment. The work is therefore directly relevant to the application area, with implementation and technology transfer issues are always being considered.

The control systems which most directly affect the power transmission system are the Governor and Automatic Voltage Regulator. It has been found that these have a considerable effect on transient stability of the generator, and also improve network stability. The following chapter describes the work carried out, and results obtained in the application of advanced control for governor and AVR functions.

Since deregulation and competition, power utilities are beginning to focus more on performance and profit. Many older plants have been closed, and those remaining are frequently being operated under load cycling conditions. Existing control systems were designed for stability at rated temperature, pressure and load. Heat rate, regulation, and low-load operation was not considered as primary objectives.

Economical, technological, and environmental constraints have also changed over the last few years. Emissions regulations are also set to become more stringent. A significant proportion of fossil fuel power plant is reaching the scheduled lifetime limit, and there is a desire to extend the life of these units by reducing component stress.

In a thermal power plant, fuel constitutes about 85% of the total running cost. As a typical example, if a 500 MW generating unit is continuously operated at an efficiency of half a percentage point below optimum, the extra fuel bill be £1 million a year [1]. Thus it is worth going to considerable effort and expense to reduce the losses to a practical minimum. Electric Power Research Institute (EPRI) studies have shown that typically a 1% to 5% improvement in heat rate of existing fossil power plant units is feasible.

The introduction of improved control and instrumentation within digital systems has brought tighter control of critical variables leading to improved plant efficiency and extension of plant life. Digital loop control can be continuously adapted to compensate for non-linearity and plant ageing. Improved instrumentation and signal processing enable critical variables to be maintained close to constraint limits. It is now established, however, that the most significant improvements in control of power plant are in the area of supervisory control, and management and operator information systems. Distributed control systems (DCS) provide rich information on the status of the plant, at a central accessible location.

The remainder of this section describes the main components of a boiler-turbine-generator. Current methods of control are shown, and recent advances highlighted.

Section 2 of this chapter describes single phase and three phase simulation of the turbine / generator system.

Section 3 introduces self-tuning control as a solution to non-linear AVR / Governor control. Single input / single output control is described, followed by experiences with multi input / output systems. Details of practical robust implementation of the controllers is considered.

Section 4 outlines a facility for testing proposed control schemes on a real generator. This introduces many implementation problems, such as computational loads, and noisy measurements.

The application of self-tuning control within a commercial industrial hardware is covered in section 5.

Section 6 describes recent breakthroughs using neural network techniques for boiler modelling, and deals with the particular problems of network training and topology, together with a practical approach to gathering training data from live plant

2.1.1. The Turbo-generator

The analysis in this chapter assumes that the boiler provides steam at a constant temperature and pressure to the high pressure main control valves

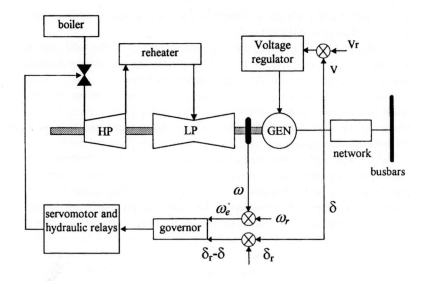

Figure 1. Turbo-generator control.

of the turbine. This is not the case however, as variations in valve main steam valve position will have an effect on the boiler which has a finite energy "inertia". For transient studies however, the stored energy within superheater pipes will justify this assumption.

Fundamental controls for the turbine / generator are shown in Figure 1. Current practice is to treat the excitation voltage control, and speed / load control as separate systems. These are however strongly coupled, and advantages in integrating control may be gained. This is discussed in section 3.

2.1.1.1. Governor control systems

The main functions of a governing system are [2]:

- To contain the speed rise within acceptable limits, should the unit become disconnected from the load.
- To control generated load from demanded setpoints.
- Control of the initial run-up and synchronisation.
- Assist in matching power generated to demanded power, in response to network frequency changes.

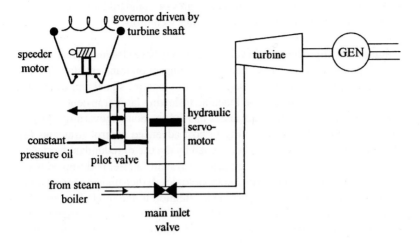

Figure 2. Mechanical Turbine Governor System.

The frequency of the power system continuously varies, as generated power rarely matches exactly load demand. Each turbine on the system therefore has a speed governor to regulate overall system frequency. Initially, a mechanical system was used. This simple feedback mechanism was a Watt centrifugal governor, where two weights move radially outwards as their speed of rotation increases. A sleeve on the central spindle moves a pilot valve piston, which in turn operates the servo-motor (Figure 2).

This technique of control exhibited significant dead band due to friction and mechanical backlash.

Developments in governor technology have seen mechanical systems replaced with fast acting electro-hydraulic systems. Speed measurement is now made electrically via a toothed wheel on the turbine shaft. The control function was initially carried out by analogue electronics, with an electro-hydraulic conversion process. Conventional PID control technology was used, and significant improvements in response were gained.

Digital control offers a variety of benefits [3]. As with any microprocessor application, the primary advantages are versatility using standard hardware components. In turbo-generator control, the use of microprocessors enables many other benefits some of which are:

- Provision of performance information.
- Simplicity in tuning controller gains.
- Eradication of thermally induced drift.

- Application of advanced control strategies on standard hardware.
- Implementation of descrete supplementary controls.

Significant improvements in valve technology have also made. Fast closure and re-opening of valves have had a major impact on transient stability. The technique of Fast Valving as a supplementary control signal can enable the retention of synchronism of the machine following large transients.

The culmination of these advances has led to robust, governing systems which have the inherent flexibility to host advanced control strategies [4]. The concern for reliability has been eased by the use of redundant hardware. Software, often blamed as the culprit for system failure, is designed using recognised software engineering methodologies. It then undergoes a rigorous dynamic test, to ensure reliability. This attention to the reliability issue has now broken down the scepticism of digital control of power generating equipment which is now widely accepted. The doors to advanced control have thus been opened.

2.1.2. Excitation of Synchronous Machines

The first generation of excitation systems were of the DC type. These were used as main and pilot exciters, and had the capability of acting as an amplifier. The DC exciter suffered form brushgear and commutator problems, and these systems essentially became obsolete following the advances and availability of high current semi-conductor devices. Rectified AC exciters in static or brushless form, now provide the fast response times required for power system transient stability (Figure 3). Static excitation systems, which supply the main generator field via sliprings, are usually specified for generators requiring high responses.

The generator terminal voltage is compared with a setpoint. The error signal is acted on by the AVR. Other main functions are provided, namely, MVAr limitation, and overflux limiting.

The advantages of microprocessor control that have been outlined in section 2.1.1.1 apply equally to AVR functions. Triplex systems are currently available, providing a high degree of functionality with reliability [5].

2.1.3. Boiler Systems

The basic function of a boiler employed in a power station is that it transforms water into superheated steam by absorbing energy from fuel burned.

Figure 4 shows the steam raising process for a fossil fuel boiler. The heating energy is released by burning oil or coal. The Second Law of

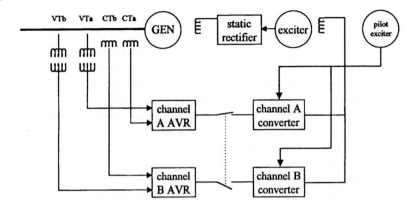

Figure 3. Dual channel AVR.

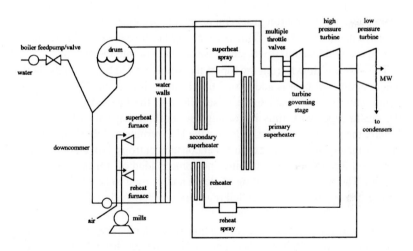

Figure 4. Boiler steam cycle.

Thermodynamics states that heat transfer will occur of its own accord down a temperature gradient, as a natural phenomenon. Hence the heat released from the burning fuel is absorbed by the water in the walls of the boiler. The heat flows by virtue of a temperature difference between the hot source (the burning fuel) and a cold sink (the cooler water in the walls of the boiler).

2.1.3.1. Steam flow

The process of generating steam begins with water pumped from the condenser, where it is stored slightly above atmospheric pressure at 30°C approximately. This pressure is maintained to prevent the contamination of the feed water, which would result from leaks in the pipework.

The feed water from the condenser is pumped through a series of feed heaters which raise the temperature to approximately 250°C before it enters the economiser in the boiler. The pressure of the feed water has also been raised to about 220 bar due to the effects of the various boiler feed pumps. The temperature of feed water continues to rise as it passes through the economiser into the drum.

The role of the drum is to separate the steam from the water. Due to the natural circulation set up in the boiler, the cooler water from the bottom of the drum, is forced along the downcomers into the base of the boiler. Here the water begins to absorb heat from the furnace. As the temperature of the water increases it rises up the tubes in the side walls of the boiler, and emerges at the top as a steam-water mix which is fed back into the drum. Inside the drum the steam goes through several devices to separate off the moisture and improve the dryness factor. This is necessary because the rotor blades in the can be easily damaged, if wet steam is carried over into the turbine.

The steam is drawn from the top of the drum and fed back into the boiler where additional heat is added so 'superheating' the steam. This allows the temperature of the steam to be raised close to the metallurgical limits of the steam pipe work thus increasing the working capacity of the steam. Finally the steam leaves the boiler for the H.P. turbine at 450°C and 160 bar approximately, where it is expanded at constant entropy. Not all the steam is expanded through the H.P. turbine and some is bled off for the feedheating system. The process of bleeding of steam and using it to heat the feed water before it enters the boiler is called regenerative feedheating, and it increases the thermal efficiency of the plant.

The steam returns to the boiler for a final time after leaving the H.P. turbine. The temperature of the steam is increased as it passes through the reheaters, which feed the I.P. then the L.P. turbines, and finally the steam returns to the condenser where it is turned back into liquid, by the cooling effect of the sea water.

2.1.3.4. Airflow

As part of the combustion process the boiler also requires a large supply of air which will feed the burning fuel with the necessary oxygen to maintain combustion.

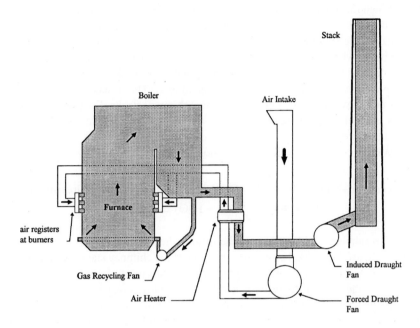

Figure 5. Simplified boiler air flow diagram.

Typically, 8 kg of air is required to burn 1kg of fuel, assuming the combustible component of the fuel is 80%. This works out at a massive 650 m^3 of air required to burn 1 kg of fuel. For an average 200 MW generating unit the boiler will require a combustion air supply of 105 kg/s. It is easy to see from these figures that the plant required to move these quantities of air will be large and slow moving. Hence it is usually the air supply loop which is the limiting factor, as to how quickly the boiler can respond to a load change.

Illustrated in Figure 5 is the flow of air through the boiler. The air is initially sucked in from atmosphere by large forced draught fans, which force it out through air heaters and into air register in the boiler. Here the air mixes with the fuel and is fed into the furnace.

The hot combustion gases rise up and transfer heat to the steam pipe work at the top of the boiler. The gas continues along its exhaust route through the air heaters, and on to the main chimney stack, where it is cleaned before being released to the atmosphere. The exhaust gas is forced out by the use of large induced draught fans, which maintains a lower than atmospheric pressure in the exhaust ductwork. This is to reduce the risk of harmful exhaust gases being released into the power station through leaks in the exhaust ductwork.

When the amount of air supplied matches exactly the oxygen requirement for the fuel burned, to support complete combustion, that ratio of air to fuel is called the stoichiometric ratio. Normally the power station operators run the boiler with approximately 1% excess O_2. However, this can be difficult to maintain as fuel supplies usually block up with age and so running the boiler with excessive amounts of O_2 is not uncommon. This is easily seen by the plumes of white smoke coming from the chimney stacks. Environmentally this is much more damaging than the unsightly black smoke which is the result of excess fuel.

With excess O_2 the following reaction takes place:

Excess Air + Sulphur Dioxide $\Rightarrow$ Sulphur Trioxide + Water $\Rightarrow$ Sulphuric Acid

$$O_2 + SO_2 \Rightarrow 2(SO_3 + H_2O) \Rightarrow 2(H_2SO_4)$$

The sulphuric acid produced is harmful to the power station as it corrodes the exhaust ductwork, but when carried out of the chimney stack it is a major contributor to the acid rain problem, which affects most of northern Europe.

As in the case of the turbo-generator, the control loops are designed and tuned independently. The various sub-systems, however, are strongly cross-coupled. Compounding this problem is the introduction of new operation regimes, in which plant that was designed for steady operation, is being load-cycled, or even two shifted (removed from the grid when demand is low). This imposes new demands on control of the process. The system is also highly non-linear, and conventional control does not remain optimal throughout the range of operation. These are obvious areas for application of advanced control techniques.

2.2. 3-PHASE AND PARK'S *D-Q* TURBOGENERATOR SIMULATIONS

2.2.1. Introduction

In most power system studies, the turbogenerator is usually represented by a voltage source, E, behind a synchronous reactance, X_S [6]. This over-simplified model, however, takes no account of such things as saliency, changes in flux linkages and the effect of damper windings. To afford a detailed study of real power systems and more particularly for the purposes of individual unit controller design, a much more complex description is therefore required. This usually takes the form of a set of non-linear equations for the generator, exciter and transmission systems, and also for the boiler and turbine [7].

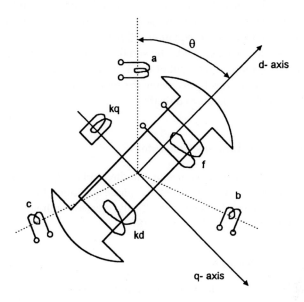

Figure 6. Idealized synchronous machine.

If one wishes to thoroughly test any measurement system that will eventually be implemented on a real machine, then the availability of a three-phase simulation is a desirable if not essential asset. For this reason, a simulation program of a three-phase synchronous generator in direct-phase quantities was developed. This program will be used to test the measurement system which is outlined in section four, and will provide a means to investigate the robustness of the scheme under virtually any conditions. Coupled with a suitable measurement program, this simulation system can then interact with a particular control program, and as such can reproduce the action of the entire controller implementation without having to resort to lengthy tests on a physical machine.

The conventional method for writing turbogenerator system equations is based on Park's d-q reference frame. Such a representation is unsuitable for unbalanced operating conditions, however, since the equations then require further transformation. The d-q model was originally formulated due to the excessive amount of computation involved in a 3-phase simulation. Modern computers are much more powerful, and with the possibilities of parallel processing, simulation of a generator in direct phase quantities is a feasible possibility.

The development of both 3-phase and d-q simulation are outlined. The main difference between the two environments clearly lies in the complexity of the synchronous generator equations. However, a common turbine and governor model has been adopted. The simulation outputs only differ in the manner of presentation. The output of the 3-phase simulation is in the form of voltage and current waveforms, while the Park's d-q approach calculates the electrical terminal quantities directly.

2.2.2. 3-Phase Synchronous Generator

A simulation of a synchronous machine in direct-phase quantities is developed. As the equations are formulated in 3-phase quantities, both unbalanced and balanced fault conditions can be easily created, without recourse to complex mathematical transformations [8].

2.2.2.1. Electrical relationships

An idealized synchronous machine, comprising damper windings, is depicted in Figure 6 [9].

In per unit quantities, the relationship between the applied voltage, $\mathbf{V}$, and the input current, $\mathbf{I}$, for a synchronous machine can be expressed as [10],

$$\mathbf{V} = -\mathbf{RI} + \frac{d\Psi}{dt} \tag{2.1}$$

where,

$$\mathbf{V} = [v_a, v_b, v_c, v_f, 0, 0]^T \qquad \text{voltage vector}$$
$$\mathbf{R} = \text{diag}[R_a, R_a, R_a, -R_f, -R_{kd}, -R_{kq}] \quad \text{resistance matrix}$$
$$\mathbf{I} = [i_a, i_b, i_c, i_f, i_{kd}, i_{kq}] \qquad \text{current vector}$$
$$\Psi = [\psi_a, \psi_b, \psi_c, \psi_f, \psi_{kd}, \psi_{kq}]^T \qquad \text{flux vector}$$

If the machine is connected to an infinite busbar through a transformer and transmission line, as Figure 7, the phase voltages then follow as,

$$v_a = \sqrt{2}V_b \sin(\omega_0 t) + R_E i_a + L_E \frac{di_a}{dt}$$
$$v_b = \sqrt{2}V_b \sin(\omega_0 t - 2\pi/3) + R_E i_b + L_E \frac{di_b}{dt} \tag{2.2}$$
$$v_c = \sqrt{2}V_b \sin(\omega_0 t + 2\pi/3) + R_E i_c + L_E \frac{di_c}{dt}$$

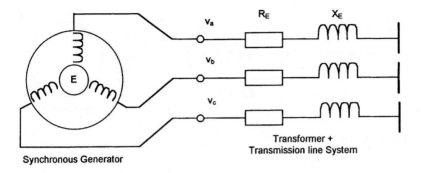

Figure 7. Single machine to infinite Busbar system.

where V_b is the busbar voltage, and R_E and L_E are the combined transformer and transmission line resistances and inductances respectively. ω_0 is the synchronous machine speed of $100\,\pi$ rad/s.

2.2.2.2. Mechanical relationships

The equation of motion for the rotor is given by,

$$\frac{d^2\delta}{dt^2} = \frac{\omega_0}{2H}\left(T_m - T_e - K_d\frac{d\delta}{dt}\right)\ \text{rad/s}^2 \tag{2.3}$$

where the electrical torque, T_e, is expressed as,

$$T_e = \frac{1}{3\sqrt{3}}\{\psi_a(i_b - i_c) + \psi_b(i_c - i_a) + \psi_c(i - i_b)\}\ \text{pu} \tag{2.4}$$

and, H represents an inertia constant, K_d is a damping coefficient and d is the machine rotor angle w.r.t the infinite busbar. The mechanical torque, T_m, provided by the turbine and boiler system (see section 2.2.4), is given by the following,

$$T_m = F_H Y_H + F_I Y_I + F_L Y_L\ \text{pu} \tag{2.5}$$

where Y_H, Y_I, and Y_L represent the output of the high, intermediate, and low pressure stages of the turbine. Similarly F_H, F_I, and, F_L indicate the relative contributions of the individual stages to the total shaft torque.

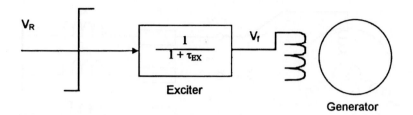

Figure 8. Thyristor exciter.

2.2.2.3. Excitation system

The exciter can be modelled fairly simply as a 1st order lag with time constant τ_{EX}, Figure 8. This model is representative of modern fast acting thyristor exciters [11],

The exciter output can therefore be calculated as,

$$\frac{dV_f}{dt} = \frac{V_R - V_f}{\tau_{EX}}, \quad -5.0 \leq V_R \leq 5.0 \, \text{pu} \tag{2.6}$$

where V_f is the field voltage, and V_R the controller excitation signal.

2.2.3. Park's $d - q$ Synchronous Generator

Using Park's d-q transformation, the properties of individual components are simulated along the direct and quadrature axes [11,12]. The flux linkage relationship for a synchronous generator connected to an infinite busbar, through a combined transformer and transmission line, as before, can be written as,

$$\begin{bmatrix} \psi_f \\ \psi_d \\ \psi_{kd} \\ \psi_q \\ \psi_{kq} \end{bmatrix} = \begin{bmatrix} L_f & -M_{ad} & M_{ad} & 0 & 0 \\ M_{ad} & -(L_d + L_E) & M_{ad} & 0 & 0 \\ M_{ad} & -M_{ad} & L_{kd} & 0 & 0 \\ 0 & 0 & 0 & -(L_q + L_E) & M_{aq} \\ 0 & 0 & 0 & -M_{aq} & L_{kq} \end{bmatrix} \begin{bmatrix} I_f \\ I_d \\ I_{kd} \\ I_q \\ I_{kq} \end{bmatrix} \tag{2.7}$$

where, the quantities are defined as the 3-phase representation. It is also important to calculate the differential flux vector as,

$$\frac{d\psi_f}{dt} = V_f - R_f I_f$$

$$\frac{d\psi_d}{dt} = V_b \sin\delta + (R_a + R_E)I_d + \Psi_q + \Psi_q\frac{d\delta}{dt}$$

$$\frac{d\psi_q}{dt} = V_b \cos\delta + (R_a + R_E)I_q - \Psi_d - \Psi_d\frac{d\delta}{dt}$$

$$\frac{d\psi_{kd}}{dt} = R_{kd}I_{kd}, \quad \text{and,} \quad \frac{d\Psi_{kq}}{dt} = R_{kq}I_{kq}$$

(2.8)

The terminal voltage, V_T, can then be calculated as,

$$V_T = \sqrt{V_{Td}^2 + V_{Tq}^2}$$

where, V_{Td} and V_{Tq} are the d- and q-axis components of voltage,

$$V_{Td} = V_b \sin\delta + R_E I_d - L_E\frac{dI_q}{dt}, \quad \text{and,}$$

$$V_{Tq} = V_b \cos\delta + R_E I_d - L_E\frac{dI_q}{dt}$$

(2.9)

The electrical torque, T_e, is similarly given as,

$$T_e = \psi_d I_q - \psi_q I_d$$

(2.10)

2.2.4. Turbine Model

A 3-stage turbine with reheater drives the synchronous generator [11]. The steam is produced by a conventional coal or oil-fired boiler. The boiler, however, is not modelled due to its slowly varying dynamics with respect to the turbine and generator system. The steam pressure at the turbine inlet valves can therefore be considered to be constant. The inertia of steam flow in each stage of the turbine is described by a 1st order transfer function, and the outputs from each stage are weighted according to their contributions to the total shaft torque. The reheater is also described by a 1st order transfer function.

Turbine losses are ignored, and steam flow is controlled by both main and intercept valves. The governors are assumed to be of the fast-acting electrohydraulic type, and are represented by a 1st order transfer function. Limits are also imposed on valve travel, and rates of movement. The system is illustrated in Figure 9.

E. SWIDENBANK *et al.*

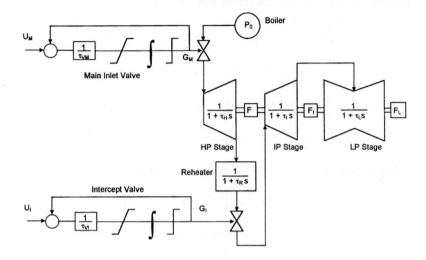

Figure 9. 3-Stage turbine with reheater and governor.

The equations for the simulation follow as,

$$\frac{dY_H}{dt} = \frac{G_M P_0 - Y_H}{\tau_H} \qquad \frac{dY_L}{dt} = \frac{Y_I - Y_L}{\tau_L}$$

$$\frac{dY_R}{dt} = \frac{Y_H - Y_R}{\tau_R} \qquad \frac{dG_M}{dt} = \frac{U_M - G_M}{\tau_{VM}} \qquad (2.11)$$

$$\frac{dY_I}{dt} = \frac{G_I Y_R - Y_I}{\tau_I} \qquad \frac{dG_I}{dt} = \frac{U_I - G_I}{\tau_{VI}}$$

where P_0 represents the constant boiler pressure. G_M and G_I indicate the positions of the main inlet and intercept valves, while U_M and U_I are the corresponding actuation signals. Similarly, τ_{VM} and τ_{VI} are the inlet and intercept valve time constants. τ_H, τ_I, τ_L, and, τ_R represent like quantities for the high pressure, intermediate pressure, low pressure and reheater stages of the turbine. Limits are imposed on the rate and position of valve movement as follows,

$$-6.7 \leq \frac{dG_M}{dt}, \frac{dG_I}{dt} \leq 6.7, \quad \text{and,} \quad 0.0 \leq G_M, G_I \leq 1.0 \qquad (2.12)$$

2.2.5. Digital Simulation

The 3-phase and Park's d-q equations given previously constitute a highly complex system of non-linear equations, and as with most power system models, the equations are also classified as *stiff* [11]. Special attention must, therefore, be given to the method of solution if accurate results are to be achieved.

The system of equations can be considered to be of the form,

$$\frac{d\mathbf{Y}}{dt} = \mathbf{F}(t, \mathbf{Y}) \tag{2.13}$$

where t is the independent variable, $\mathbf{Y}$ the dependent variable, and $\mathbf{F}(t, \mathbf{Y})$ a function of both $\mathbf{t}$ and $\mathbf{Y}$. In computing terms, $\mathbf{F}(t, \mathbf{Y})$ is known as the right-hand side expression. If it is assumed that the solution of Equation (2.13) is known at one point $\mathbf{Y}_0$, then the solution at the next point $\mathbf{Y}_1$ can be predicted using $\mathbf{F}(t, \mathbf{Y})$, the rate of change of $\mathbf{Y}$.

The Runge-Kutta-Merson 4th order Fixed Step Method has been selected for integration of the system equations [13]. This method involves 5 right-hand side function evaluations per step, giving improved accuracy over the classical 4th order Runge-Kutta method. The algorithm is outlined below, where h is the step width,

$$\mathbf{L}_1 = \mathbf{Y}_n + \frac{h}{3}\mathbf{F}(t_n, \mathbf{Y}_n)$$

$$\mathbf{L}_2 = \mathbf{Y}_n + \frac{h}{6}[\mathbf{F}(t_n, \mathbf{Y}_n) + 3\mathbf{F}(t_n + \frac{h}{3}, \mathbf{L}_1)]$$

$$\mathbf{L}_3 = \mathbf{Y}_n + \frac{h}{8}[\mathbf{F}(t_n, \mathbf{Y}_n) + 3\mathbf{F}(t_n + \frac{h}{3}, \mathbf{L}_2)] \tag{2.14}$$

$$\mathbf{L}_4 = \mathbf{Y}_n + \frac{h}{2}[\mathbf{F}(t_n, \mathbf{Y}_n) - 3\mathbf{F}(t_n + \frac{h}{3}, \mathbf{L}_2)] + 4\mathbf{F}(t_n + \frac{h}{2}, \mathbf{L}_3)]$$

$$\mathbf{Y}_{n+1} = \mathbf{Y}_n + \frac{h}{6}[\mathbf{F}(t_n, \mathbf{Y}_n) + 4\mathbf{F}(t_n + \frac{h}{2}, \mathbf{L}_3)] + \mathbf{F}(t_n + h, \mathbf{L}_4)]$$

2.3. SELF-TUNING CONTROL

2.3.1. Introduction

The characteristics of a power system change significantly between heavy and light loading conditions, with varying numbers of generating units and transmission lines in operation at different times. These affect voltage levels and system damping, causing the modes of oscillation to change continuously. The magnitude of plant disturbances can also vary from minor to large imbalances in mechanical and electrical generated power. The result is a highly complex and non-linear system, subject to varying loads and generation schedules.

Although conventional regulators have proven remarkably effective at tackling practical control problems, tuning and integration of the algorithms, in the case of plants with many control loops, can prove to be a costly and time consuming exercise [14]. Fixed parameter controllers and stabilizers, designed at a particular operating point, are therefore unable to provide the most effective plant and system control over the full operating range. However, research has demonstrated that adaptive control can offer a solution to such problems through allowing the controller parameters to adjust as the operating conditions change, automatically coping with plant ageing [15]. It has been shown to improve overall control in turbogenerator systems [16], with the objective of extending operational stability margins.

Adaptive control was first investigated in the early 1950s, but interest diminished due to a lack of sufficient computational power, and failings in the actual control theory. It was not until the 1970s that practical control schemes began to emerge. Today, there are essentially two approaches to adaptive control — model reference and self-tuning control. In power systems, early work centred on the application of model reference adaptive control, but difficulties arose in selecting a suitable reference model for such a highly non-linear system [17]. By comparison, self-tuning control has proved much more successful in the areas of turbogenerator excitation and field control. These techniques have performed acceptably in power stations [18]. Indeed, commercial power system control manufacturers are beginning to apply self-tuning strategies to their own automatic voltage regulators.

Self-tuning control is based on the principle of separating the estimation of the unknown process parameters from the design of the controller [19]. The scheme can be thought of as consisting of two loops, Figure 10. The outer loop incorporates the process and a feedback regulator, while the inner loop comprises a recursive parameter estimator and a design calculation. If it is assumed that the estimated parameters represent the true parameters then a selection of methods becomes available to design the self-tuning controller itself.

An important aspect of adaptive control, is the need for an estimated model of the plant. System identification deals with the problem of building mathematical models of dynamical systems, based on observed data from the system. The models obtained attempt to link observations together into some sort of pattern. Many methods have been suggested for the identification of suitable plant models [20]. Techniques including instrumental variables, maximum likelihood, stochastic approximation, and recursive least squares have previously been applied.

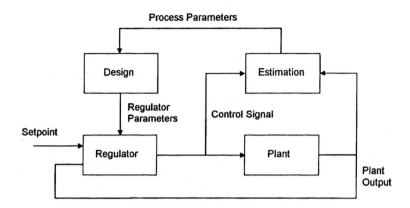

Figure 10. Self-tuning adaptive control.

2.3.2. Generator Excitation Control

Having obtained a suitable model of the system under investigation, a wide range of self-tuning control strategies become available. A simple regulator with few parameters is likely to do well if only replacement of a PID scheme is intended. On the other hand, in demanding applications such as generator control, there is a need to increase the regulator complexity to achieve the desired performance.

2.3.2.1. Fixed gain automatic voltage regulator

The fixed gain controller consists of an automatic voltage regulator (AVR), coupled with a power system stabilizer (PSS). Under some circumstances, the voltage regulator can introduce negative damping into a power system, with almost all the negative damping for a regulated machine originating in the AVR. A PSS stabilizes the system by introducing damping torque through regulating the field flux linkage, in phase with variations in shaft speed [21].

A typical controller with fixed gain proportional feedback, coupled with a speed stabilizer, is shown in Figure 11. The controller parameters were obtained using eigenvalue analysis with a linearized 10th order state-variable model of the turbogenerator [22].

Fault studies and long term operation tests, through simulation and on a microalternator, have proved the acceptable performance of this controller over a wide range of operating conditions and environments. It can be employed as the basis for comparison with more advanced adaptive control

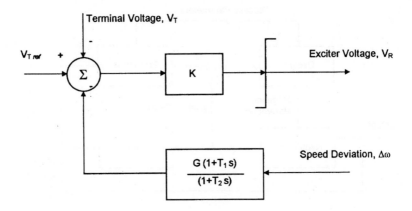

Figure 11. Fixed gain automatic voltage regulator.

algorithms, presented subsequently. It should be noted though that the derivation of the fixed gain controller gains is based on an analytical model of the generator system. While this is readily available for a laboratory machine, such models are difficult to obtain in practice, and consequently the selection of individual gains is not a trivial exercise.

2.3.2.2. Generalized minimum variance control

A generalized minimum variance (GMV) self-tuning strategy is proposed, due to the excitation control loop's predominantly regulatory requirements [23]. In comparison with generalized predictive control (GPC) it is more computationally efficient, with fewer *a priori* parameters to be selected. The synchronous generator can be considered as a single-input single-output (SISO) system, described by the following equation [24],

$$A(z^{-1})y(t) = B(z^{-1})u(t - t_d) + C(z^{-1})\xi(t) \qquad (3.1)$$

where $u(t)$ and $y(t)$ represent the input and output of the system at the sample instant t, respectively. These quantities are obtained as,

$$u(t) = \Delta V_R(t) \quad y(t) = \Delta V_T(t) + \gamma \Delta \omega(t); \quad -1 < \gamma < 0$$

where $\Delta V_T(t)$ is the deviation in the terminal voltage, $\Delta \omega(t)$ the rotor shaft speed deviation, γ a factor that determines how much weight is placed on the speed deviation signal, and $\Delta V_R(t)$ the deviation in the exciter voltage signal applied to the generator. The inclusion of an auxiliary signal, $\Delta\omega(t)$, in $y(t)$ introduces an adaptive power stabilization function to enhance system damping [25] $A(z^{-1})$, $B(z^{-1})$ and $C(z^{-1})$ are polynomials in the backward shift operator z^{-1}, while $\xi(t)$ is a zero mean, white noise sequence disturbing the system. t_d is the system time delay in an integral number of sample intervals.

A GMV control strategy is implemented by minimizing the cost function,

$$J = E\{[y(t + t_d) - w(t + t_d)]^2 + Ru^2(t)\}$$

where the sequence $W(t+t_d)$ defines the desired track that the system should follow, rather than the output setpoint itself. This has the effect of reducing the normally excessive control signals of minimum variance control. The control law will act to minimize the variance between the actual and predicted outputs, rather than that between the actual output and the system setpoint. $W(t + t_d)$ is obtained by designing an auxiliary, one step ahead, predictor as below,

$$w(t + t_d) = P(z^{-1})y(t)$$

The $P(z^{-1})$ coefficients describe the relationship between the desired outputs of the optimal predictor and the outputs of the dynamic system. They may be selected according to the qualitative requirements of the controlled turbogenerator system [24]. The weighting factor, R, where $R > 0$, permits *detuning* of the control signals and becomes necessary when dealing with non-minimum phase systems. The selection of R trades closeness of desired output reference following against control effort. After rearrangement, minimization of the above cost function leads to the following control law,

$$u(t) = -\frac{g_0}{R}\left[\frac{F(z^{-1})y(t) + G(z^{-1})u(t)}{C(z^{-1})}\right] \qquad (3.2)$$

where g_0 is the first coefficient of the $G(z^{-1})$ polynomial. The parameters of the polynomials $F(z^{-1})$, $G(z^{-1})$ and $C(z^{-1})$ are determined by extended least squares (ELS) identification [33]. If the order of $C(z^{-1})$ is selected to be zero, then Equation (3.2) reduces to

$$u(t) = -\frac{g_0}{R}[F(z^{-1})y(t) + G(z^{-1})u(t)] \qquad (3.3)$$

where the parameters are estimated by recursive least-squares (RLS) identification [33].

2.3.2.3. Self-tuning PID controller

Self-tuning regulators may be configured to provide PID control actions. An approach based on eigenvalue placement permits the gain settings of the controller, K_P, K_I and K_D, to be adjusted such that the eigenvalues of the system are maintained at specified locations, in spite of variations in operating conditions.

The control signal, $u(t)$, for a general linear regulator is described by [26],

$$u(t) = \frac{T(z^{-1})}{R(z^{-1})} y_{\text{ref}}(t) - \frac{S(z^{-1})}{R(z^{-1})} y(t) \tag{3.4}$$

where $y_{\text{ref}}(t)$ is the reference signal of y, and $R(z^{-1})$, $S(z^{-1})$ and $T(z^{-1})$ are polynomials in z^{-1}. Combining Equations (3.1) and (3.4), the closed-loop transfer function for the system is obtained as,

$$\frac{y(t)}{y_{\text{ref}}(t)} = \frac{z^{-1} B(z^{-1}) T(z^{-1})}{A(z^{-1}) R(z^{-1}) + z^{-1} B(z^{-1}) S(z^{-1})} = \frac{z^{-1} B_m(z^{-1})}{A_m(z^{-1})}$$

where $A_m(z^{-1})$ and $B_m(z^{-1})$ correspond to the desired transfer function poles and zeros. The denominators can be compared after pole-zero cancellation, and substituting for $B(z^{-1})$, to give

$$A(z^{-1}) R(z^{-1}) + z^{-1} B(z^{-1}) S(z^{-1}) = P_1(z^{-1}) B_2(z^{-1}) \tag{3.5}$$

where the term $P_1(z^{-1})$ is derived from pole-zero cancellation. $B_2(z^{-1})$ corresponds to the zeros of $B(z^{-1})$ lying inside the unit circle. In order to obtain a self-tuning PID structure it is assumed that

$$R(z^{-1}) = (1 + r_1 z^{-1})(1 - z^{-1}) \qquad T(z^{-1}) = s_0 + s_1 + s_2$$
$$S(z^{-1}) = s_0 + s_1 z^{-1} + s_2 z^{-2} \qquad P_1(z^{-1}) = (1 + p_1 z^{-1} + p_2 z^{-2})$$

With p_1, p_2 selected, the 4 parameters r_1, s_0, s_1, s_2 may be obtained by solving Equation (3.5). The self-tuning PID control signal, $u(t)$, can then be calculated as,

$$u(t) = \frac{s_0 + s_1 + s_2}{(1 + r_1 z^{-1})(1 - z^{-1})} y_{\text{ref}}(t) - \frac{s_0 + s_1 z^{-1} + s_2 z^{-2}}{(1 + r_1 z^{-1})(1 - z^{-1})} y(t)$$

Comparing this with the PID controller described by [27],

$$u(t) = K_P y(t) + \frac{-T_S K_I}{(1 - z^{-1})(1 + r_1 z^{-1})} (y_{\text{ref}}(t) - y(t)) + \frac{K_D(1 - z^{-1})}{T_S(1 + r_1 z^{-1})} y(t)$$

where the proportional and derivative actions operate only on the plant output, the PID controller gains follow as,

$$K_P = \frac{s_1 + 2s_2}{1 + r_1}, \quad K_I = -\frac{s_0 + s_1 + s_2}{T_s}, \quad K_D = \frac{r_1 s_1 - (1 - r_1)s_2}{1 + r_1} T_s$$

Since the output signal, y, is formed from a combination of ΔV_T and $\Delta \omega$, the reference signal, y_{ref}, is zero. The PID controller output simplifies to,

$$u(t) = -\frac{s_0 + s_1 z^{-1} + s_2 z^{-2}}{(1 + r_1 z^{-1})(1 - z^{-1})} y(t)$$

Having estimated the plant model coefficients using RLS identification, the control signal can be generated by updating the PID gain settings.

2.3.3. Integrated Control Strategies

The self-tuning techniques developed for the generator AVR loop may similarly be applied to the governing systems. Two approaches may be adopted. One employs a separate self-tuner for both the governor and AVR, and is referred to as *multi-loop control*. The second method integrates both the governor and AVR systems into a single multivariable controller [28]. Figures 12 and 13 illustrate the two procedures, where U_G is the governor input signal.

In the multi-loop case, the governor and AVR control laws are the same as for the SISO system. The multivariable controller has the same form as the SISO algorithm, except that the associated vectors are replaced by matrices, and multivariable least squares is used to estimate the system parameters. Although the multi-loop system is easier to design and requires much less computation time to implement, it is expected that the multivariable controller will better exploit the inherent coupling in the turbogenerator system [29]. However, as outlined in Section 2.4.6 these MIMO (multi-input multi-output) strategies can significantly increase the computational burden.

2.3.4. Supervision Schemes

Previous investigations have illustrated that self-tuning schemes will work well if the preconditions for stability and convergence are satisfied [8]. In practice, however, such stipulations may be violated. The non-linear nature of power systems implies that the model of Equation (3.1) is only valid for a small region about a given operating point. Significant deviations from this region will cause the estimator to operate outside its linear range, and the resulting model will be invalid.

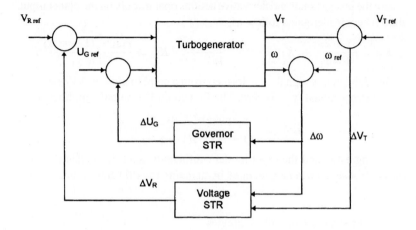

Figure 12. Multi-loop self-tuning control.

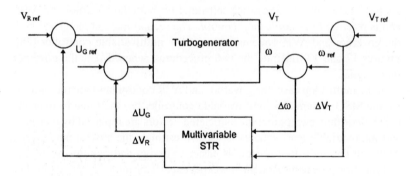

Figure 13. Multivariable self-tuning control.

A power system is frequently subjected to various disturbances such as transformer tap-changing, line switching and occasional major disturbances such as short-circuits or lightning surges. All of these may cause excessive variations in the outputs of synchronous generators, leading to abrupt changes in operating conditions, and possibly hunting of the machines over a linear range. Such occurrences are violations of the preconditions for parameter estimation. Therefore, if a self-tuning controller is to work safely in practice, it must incorporate a reliable and robust supervision scheme. Indeed, individual controllers may be coordinated by an expert system to provide improved system performance and increased robustness of the self-tuning regulators.

Besides determining the most appropriate control strategy, the expert system can monitor estimator parameters, detect transducer failures, introduce external excitation, etc [30].

A number of methods have been developed to ensure satisfactory operation of self-tuning controllers. These usually take the form of protection algorithms for the parameter estimator, and are commonly referred to as jacketing software. A selection of four such methods will be discussed, namely enhancing process excitation, modification of the identification algorithm to deal with time varying systems, moving boundaries for the estimated parameters, and switching the estimator off during transient conditions.

2.3.4.1. Excitation of estimator inputs

For the process of identification it is essential that the dynamics of the process are persistently exciting. However, under normal circumstances, the excitation present on a system is not sufficiently rich in frequency, and artificial input signals must be introduced. Their inclusion eliminates ambiguity in the relationship between plant input and output signals, through introducing an independent signal into the feedback loop. For industrial applications, consideration must be given to the properties of the actuator, limitations on the input signal, the permissible deviation from the operating point, and the maximum sampling rate [24].

A pseudo random binary sequence (PRBS) is often selected as a test signal. The autocorrelation function of this sequence approximates a Dirac delta function, simulating a white noise process. Practical experience, and suggestions from other authors has resulted in a 7th order PRBS signal being selected, with a width of 100 ms and an amplitude set at 4% of the maximum excitation voltage [32]. To ensure that the estimator inputs are persistently exciting the energy of the control signal, $u(t)$, is monitored. If the variance of the input signal, or the trace of the estimation covariance matrix, are beyond selected thresholds it can be concluded that there is *sufficient* frequency content present to permit parameter estimation to proceed. However, if there is not, then a pseudo random binary sequence input will need to be injected to improve system excitation. The threshold level is selected depending on the amount of noise present in the input signal.

As an alternative to injecting PRBSs to improve system excitation, deadbands or dead-zones may be introduced to protect the parameter estimator. In these techniques the estimator is switched off when the excitation level is very low, or the estimation error has converged within a predetermined zone. Such approaches, however, require appreciable signal-to-noise ratios. In turbogenerator systems the magnitude of the dead-zone can be difficult to select due to unmodelled plant disturbances, e.g. increasing the governor input smoothly and slowly can induce significant changes in operating conditions, with only minimal changes in the system output, while the parameter estimator remains frozen. This can give rise to slow drift instability [24].

2.3.4.2. Time varying parameters

An important aspect of reliable operation is that the parameter estimator should be able to track slowly varying process parameters, while at the same time not discarding important information too rapidly. This leads to a scheme involving a variable forgetting factor [24]. In a generator system there may be long periods at constant operating conditions, and this may result in insufficient excitation signals being available. In this case, the estimator will discard old information and uncertainties in the parameters will rise, leading eventually to estimator wind-up.

Increasing trends in the elements of the estimator covariance matrix can be counteracted by invoking a random walk, introducing a constant trace algorithm, increasing the forgetting factor towards unity, etc. [31]. An alternative solution to this problem lies in monitoring the Kalman gain vector, $\mathbf{K}(t)$, of the estimator as follows,

$$v(t) = \sum_{\tau=1}^{T} \sum_{i=1}^{m} k_i(t - \tau)$$

where $v(t)$ is a function analogous to the power spectrum of $\mathbf{K}(t)$. m is the number of estimated parameters, while T equates to the summation level. If $v(t)$ exceeds a preset level, σ, then the forgetting factor, λ, will be reset [8].

2.3.4.3. Moving parameter boundaries

Transient disturbances on a power system may give rise to abrupt changes in the estimated parameters, which are not due to a change in the process dynamics. Individual moving boundaries are therefore introduced to protect each of the parameters against such disturbances [8]. Such an approach is preferable to the imposition of absolute boundaries, which require prior knowledge of the parameter estimates [24]. The mean values, $\beta(t)$, of the estimated parameters at time t, are given by,

$$\beta_i(t) = \frac{1}{T'} \sum_{\tau=1}^{T'} \theta_i(t - \tau) \tag{3.6}$$

where $\theta_i(t)$ is the ith element in the parameter vector $\theta(t)$. T' determines how adaptable the parameters are – the larger the value of T', the more stable the parameters become. The high and low parameter boundaries, β_{iH} and β_{iL}, are defined as,

$$\beta_{iH} = \beta_i(t) + \eta|\beta_i(t)| \quad \beta_{iL} = \beta_i(t) - \eta|\beta_i(t)| \tag{3.7}$$

The factor η, where $0 < \eta < 1$, specifies how quickly the parameters can adapt — the larger the value of η, the more likely it is that the parameters will change. If a parameter $\theta_i(t)$ moves outside the range defined by Equation (3.7) then the appropriate limiting range will be substituted for the parameter. If $\theta_i(t)$ lies within the range then no action is taken.

2.3.4.4. *Estimator activation*

One of the most important features of the supervision scheme is deciding when the estimator should be used. During a transient condition on the power system, the synchronous machine outputs may vary to an excessive degree. Non-stationary and periodic disturbances, faults at transmission lines and emergency operation in power systems, etc., would lead to a sudden change of the estimated parameters. This may lead to ill-conditioning of the estimator resulting in a model that does not represent the process behaviour. The estimator should, therefore, only be introduced when it is likely to produce acceptable estimates, and should be switched off or deactivated during transient disturbances.

As the purpose of the control schemes is to regulate the terminal voltage, the deviation of this signal from its setpoint has been selected as an estimator deactivation indicator. If the terminal voltage deviation exceeds a preset limit the estimator will be switched off, and will only be switched on again once the terminal voltage returns to its preset level, remaining there for a fixed time. This ensures that the estimator will remain deactivated during severe oscillations and generator hunting. A typical terminal voltage pattern during a transient short period illustrates the operation of this method, Figure 14.

2.3.5. Comparison of Individual Controllers

This section will present some typical results obtained when using the fixed gain, PID, and self-tuning regulators to control various transient disturbances on a simulated power system. The estimator supervision schemes described in Section 2.3.4 form an integral part of the self-tuning regulator.

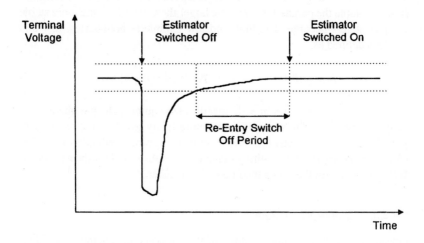

Figure 14. Estimator deactivation during a voltage transient.

The controllers are examined by simulating a line-line short circuit, duration 200 ms, at the sending end of the transmission line system, at an operating point of $P_T = 0.5$ and $Q_T = 0.1$ pu. Figure 15 depicts the terminal voltage response and rotor angle oscillations for the fixed gain, PID, and, 5-parameter RLS & 7-parameter ELS GMV self-tuning controllers. For these tests, the fixed gain controller is employed as a reference, facilitating comparison between the individual schemes. While at first sight the main purpose of an automatic voltage regulator should be to minimize deviations of the terminal voltage, its main role is actually to maintain machine rotor angle, and, therefore, to assist in preserving steady-state stability [34]. Paradoxically, reducing the rotor angle oscillations is more important than minimizing voltage deviations after a fault condition. The results illustrate that the fixed gain, PID and self-tuning algorithms provide satisfactory performance for the various fault conditions, with well damped responses achieved. However, it is observed that the actions of the self-tuning controllers are superior, with a significant reduction in the second rotor swing and improved damping of the subsequent oscillations, in comparison with the fixed gain controller response.

A more stringent short-term operating sequence encompassing large transient disturbances, and changes in operating conditions, may be developed. At an initial operating point of $P_T = 0.8$ and $Q_T = 0.2$ pu, a series of external faults are applied, as follows,

Terminal Voltage (pu)

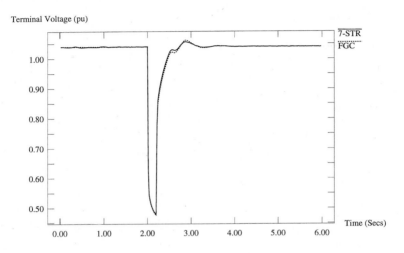

Rotor Angle (Deg.)

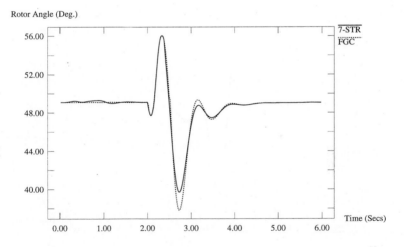

Figure 15a. GMV-ELS self-tuning controller — line-line short circuit, duration 200 ms.

Terminal Voltage (pu)

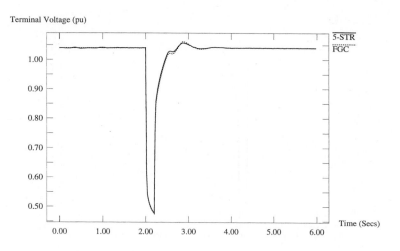

Rotor Angle (Deg.)

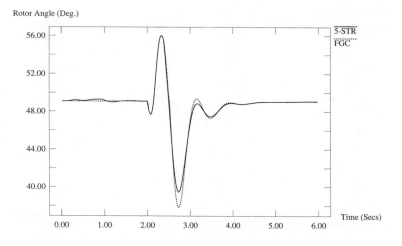

Figure 15b. GMV-RLS self-tuning controller — line-line short circuit, duration 200 ms.

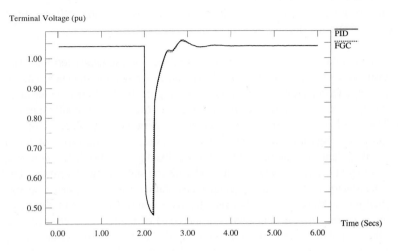

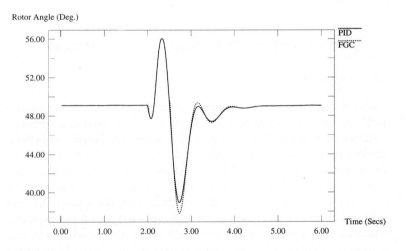

Figure 15c. Self-tuning PID controller — line-line short circuit, duration 200 ms.

- 3-phase-earth short circuit after 2.0 seconds, duration 200 ms, a single transmission line switched out
- 3-phase-earth short circuit after 10.0 seconds, duration 200 ms, both transmission lines switched in

The initial fault is cleared by opening the faulted transmission line. The resultant increase in system impedance causes a reduction in synchronizing power, and decreases the stability limits. Figure 16 illustrates the terminal voltage and rotor angle responses for the four control configurations. The fixed gain controller response again permits an evaluation of the individual algorithms. As previously, the responses to the series of faults are well damped. The performance of the fixed gain controller is, however, significantly inferior, when compared with the self-tuning algorithms. The 7-parameter ELS self-tuning controller exhibits a more damped response than that achieved by the 5-parameter RLS controller and PID controller after the second transient fault. On the basis of the results presented, and other simulation tests, the four controllers can be ordered in increasing quality of performance as ELS-GMV self-tuning regulator (*best*), RLS-GMV self-tuning regulator, PID controller, and fixed gain regulator (*worst*).

2.4. QUEEN'S UNIVERSITY OF BELFAST TEST FACILITY

2.4.1. Introduction

A laboratory micromachine provides a practical test bed for both measurement and control algorithms under an industrial environment. Although it cannot be realistically expected to give results comparable with those obtained on a full scale power station, it does provide a means of verifying the behaviour of controllers which have shown themselves to be successful in simulation. As well as permitting particular control systems to be tested under real-time constraints, the micromachine provides a more stringent testing environment, incorporating many features not present in a computer simulation. These effects include non-ideal transducer characteristics leading to limited resolution and noise, computational delays, variations in busbar voltage and frequency, saturation, hysteresis, and other non-linearities present on a real machine.

The performance of any control system depends almost entirely on the quality of information presented to it. If a controller is to be implemented reliably, accurate measurements of feedback variables are essential. The availability of auxiliary variables for display purposes and control performance monitoring is a further requirement. Reliable measurements of

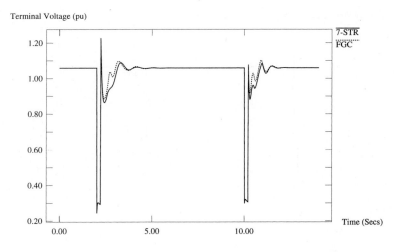

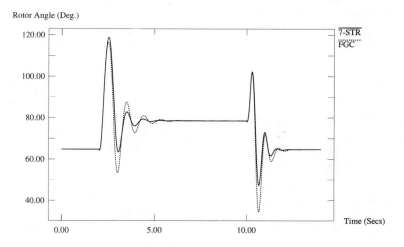

Figure 16a. GMV-ELS self-tuning controller — short-term operating sequence.

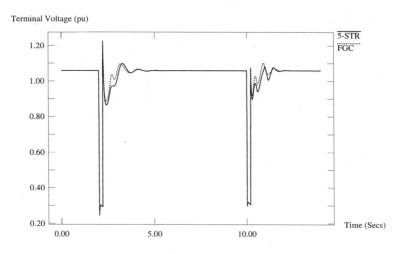

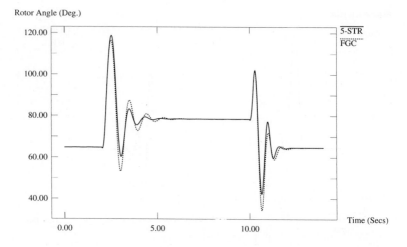

Figure 16b. GMV-RLS self-tuning controller — short-term operating sequence.

Terminal Voltage (pu)

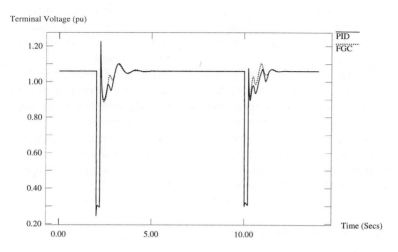

Rotor Angle (Deg.)

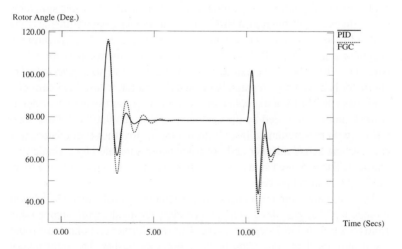

Figure 16c. Self-tuning PID controller — short-term operating sequence.

generator terminal quantities are difficult to achieve in the electrically noisy environment of power stations. Harmonic interference and unbalanced generator operation inevitably lead to distortion and ripple, while any subsequent filtering may further degrade the information content, especially during transient conditions. Therefore, improved control can only be realized if enhancements in measurement strategies are introduced.

This section describes an algorithm, based on a finite Fourier series, that is used to extract the terminal quantities from the complex periodic waveforms obtained from the synchronous machine. Measurements of machine speed and rotor angle are achieved digitally using signals derived from an optical transducer mounted on the rotor shaft. The entire implementation makes use of industry standard VME bus hardware, employing both Motorola 680×0 microprocessors and Inmos transputers. Practical test results are then given for the SISO and MIMO controllers outlined in section 2.3.

2.4.2. Micromachine System

The Queen's University of Belfast micromachine system consists of a specially designed synchronous generator, with an associated turbine simulator, tied to the busbar through a transformer and artificial transmission lines [35]. The entire system is shown schematically in Figure 17. The synchronous generator is a 3 kVA, 220 V, 50 Hz, 1500 rpm 4 pole microalternator, whose parameters have been selected to match those of a full size generator. The alternator is driven by a separately excited d.c. motor, whose field current is held constant. The torque supplied by the motor is proportional to its armature current, and is controlled by the analogue turbine simulation. A three stage turbine with reheater and a fast electrohydraulic governor is emulated, with each turbine stage, reheater and governor being simulated by a single time constant. The weighted sum of signals from each stage is proportional to the turbine mechanical power.

The alternator is directly connected to a delta-star transformer which has an on-load tap-changing device on the secondary terminals, with tapping ratios from 65% to 116% in 7 steps available. This transformer is connected through a transmission line simulation to the laboratory busbar. The transmission system is simulated by lumped parameter Π-networks, representing a typical double line transmission system. Provision is made for the application of short circuits at the secondary terminals of the transmission transformer or half way through the line. It is also possible to switch out one of the transmission lines. A linear power amplifier with a time constant regulator supplies the field current for the alternator.

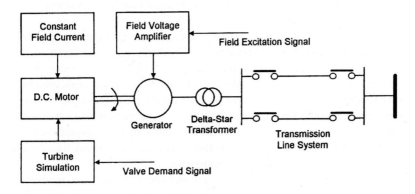

Figure 17. Laboratory Micromachine System.

2.4.3. Fourier Measurement Algorithm

Deficiencies in existing RMS techniques for measuring the generator terminal quantities of voltage, current, real power and reactive power, has resulted in the development of alternative strategies. To cope with signals contaminated by harmonics and noise, an advanced algorithm based on a finite Fourier series has been adopted [8] The Fourier algorithm effectively acts as a band pass filter, centred around the main power frequency of 50 Hz. High frequency noise, d.c. offsets, and low frequencies are completely rejected. The harmonic content, which is always present in the raw waveforms generated by the synchronous machine, will also have no effect on the final calculation of the terminal quantities.

If a 3-phase system is perfectly balanced, the harmonic content in the signals will cancel out. It is then valid to use the simpler, traditional RMS method for calculating the terminal quantities. The individual phase quantities can be combined directly to form the terminal quantities without the need for complex filtering algorithms. In practice, however, physical systems are never fully balanced and any unsymmetrical behaviour, e.g. unbalanced faults/loads, will invalidate the measurements obtained by this method.

The Fourier algorithm, on the other hand, suffers from no such drawbacks. The only requirement is that the waveforms are periodic, so that an accurate picture of the terminal quantities will continue to be provided during the most severe transient disturbances. Under such conditions, the RMS measurement may cause violent fluctuations in the controller signal due to the highly oscillatory nature of the measured feedback signals. This is clearly undesirable, as rapid movements in control signals can lead to excessive heating in field excitation coils. The Fourier algorithm supplies continuous feedback signals, permitting smooth control.

Traditional methods for determining machine terminal quantities depend upon the rectification and summation of individual phase quantities to form an average RMS value. As the voltage and current waveforms generated by a synchronous machine are rarely perfectly balanced, invariably contaminated by harmonics and noise, they often produce poor representations of the true terminal quantities. This is particularly evident during periods of transient disturbances. Inaccurate measurements inevitably lead to a degradation in control performance, and often mean that more advanced control algorithms work no better than their simpler counterparts.

Fourier analysis of the 3-phase voltage and current waveforms, produced by a synchronous generator, provides one solution. By extracting the fundamental components of the waveforms, very accurate measurements of terminal quantities under both symmetrical and unsymmetrical transient, and steady-state conditions can be produced.

2.4.3.1. *Time series harmonic filter*

The applied Fourier analysis algorithm is based on an N sample point, moving window approximation to the general Fourier series for a periodic waveform [36]. N is selected to be 12, being a compromise between accuracy and computational burden. The 12 point time series harmonic filter can be derived as follows,

Any periodic waveform, $F(t)$, can be expressed by its Fourier series as,

$$F(t) = \frac{a_0}{2} + \sum_{n=1}^{\infty}(a_n \cos(nt) + b_n \sin(nt)) \tag{4.1}$$

Through approximating the series as,

$$F(t) = \frac{a_0}{2} + a_1 \cos t + \cdots + a_6 \cos 6t + b_1 \sin t + b_2 \sin 2t + \cdots + b_5 \sin 5t \tag{4.2}$$

an expression for $F(t)$ with 12 unknown coefficients is obtained. The waveform fundamental components are given by the coefficients a_1 and b_1. If a periodic waveform is sampled 12 times per cycle as in Figure 18, then, applying Equation (4.2) for the sampled point U_0,

$$\begin{aligned} U_0 = F(t_0) &= \frac{a_0}{2} + a_1 \cos t_0 + a_2 \cos 2t_0 + \cdots + a_6 \cos 6t_0 \\ &+ b_1 \sin t_0 + b_2 \sin 2t_0 + \cdots + b_5 \sin 5t_0 \end{aligned} \tag{4.3}$$

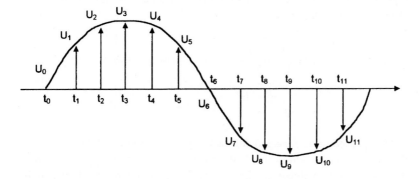

Figure 18. Sampled periodic waveform.

Repeating for the remaining sample points, a system of 12 equations in the 12 unknown coefficients $[a_1, \ldots, b_5]$ is created. A solution for the fundamental components, a_1 and b_1, is then obtained as an algebraic sum of past samples as follows,

$$
\begin{aligned}
a_1 = \frac{1}{6}\Bigg[& (U_0 - U_6) + \frac{\sqrt{3}}{2}(U_1 - U_5 - U_7 + U_{11}) \\
& + \frac{1}{2}(U_2 - U_4 - U_8 + U_{10}) \Bigg] \\
b_1 = \frac{1}{6}\Bigg[& (U_3 - U_9) + \frac{\sqrt{3}}{2}(U_2 + U_4 - U_8 - U_{10}) \\
& + \frac{1}{2}(U_1 + U_5 - U_7 - U_{11}) \Bigg]
\end{aligned}
\tag{4.4}
$$

The time series filter Equations (4.4), are executed at every sample interval to provide a moving average of the fundamental components of the periodic waveform.

2.4.3.2. Calculation of electrical terminal quantities

Having determined the individual waveform components, the electrical terminal quantities of voltage (V_t), current (I_t), real (P_t) and reactive power (Q_t) can be calculated. If c_1 and d_1 are the fundamental components of the *a* phase current, then expressions for V_T, I_T, P_T and Q_T in terms of a_1, b_1, c_1 and d_1 can be derived as:

$$V_T = \sqrt{\frac{a_1^2 + b_1^2}{2}} \qquad P_T = \frac{1}{2}(a_1c_1 + b_1d_1)$$

$$I_T = \sqrt{\frac{c_1^2 + d_1^2}{2}} \qquad Q_T = \frac{1}{2}(a_1d_1 - b_cd_1)$$

(4.5)

This procedure is repeated for each phase. An average RMS value can then be calculated across the 3 phases. Equation (4.5) is applicable when the phase-neutral voltages are measured. However, the inaccessibility of the neutral point on the micromachine means that phase-phase or line voltages must be used instead. The fundamental components of the phase-phase voltages are thus calculated, and combined to form expressions similar to Equation (4.5) as follows,

$$V_T = \sqrt{\frac{a_1^2 + b_1^2}{6}} \qquad P_T = \frac{\sqrt{3}(a_1c_1 + b_1d_1) + (a_1d_1 - b_1c_1)}{4\sqrt{3}}$$

$$I_T = \sqrt{\frac{c_1^2 + d_1^2}{2}} \qquad Q_T = \frac{\sqrt{3}(a_1d_1 - b_1c_1) - (a_1c_1 + b_1d_1)}{4\sqrt{3}}$$

(4.6)

Again, average values are taken across the 3 phases.

2.4.4. Measurement of Machine Speed and Rotor Angle

Measurement of both machine speed and rotor angle is achieved through the introduction of an optical transducer mounted on the rotor shaft [35], as shown in Figure 19. The aluminium disc is attached to the non-drive end of the alternator rotor by a flexible coupling, with four slots cut at approximately 90° intervals [37]. One slot is slightly longer than the others, permitting it to be used as a reference slot. The disc rotates through a fixed head that contains an optical transducer consisting of three lamps, and associated light detectors and circuitry.

Each time a slot passes through the fixed head a pulse is generated, negative edge triggering the reading of a 1 MHz counter, representing the time period from one slot to the next. The machine speed is inversely proportional to the measured count. This operation takes place each time a slot passes through the fixed head, and hence a measurement of speed is available every 10 ms, for a 4 pole synchronous machine. As it is not possible to space the slots

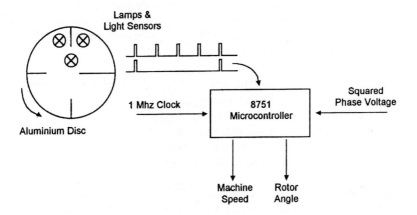

Figure 19. Speed and rotor angle measurement.

exactly 90° apart, a moving average of the last four counts is calculated — corresponding to a complete revolution of the aluminium disc — every 10 ms, as a measure of the machine speed.

The machine rotor angle is measured in a similar manner to that for speed. A phase-phase voltage signal, v_{ab}, is squared and similarly triggers a further read of the 1 MHz counter. Since the mains signal cycles every 20 ms, a measure of rotor angle — which is again proportional to the count — is available at this time.

It should be noted that the value of rotor angle calculated by this method does not represent the true transmission angle between the infinite busbar and the generated EMF of the machine. Rather, it is the angle between the terminal voltage and the generated EMF. In a power station, the infinite busbar voltage is difficult to measure and so the terminal voltage of the machine would be used instead. This policy has been adopted for measurement of the rotor angle on the micromachine, although the mains signal is available in the laboratory environment.

2.4.5. VME Transputer Hardware System

In order to create a structurally open-ended environment, a VME bus based system was selected as the host hardware system for the implementation of both control and measurement algorithms [8]. This standard bus system is already being used by power system control manufacturers [38], and is well established in many industries. In fact, the measurement system has been designed to be compatible with existing industrial implementations. The VME system is connected to a personal computer, providing additional facilities for communication and data storage. The specification for each of the slot-in boards on the VME rack is as follows,

Master	20 MHz 68020, 68881 FPU, 1 MB DMA, Monitor
Transputer	T800, slave/master
Analogue I/O	12 bit, 50 input, 4 output
Digital I/O	48 bit I/O, interrupt, clock
Sample Control	8751 microcontroller (custom board)
Serial I/O	8 RS232 channels, 19.2 kBaud

Raw values of voltage and current from the 3 phases are taken from instrumentation already installed on the micromachine system. The waveforms from these voltage and current transformers are directed through signal conditioning circuits, which adjust their levels to be compatible with the analogue to digital converters on the hardware system [35]. The resultant signals are then passed through anti-aliasing low pass filters on a custom designed board, before reaching the analogue I/O board. The custom board also contains an 8751 programmable microcontroller, which calculates values of machine speed and rotor angle.

The microcontroller generates an interrupt signal at 12 times the system frequency, i.e. 600 Hz. The interrupt triggers the master 68020 board to read in samples of the filtered electrical waveforms from the analogue I/O board, and to record measurements of speed and rotor angle calculated by the 8751 microcontroller. On completion of a read sequence, the raw values are passed to shared Direct Memory Access (DMA) memory for retrieval by the IMS BO11 T800 transputer. A handshaking mechanism safeguards data communications between the 2 processors.

The Fourier measurement algorithm is then performed on the transputer, to produce the 4 electrical terminal quantities. As the values obtained for speed and rotor angle are subject to spurious noise, the input signals are passed through a digital 1st order RC filter, with a cut-off frequency of 10 Hz. The entire sequence of operations can be seen in Figure 20.

The BO11 transputer card is connected to PC based transputer modules (TRAMs), mounted on a TMB04 motherboard, via an INMOS hardware link wire. The calculated terminal values are transmitted along with speed and rotor angle, through the hardware link, to a program running on the TMB04 transputer board.

All the software for the measurement system has been sourced in C, apart from the code that calculates machine speed and rotor angle, which is written predominantly in 8751 assembly language. The selection of C for the majority

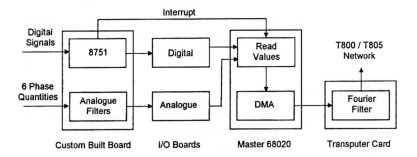

Figure 20. Hardware measurement process.

of the software permits debugging and system testing using standard C compilers. Code can also be easily ported from a simulation environment to the test system, with little or no modification. The source code for the transputer system is written in 3L Parallel C. This is a variant of the standard Kernighan & Ritchie C language, containing additional constructs that allow programs to operate in parallel.

The multi-tasking parallel implementation of the measurement and control algorithms is facilitated by the use of transputers. The VME transputer can be linked, as in this application, to an external transputer system containing any number of processors. This creates a system with vastly increased computing power, and with the advent of the next generation of processors will provide the control systems designer with the opportunity to implement virtually any advanced control strategy.

2.4.6. Control Implementation

Industrial standards set a maximum control interval of 10 ms for digital voltage regulation. All control programs must therefore operate within this specification. The transputer-based control system developed here is capable of providing the necessary computing power to achieve the required standard for all the controllers described in section 3. Program execution strategies are outlined in the following sections.

It should be mentioned that all control software is written in C using double-precision arithmetic. Using a medium-level language such as C not only reduces the development time of the software, but also means that the control programs can be moved from simulation to the physical system with little or no changes. The numerical robustness of the algorithms is also greatly improved by the use of double-precision arithmetic. This represents an improvement over more conventional implementations where a substantial part of the software may be sourced in some form of assembler due to hardware limitations.

2.4.6.1. SISO controllers

Benchmark tests on the SISO STRs gave computation times of 6.0 ms for the 5-parameter RLS controller (Eqn. 3.3) and 8.0 ms for the 8-parameter ELS controller (Eqn. 3.2). No further decomposition or multi-transputer implementation was required for these SISO controllers since they both met the required specification.

2.4.6.2. Multi-loop controller

The self-tuning AVR and governor both require 6.0 ms to run since they essentially execute identical code. If both controllers were to run sequentially on a single transputer, this would mean that the minimum sample rate would be 12 ms, which is outside the 10 ms specification. The parallelism of the multi-loop controller is obvious since the AVR and governor loops are distinct. Two transputers can therefore be used to implement the controller, with no associated increase in the sequential computation, leading to an overall control interval of 6.0 ms. The mapping technique involved here is termed functional decomposition since the multi-loop controller is composed of 2 functions (AVR and governor).

2.4.6.3. Multivariable controller

Benchmark tests carried out on the multivariable control algorithm showed that the execution time of the software was 85.0 ms. This is clearly unacceptable and it was necessary to examine procedures to accelerate the algorithm.

Close analysis of the multivariable equations showed a large amount of inherent matrix redundancy. Restructuring of these equations led to a non-redundant multivariable algorithm and a quite marked reduction in overall execution time, from 85.0 ms to 11.4 ms. This is still outside the required specification, however, and so further decomposition is necessary.

Examination of the equations revealed that the control algorithm could be split up into four semi-independent tasks as follows:

(a) calculation of model error, supervision and variable forgetting factor;
(b) calculation of Kalman gain;

Table 1. Controller execution times.

controller	1 transputer	2 transputers
SISO (Eq. 3)	6.0 ms	—
SISO (Eq. 2)	8.0 ms	—
multi-loop	12.0 ms	6.0 ms
multivariable (full matrix form)	85.0 ms	—
multivariable (non-redundant)	11.4 ms	6.6 ms

(c) updating of parameters, supervision, control law;

(d) updating of covariance.

These tasks were formulated to execute in parallel on two transputers, reducing the overall computation time to 6.6 ms, well within the required specification. The mapping technique used here is termed algorithmic decomposition, since the control algorithm itself is decomposed into different tasks. The execution time for all control algorithms are summarized in Table 1.

2.4.6.4. Test results

This section presents results obtained when the concurrent adaptive control system was used to control the behaviour of the synchronous generator during various transient disturbances on the laboratory power system.

In this application, the controllers sampled the system at 10 ms intervals and ran on the PC-based T800 transputer in parallel with the VME system. A 7th order Pseudo Random Binary Sequence (PRBS) was superimposed on the exciter/governor inputs for an initial period of 8 seconds to ensure proper convergence. This input was subsequently removed, leaving noise on the system to provide sufficiently exciting signals.

The figures show the machine response to a three-phase to ground short circuit, which occurs at the sending end of one transmission line, when the generator is controlled by the multivariable STR. The fault is removed by switching out this line after 120 ms, and then the generator then supplies power through the one remaining line. The initial operating conditions are 0.8 p.u. real power and 0.2 p.u. reactive power (nominal operating point). This sequence constitutes a very severe disturbance, immediately followed by a change of system configuration, and consequently it provides a stringent evaluation of controller performance.

Figure 21(a–d) presents a comparison between a conventional fixed-gain AVR, the SISO RLS STR, SISO ELS STR with noise parameters, a conventional multi-loop fixed-gain controller, multi-loop STR, and the multivariable STR under the same transient conditions as the previous test.

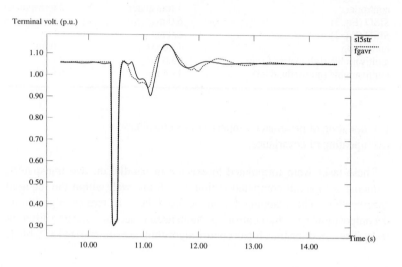

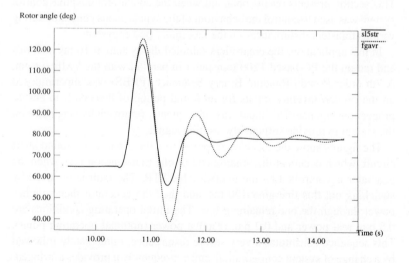

Figure 21a. GMV-RLS self-tuning — 3-phase line out short circuit, duration 120 ms.

Terminal volt. (p.u.)

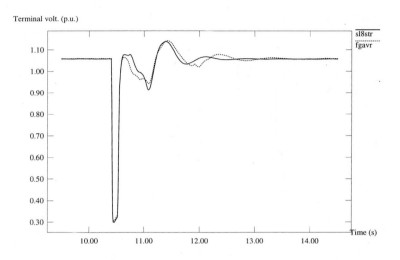

Rotor angle (deg)

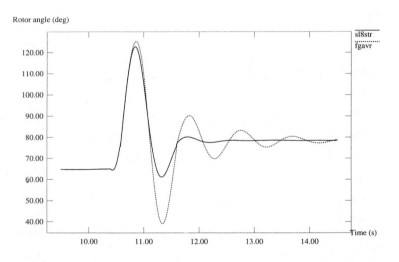

Figure 21b. GMV-ELS self-tuning controller — 3-phase line out short circuit, duration 120 ms.

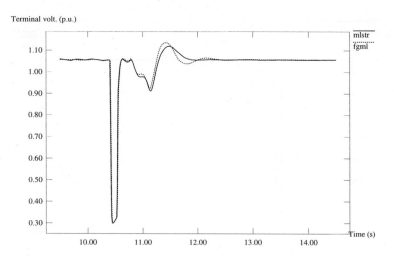

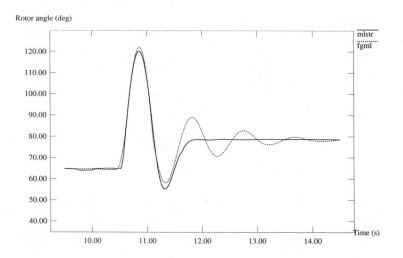

Figure 21c. Multi-loop self-tuning controllers — 3-phase line out short circuit, duration 120 ms.

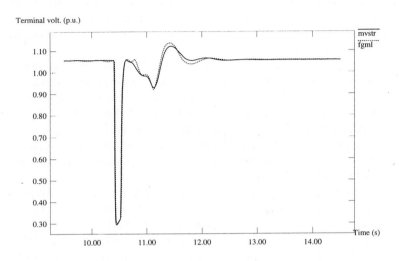

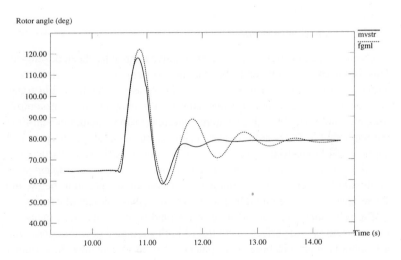

Figure 21d. Multivariable self-tuning controller — 3-phase line out short circuit, duration 120 ms.

As can be seen, the MIMO controllers provide a better transient response than the SISO variants, and that the self-tuning controllers perform consistently better than their fixed-gain counterparts, with a reduction in first rotor angle swing and post fault power oscillations (two important criteria in overall power system stability).

Figure 22(a–d) shows the machine response to a three-phase to ground short circuit, which occurs, as before, at the sending end of one transmission line. Again, comparisons are made between the controllers mentioned previously. The fault is removed by switching out the line after 120 ms, and then the generator then supplies power through the one remaining line. The initial operating conditions are 0.8 p.u. real power and 0.0 p.u. reactive power. In this instance, the generator is open loop unstable and requires stringent control to sustain safe operation.

As can be seen from Figure 22, the fixed-gain AVR is unable to maintain machine stability, with the generator losing synchronism shortly after the fault has occurred. The SISO self-tuning controllers are able to control this major disturbance, but the initial rotor angle swing is considerable. The MIMO controllers reduce this rotor angle excursion (due to the active control of the steam governoring valve), with the multivariable controller providing the best overall response.

2.4.6.4. Discussion of results

A laboratory demonstrator for parallel adaptive control has been developed. This is based on a model turbogenerator system, but the control techniques, software and hardware systems are generic. The work illustrates the complete systems engineering required for parallel adaptive control, from measurement, through algorithms and supervision schemes, to real-time implementation. The benefits of parallel processing technology, in a practical application of industrial significance, have also been established, particularly speed of implementation since it has proved possible to implement a multivariable, self-tuning regulator with a control sample time below the 10 ms specification required by power system equipment manufacturers.

The self-tuning regulation strategy, coupled with an advanced measurement system provide effective control of both simulated power systems and a laboratory micromachine. Transputers are used to calculate the terminal electrical quantities and to perform the self-tuning control law. Extensive computer simulation studies, performed prior to real-time implementation on the laboratory micromachine, using a multi-task parallel model of the control architecture running on a network of transputers, have shown that improved control performance can be gained by using self-tuning regulators instead of a more conventional fixed gain regulator. The self-tuning regulators exhibited

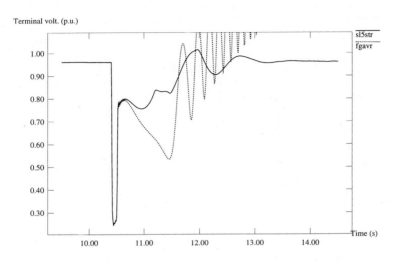

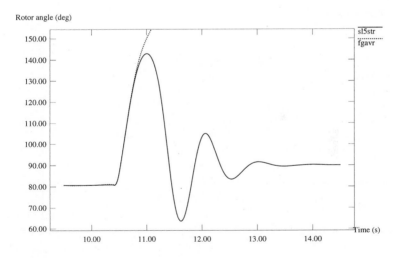

Figure 22a. GMV-RLS self-tuning controller — 3-phase line out short circuit, duration 120 ms.

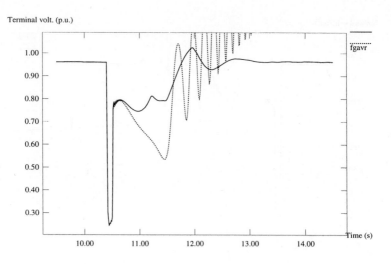

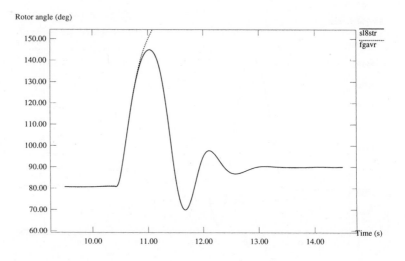

Figure 22b. GMV-ELS self-tuning controller — 3-phase line out short circuit, duration 120 ms.

Terminal volt. (p.u.)

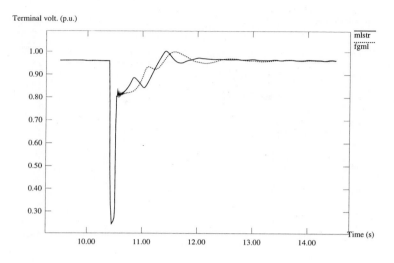

Rotor angle (deg)

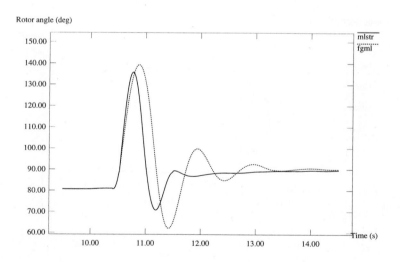

Figure 22c. Multi-loop self-tuning controllers — 3-phase line out short circuit, duration 120 ms.

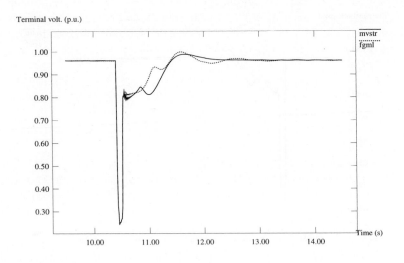

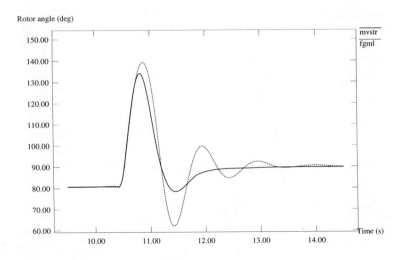

Figure 22d. Multivariable self-tuning controller — 3-phase line out short circuit, duration 120 ms.

better damping, less power oscillations and quicker voltage stabilisation than the fixed gain AVR. This not only confirms the viability of the self-tuning algorithm but also validates the robustness of the supervision scheme. This should be viewed in the light of the fact that the self-tuning algorithms have no prior knowledge of the plant and are based on a model which could not be derived by any analytical means.

Although the results achieved with the micromachine studies cannot be reasonably expected to match those obtainable on a full size generating set, they do serve to demonstrate the practical viability of both the measurement and control systems, and to point to the benefits which may accrue form adaptive self-tuning governors and AVRs.

In this context it is important to realise that self-tuning controllers demand significant commissioning effort in that a number of tuning parameters like fixed estimates and predictor polynomials (section 3) must be selected. Although off-line simulation proved to be an important tool for this task, the resultant performance in both simulation and on the micromachine system could be significantly altered by an incorrect choice. In particular, the multivariable adaptive controller, which compensates for the inevitable cross-coupling between governor and voltage loops, was difficult to set up in practice since 22 parameters had to be chosen. For this reason, the multi-loop approach to independent voltage and governor control is likely to remain the preferred solution.

2.5. INDUSTRIAL IMPLEMENTATION

2.5.1. Introduction

The introduction of adaptive control schemes on the Queen's University of Belfast micromachine system suggested the deployment of complementary advanced measurement techniques [39]. As discussed in section 2.4, in order to achieve this aim multiple voltage and current transformers, as well as sophisticated filtering algorithms, were employed. However, for an industrial installation a minimalist hardware arrangement is normally adopted, employing perhaps one or two voltage transformers and current transformers, and corresponding comparatively simple measurement schemes. The Parsons Power Generation Systems (PPGS) measurement system, for a single channel digital automatic voltage regulator (AVR) follows these principles [40].

It is of interest to evaluate the performance of the adaptive controllers developed at QUB on an industrial test-bed, utilizing a standard digital AVR and associated measurement systems.

2.5.2. Industrial VME Hardware System

The PPGS VME hardware has been designed specifically for industrial installation. Originally, a triplex modular redundant (TMR) architecture was used, in conjunction with majority voting of the outputs of three control channels. However, a recent trend has been towards the design of much cheaper single channel systems [43]. This has arisen due to an increased user confidence with digital systems, and improved reliability of digital hardware. The quantities measured include terminal voltage, real and reactive power, frequency and field current. For display and analysis purposes, mechanical speed and rotor angle measurements have also been made available.

2.5.2.1. Measurement of signals

The raw outputs from the synchronous machine are fed to a 3-phase transformer with three secondaries - two of them at 55 V phase-phase, and the other at 3.5 V phase-neutral. The 55 V secondaries are connected in star — delta providing a 6-phase system. A current transformer is also connected to a single phase of the microalternator. The 6-phase voltage system permits measurement of the terminal voltage, while the 3.5 V transformer output, in conjunction with a measured phase current, allows both the real and reactive power to be calculated. The six phases from the 3-phase transformer are rectified to produce a composite signal, which is then low pass filtered. The resulting d.c. signal, proportional to the terminal voltage, V_T, is sampled using a 12 bit analogue to digital converter.

The real power, P_T, and reactive power, Q_T, are measured by synchronous rectification, whereby the generator current, I_T, is resolved into direct and quadrature components, $I_T \cos \phi$ and $I_T \sin \phi$. Real power and reactive power are then determined as $P_T = V_T I_T \cos \phi$ and $Q_T = V_T I_T$, $\sin \phi$ where ϕ represents the angle between the terminal voltage and the terminal current. Resolution of the terminal current is achieved through measurement of a line voltage from the 3-phase transformer 3.5 V secondary star winding, and the current from the current transformer. The voltage signal is passed through a comparator edge triggered network, to produce a train of pulses at ≈ 10 ms intervals. The quadrature and in-phase current components are then obtained by integrating the line current between two successive pulses, and two 90° phase shifted pulses [24] The analogue measurements are then sampled, with additional hardware compensating for the alternating signs of $I_T \cos \phi$ and $I_T \sin \phi$, corresponding to the positive and negative halves of the current waveform.

During the auto excite sequence, generator frequency is monitored to prevent the generator transformer overfluxing. However, once synchronized the AVR no longer requires measurement of system frequency for generator control. The scheme employed is based around estimating the time between zero crossings of a voltage waveform. Under fault conditions, such an approach is suspect, producing incorrect measurements, due to the voltage waveforms transiently being aperiodic. For analysis purposes, a second scheme, which does not form part of the industrial implementation, has been adopted which overcomes these problems [44]. By attaching a 64 slot disc to the free end of the prime mover, rotor pulses are generated using an eddy current probe, at ≈ 0.625 ms intervals. A further pulse train is then created for every 16 eddy current probe pulses, producing a signal similar to that of the original frequency measurement method. However, as the pulses are generated mechanically, measurements remain valid under electrical faults. Arrival of each pulse triggers the reading of a 1.2 MHz counter, which represents the time between successive pulses.

The rotor angle is measured in a similar manner to that for machine speed. The derived pulse train is phase shifted through a precise angle, such that the shifted signal is in phase with a similar pulse train obtained from the busbar voltage on open circuit. As the generator is progressively loaded, the rotor advances with respect to a reference pulse train produced by the busbar voltage. The advance is measured by a phase difference timer. The accuracy of the measured rotor angle depends on the system frequency, and can be corrected for frequency variations by scaling the phase difference in proportion to the measured machine speed [40].

2.5.2.2. VME hardware system

The physical hardware of the single channel digital AVR system is based around a VME bus compatible CPU board. The specification for the industrial hardware system is as follows,

Master	25 MHz 68030, 68882 FPU, 1 MB DRAM
Analogue I/O	20 channel 12 bit A/D, 2 analogue outputs
Digital I/O	16 16 bit I/O lines
Speed Detector	(custom board)
Sample Control	(custom board)
Serial I/O	8 RS232/422 channels, 38.4 kBaud

Although a different arrangement of VME cards has been used in comparison with the Queen's University of Belfast measurement system, the two configurations are compatible. The main difference between the two systems is that the controller software for the PPGS system runs on a Motorola 68030 microprocessor, while the QUB system employs an array of transputers. A *pseudo* QUB VME system may be constructed by inserting a VME transputer card into the industrial system, Figure 23.

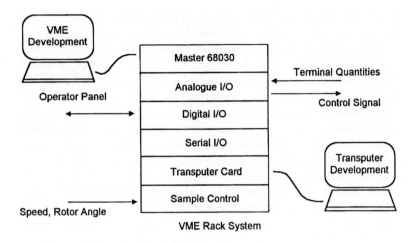

Figure 23. VME hardware system.

The measurement system is based around a 10 ms sample period, which originates from zero crossings of the v_{ac} voltage waveform. To improve noise immunity input signals are read three times in quick succession, and the median value selected. Once the AVR control signal has been calculated, it is transmitted by the analogue I/O module to a thyristor converter. For power station implementation this signal would be communicated via RS422 serial links, using the serial I/O module, to a Bridge Control Unit (BCU).

2.5.2.3. Software development

The AVR software is implemented using the OS9 real-time operating system. It is a multi-tasking environment, allowing several independent programs to be executed simultaneously through time slicing [41]. The system's CPU is interrupted at a regular rate of *usually* 100 times per second. The interrupt is generated from zero crossings of the v_{ac} waveform. At each interrupt or tick, OS9 can suspend execution of one program and begin execution of another [41]. The starting and stopping of programs is performed in a manner that does not affect an individual program's execution. Thus each second of CPU time is split up, to be shared among several concurrent processes.

The software modules themselves have been written in ISO standard Pascal, with the addition of OS9 system calls as necessary. The AVR uses the OS9 operating system to perform all routine housekeeping jobs associated with a software based system, such as monitoring switches on the user interface and updating the display.

2.5.3. Control Strategies

In simulation and QUB micromachine studies the self-tuning strategies were compared with an AVR + PSS. This fixed gain controller is replaced with a digital proportional filter, as employed on industrial digital excitation systems.

2.5.3.1. Fixed gain controller

Industrial digital excitation systems employ a digital proportional filter. For the purposes of comparison, it should be noted that the PSS module has not been included, and the controller can therefore not be expected to match the transient damping capabilities of the GMV and PID controllers. The transfer function for the controller is expressed as follows,

$$\frac{4548}{128} \frac{(1+0.7s)}{(1+7s)} \frac{(1+0.4s)}{(1+0.2s)} \tag{5.1}$$

The controller time constants selected correspond to those employed by four single channel digital AVRs at a 2000 MW coal-fired power station. The controller was tuned from open-circuit step response tests on the generators. As the controller does not incorporate integral action, a steady state control error may be anticipated.

As the AVR system is intended for industrial use, the software contains various generator protection schemes. Provision is made to restrict the AVR output under field forcing conditions to avoid overheating the rotor. VAr limiting, under leading power factor operation, is provided to maintain synchronous stability. Overflux protection during generator synchronization is incorporated. Auto-excite sequencing and de-excite facilities are also available.

2.5.3.2. Selection of power system stabilization signal

The traditional selection for a stabilizer input signal is rotor shaft speed. However, the addition of speed detectors to large steam turbine generators has emphasized that turbine shafts cannot be regarded as infinitely stiff, with the detectors having to be restricted to points along the turbine shaft corresponding to nodes of oscillation. The presence of vibrations can subsequently lead to operational difficulties of power system stabilizers [42]. Consequently, a signal derived from electrical output power is usually employed in practice. This has minimal effect on the operation of individual controllers, however it can have an unwanted effect on transient stability by preventing the beneficial action of field forcing during severe disturbances. Excessive terminal voltage excursions may also arise during mechanical power changes. These factors necessitate limiting the stabilizing signal.

2.5.4. Micromachine Test Results

For power station implementation, the AVR and governor control systems have traditionally been implemented as distinct systems. Hence, with conventional industrial hardware it is not practical to implement the MIMO control strategies, previously demonstrated in section 2.4.6. However, current research is investigating the industrial implementation of integrated AVR and governor control, with potential improvements in control performance and significant reductions in hardware cost being foreseen [44]. Therefore, the performance of SISO control algorithms alone are investigated under the following scenarios,

- 3-phase-earth short circuit after 2.0 seconds at the sending end of the transmission line system, duration 100 ms, at an operating point of $P_T = 0.8$ pu and $Q_T = 0.2$ pu lagging
- voltage setpoint change of $\Delta V_{T_{\text{ref}}} = 0.02$ pu after 7.0 seconds, and subsequent recovery after 17.0 seconds, at an operating point of $P_T = 0.5$ pu and $Q_T = 0.1$ pu lagging.

Figure 24 illustrates the terminal voltage and rotor angle signals for the GMV-ELS, GMV-RLS and PID self-tuning schemes. In each case, the fixed gain controller response is included for comparison. The machine rotor angle is introduced to assess the transient performance of the excitation control system. Examination of the 3-phase short circuit responses illustrates that the various control schemes provide satisfactory performance for the applied conditions, with well damped responses achieved. However, it is observed that the actions of the self-tuning controllers are clearly superior, with a significant reduction in the second rotor swing and improved damping of the subsequent oscillations, illustrating the transient behaviour benefits of PSS action. The GMV schemes also provide a significant reduction in the initial rotor swing, while, the self-tuning schemes, in general, offer improved voltage regulation, with minimal delay before the terminal voltage returns to its preset reference level.

Similarly, in Figure 25, the application of successive voltage setpoint changes demonstrates a faster transient response, and improved voltage regulation, for the self-tuning regulators over the fixed gain scheme. To facilitate comparison the steady state voltage offset has been removed from

Terminal Voltage (pu)

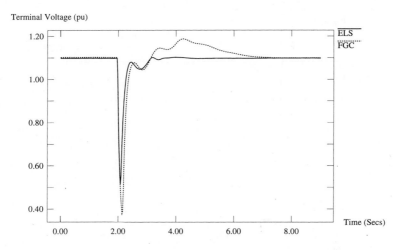

Rotor Angle (Deg.)

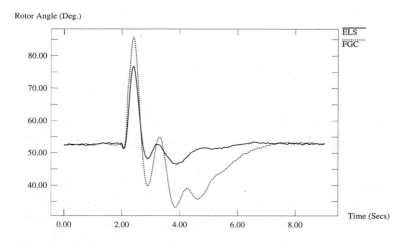

Figure 24a. GMV-ELS self-tuning controller — 3-phase earth short circuit.

Terminal Voltage (pu)

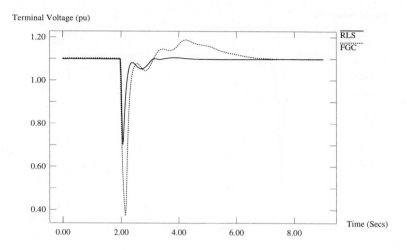

Rotor Angle (Deg.)

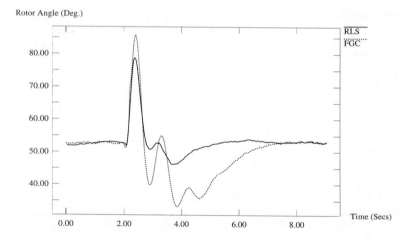

Figure 24b. GMV-RLS self-tuning controller — 3-phase-earth short circuit.

Terminal Voltage (pu)

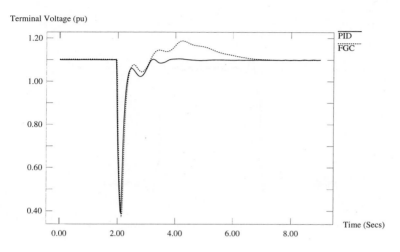

Rotor Angle (Deg.)

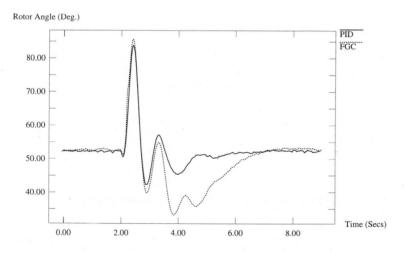

Figure 24c. Self-tuning PID controller — 3-phase-earth short circuit.

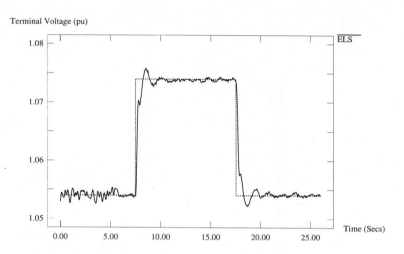

Figure 25a. GMV-ELS self-tuning controller — voltage setpoint changes.

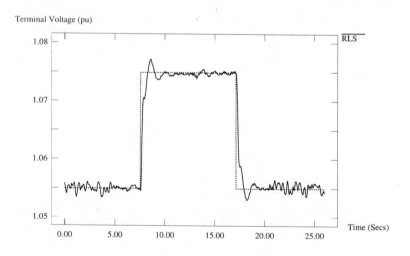

Figure 25b. GMV-RLS self-tuning controller — voltage setpoint changes.

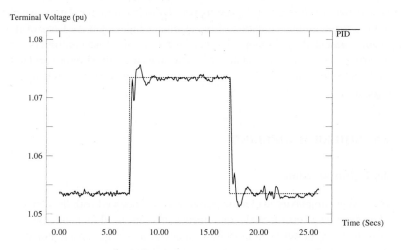

Figure 25C. Self-tuning PID controller — voltage setpoint changes.

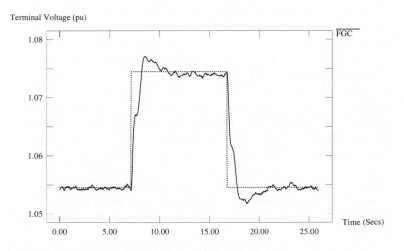

Figure 25d. Fixed gain controller — voltage setpoint changes.

the fixed gain controller response. The addition of PRBS excitation signals is also clear from some of the results with the self-tuning regulators, which help to maintain the excitation of the estimator inputs. On the basis of the results presented, and other micromachine tests, the four controllers can be ordered in increasing quality of performance as fixed gain regulator, PID controller, GMV-RLS self-tuning regulator, and GMV-ELS self-tuning regulator.

2.6. BOILER MODELLING

2.6.1. Introduction

Increasing complexity in electric power systems, coupled with the demands of economic and operational requirements, drive the need for continuing improvements in power plant modelling and control. Accurate plant modelling is most important in the assessment and prediction of performance, and in seeking improvements in design, operating procedures and control strategies. A generating unit is a highly complex nonlinear system, and extensive work has been performed on derivation and validation of models, ranging from very detailed representations based on physical laws, to low-order linear ones obtained by system identification [45,46]. There is also increasing interest in developing real-time models to run in parallel with the plant for operator training and for predicting the effects of possible control action [47].

Detailed nonlinear models, however, tend to be computationally intensive, are expensive to build, and also present difficulties in obtaining appropriate plant data. Turbine and generator systems are generally assumed to be more readily defined, but the boiler presents particular difficulties. For example, the time constants differ greatly in various parts of the boiler system, and the performance changes significantly due to variations in fuel quality and fouling. There is a clear need for accurate nonlinear models of individual units, which take account of the factors indicated above. In pratice, a selection of models of varying complexity is desirable for defined purposes, such as detailed performance evaluation, real-time simulation and control system design.

Neural networks offer a framework for nonlinear modelling and control [48] based on their ability to learn complex nonlinear functional mappings. This section reports on the application of a particular network paradigm, the multilayer perceptron, to off-line identification of a 200 MW, oil-fired, drum-type boiler unit represented by a detailed computer simulation model. The latter has been produced in collaboration with Ballylumford power station in Northern Ireland.

A description of the boiler system is given in section 2.6.2 and practical issues for plant identification, such as noise levels and allowable test conditions, are discussed. Section 2.6.3 describes the application of linear, multivariable ARX modelling leading to 4-input, 4-output models at operating points of 100 MW and 200 MW. Section 2.6.4 deals with the application of the MLP to the same training data as employed for the earlier linear modelling, highlighting the problems of training from noisy data, and confirming the ability of the network to learn local nonlinear models with comparable performance to the ARX ones. The full potential of the new technology is also made clear, where the network, trained to form a global model of the Ballylumford boiler system, proves capable of capturing the dynamics right across the whole operating range.

2.6.2. Ballylumford Boiler Model

The model of the boiler system contains all of the important control loops, together with typical subsystems of evaporation, heat exchangers, spray water attemperators, steam volume, turbines and water, and steam and gas properties [49]. This simulation contains 14 non-linear differential equations and 112 algebraic equations, and has been designed and verified using tests on a 200 MW oil-fired drum type boiler turbogenerator unit at Ballylumford power station, N. Ireland.

The model is a direct simulation of unit 5 at Ballylumford [50] and consists of a 200 MW two-pole generator supplied by GEC, directly coupled to a three-stage turbine also supplied by GEC. The turbine is driven by steam from a drum-type boiler. Figure 26 shows the layout of the complete boiler turbine unit.

2.6.2.1. Test conditions and allowable perturbation signals

The Ballylumford simulation has 6 control loops, as shown in Figures 27–30 (there are 3 superheater sections). All are digital, with a sampling interval of 30 seconds, and the plant always operates in closed loop. Suitable noise levels and deviations were chosen from plant data and previous identification tests.

Due to the critical conditions that apply within a thermal power generation unit, extreme care must be taken when applying disturbances for the purposes of identification. The power plant must always operate in closed loop. Based on previous power plant experience, PRBS (pseudo random binary sequence) test signals on the attemperator spray control input, on the main steam pressure set point, and on the governor valve control input are allowable.

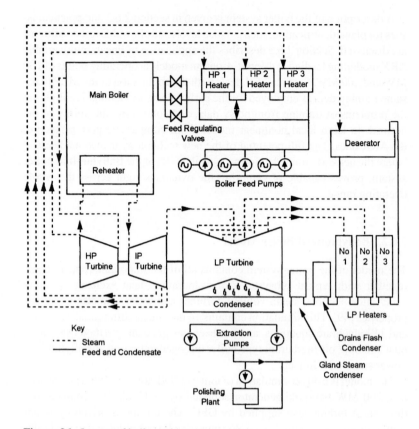

Figure 26. Layout of boiler turbine (unit no. 5).

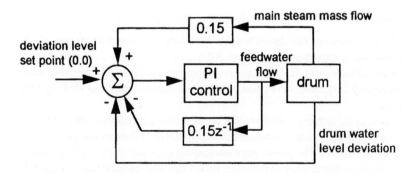

Figure 27. Steam drum water level controller.

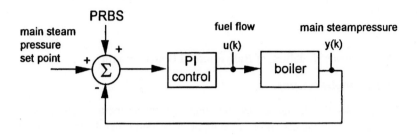

Figure 28. PRBS on main steam pressure set point.

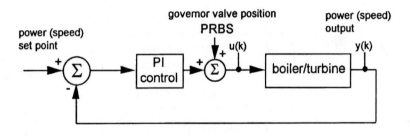

Figure 29. PRBS on governor valve input.

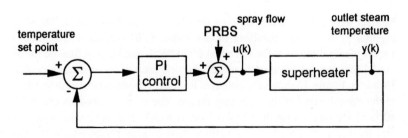

Figure 30. PRBS on attemperator spray input.

The test sequence width was set to four times the digital control sample rate and each test lasted for 1.5 h (i.e. 180 samples at a control sample rate of 30 s).

2.6.3. Linear Modelling of Ballylumford Boiler System

2.6.3.1. Model structure

A 7-th order PRBS signal was superimposed on the main steam pressure setpoint and on the governor valve input. A second order 4-input 4-output ARX (autoregressive with exogenous inputs) model was then formed from the resulting data. The structure of the model follows the usual ARX format:

$$A(z^{-1})Y(k) = B(z^{-1})U(k) + E(k) \qquad (6.1)$$

where $A(z^{-1})$ and $B(z^{-1})$ are polynomial matrices in the backward shift operator and:

$$Y(k) = \begin{bmatrix} y_1(k) & \text{(drum water level dev.)} \\ y_2(k) & \text{(electrical power output dev.)} \\ y_3(k) & \text{(main steam pressure dev.)} \\ y_4(k) & \text{(main steam temp. dev.)} \end{bmatrix} \qquad (6.2)$$

$$U(k) = \begin{bmatrix} u_1(k) & \text{(feed water flow dev.)} \\ u_2(k) & \text{(governor valve input dev.)} \\ u_3(k) & \text{(fuel flow dev.)} \\ u_4(k) & \text{(at temperature spray dev.)} \end{bmatrix} \qquad (6.3)$$

$$E(k) = [4 \times 1 \ \text{vector of uncorrelated noise sources}] \qquad (6.4)$$

2.6.3.2. Test results

Models were formed using training data both at 100 MW and at 200 MW. Figures 31 shows the resulting model output (180 step-ahead-prediction) versus plant output over the test set at 200 MW. It can be seen that the linear model matches the plant quite closely around each of the operating points. However, Figure 32 shows the output of this linear model when using test data at the opposite end of the operating region. The poorer responses obtained, particularly marked on the graph of drum water level deviation, indicates that the linear models are only valid around a small region at a particular operating point. When moving to a different operating point, the dynamics of the plant will change and the original linear model is no longer valid. This was expected and follows the usual behaviour of non-linear plant.

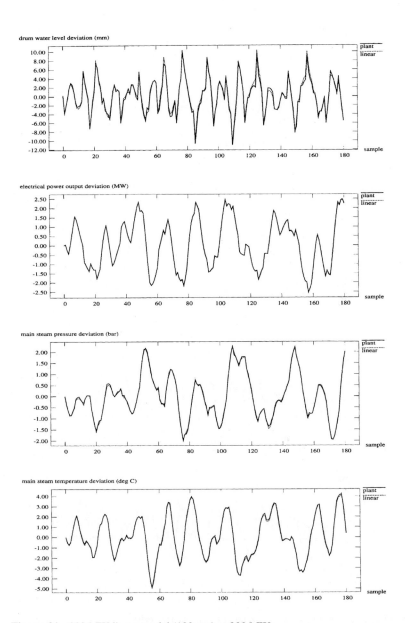

Figure 31. 200 MW linear model (180 sap) at 200 MW.

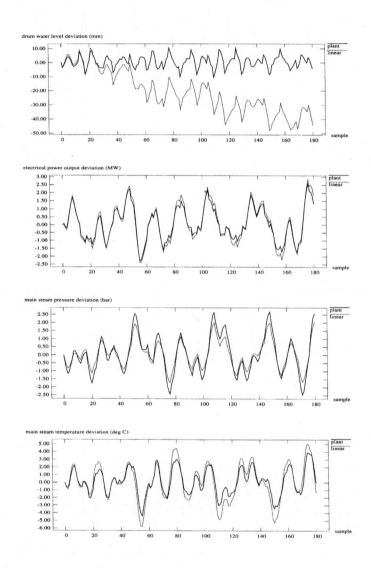

Figure 32. 200 MW linear model (180 sap) at 100 MW.

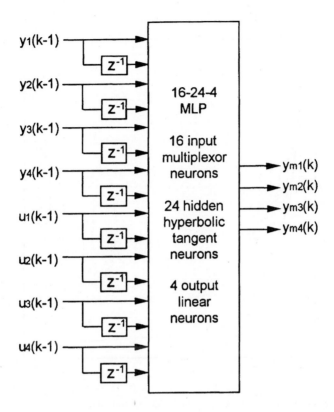

Figure 33. MLP neural network structure.

2.6.4. Neural Modelling of the Ballylumford Boiler System

2.6.4.1. Network structure

A 4-input 4-output 2nd order dynamic non-linear model was simulated using a 16-24-4 Multilayer Perceptron (MLP) as in Figure 33.

2.6.4.2. Neural network training

As in the linear modelling of section 2.6.3, two data sets were generated in the neural modelling procedure — one for training and the other for testing. The weights of the neural network were adjusted using an output (1 step ahead) prediction error square sum over the training set with the BFGS algorithm [51].

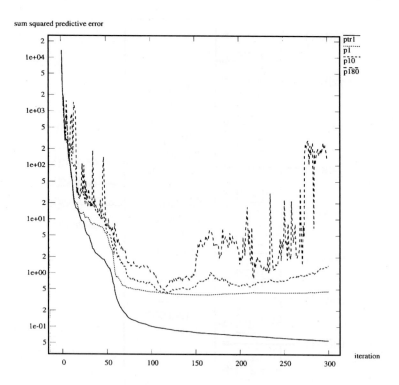

Figure 34. Evolution of predictive errors during training.

Since the data are contaminated with noise, extra care must be taken to ensure that only the dynamics contained in the data are learned and not the noise, as this lessens the predictive capability of the model and results in an over-trained network. With this in mind, the optimum number of presentations of the training data (or iterations) must be found. The effect of over-training is obvious by referring to Figure 34 which shows the evolution of predictive error versus training iteration for 1-step-ahead (1-sap) training set (the error which is used to adjust the weights), 1-sap test set, 10-sap test set, and 180-sap test set. From these graphs it can be seen that the optimum predictive capability of the network occurs around iteration 115, and that the predictive error (especially the 180-sap) increases significantly as the number of iterations increases. Notice also that the 1-sap over the training set always decreases — this is due to the minimization of 1-sap predictive error by the training algorithm.

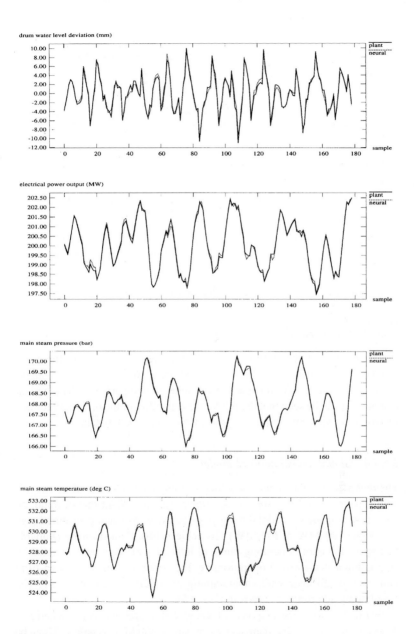

Figure 35. 200 MW neural model (180 sap) at 200 MW.

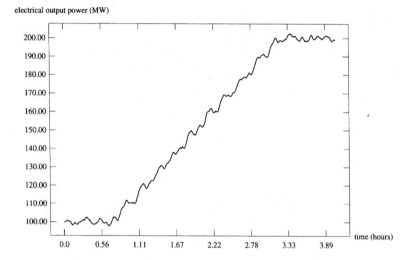

Figure 36. Two-shift load profile used for training.

2.6.4.3. *Local neural network models*

Neural models of the Ballylumford system were formed using the same training and test data as for the linear modelling exercise. Figure 35 shows the 180-sap at 200 MW for this neural model in comparison with the plant response. It can be seen that the predictive capability of the optimally trained networks is excellent, despite the presence of noise on the data.

2.6.4.4. *Global neural network model*

The real power of the neural modelling technique comes from the network's ability to represent an arbitrary non-linear function. With this in mind, an attempt was made to model the boiler system over its usual operating range (i.e. from 100 MW – 200 MW). To achieve this, data must be available that spans the entire operating range of interest. This data was obtained by simulating a two-shifting operation, i.e. the boiler system was driven from half load to full load, whilst maintaining the PRBS signals as before. The simulated load profile is shown in Figure 36.

Figure 37 shows the response of the global neural network, trained and tested using two-shifting data, using test data at 100 MW as before. It can be seen that the network is capable of representing the behaviour of the boiler system at the two extreme operating regions, indicating that the non-linear response of the system has been captured.

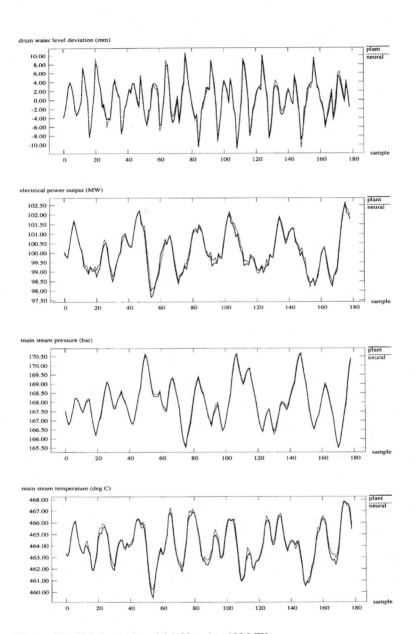

Figure 37. Global neural model (180 sap) at 100 MW.

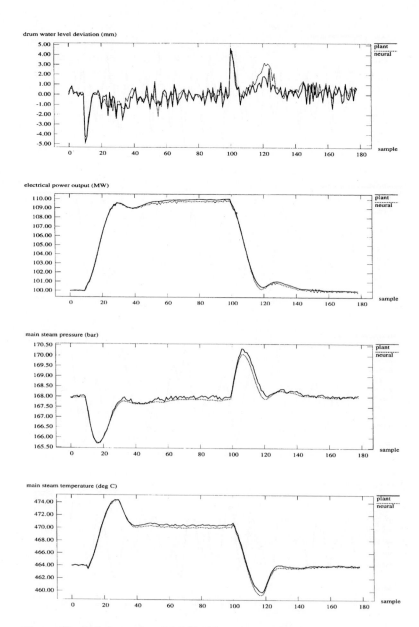

Figure 38. Global neural model (180 sap) step test at 100 MW.

Figure 38 shows the response of the network under a completely different test condition. Here, the network simulates the behaviour of the plant under a +10 MW demand in power followed by a −10 MW demand in power. Again, the response of the plant and the network are in close agreement. These results clearly illustrate the improvements in predicted plant output, which can accrue from nonlinear modelling.

2.6.5. Conclusions

This section has been concerned with the application of neural networks to off-line identification of a 200 MW, oil-fired, drum-type boiler unit at Ballylumford power station in Northern Ireland. The aim was to demonstrate the potential advantages of these new techniques for nonlinear modelling of power plant compared with conventional linear ARX approaches.

Multivariable ARX models were formed for 100 MW and 200 MW operating points. These showed good predictive capabilities at their respective operating points, but as expected, the performance deteriorated when the models were tested at opposite ends of the operating range. These results illustrate clearly the limitations of linear models for a highly nonlinear plant which are perhaps most clearly reflected in the predictived dynamic responses of drum water level deviations.

A single hidden layer (16,24,4) MLP was trained and tested on the same noisy, and under the same operating conditions used for the ARX modelling. Local nonlinear neural models were produced for both 100 and 200 MW operating points. These gave good results in predicting the plant output, demonstrating that the MLP was at least capable of matching the performance of the linear models. However, the technique is only of benefit if the linear results can be bettered. To this end a single global neural nonlinear model has been produced and was shown to perform sastisfactorily at both extremes of the operating range. This constitutes a considerable improvement over the best linear results and shows that neural networks are a powerful modelling tool, if the training data are sufficiently rich in information about the system dynamics across the operating range of interest.

REFERENCES

1. Armor, A.F. and Weiss, J.M., 1995, *IFAC Symposium on Control of Power Plants and Power Systems*, Cancun, Mexico.
2. Ham, P.A.L., 1972, *Inst. Measurement and Control. Turbine and Compressor Control.*
3. Hertzog, H. and Baumberger, H., 1990, *ABB Review*, 1/90.
4. Ham, P.A.L. and Green, N.J., 1988, *IEEE Winter Power Meeting*, Paper 88 WM 236-2.

5. Rush, P.W. and White, B.J., 1986, *21st UPEC*, 15–17th April, 1986, Imperial College, London.
6. Weedy, B.M., 1984, *Electric Power Systems*, J. Wiley & Sons.
7. Hogg, B.W., 1981, Chp. 5 in H. Nicholson, *Modelling of Dynamic Systems*, vol. 2, P. Peregrinus.
8. Brown, M.D., 1991, *Transputer Implementation of Adaptive Control for Turbogenerator Systems*, PhD Thesis, The Queen's University of Belfast, U.K.
9. Subramaniam, B.E. and Malik, O.P., 1971, *Proc. IEEE*, **118**(1), 153–160.
10. Barber, M.D. and Giannini, M., 1974, *Proc. IEEE*, **121**(12), 1512–1521.
11. Anderson, P.M. and Fouad, A.A. 1977, *Power System Control and Stability*, Iowa State University Press, USA.
12. Matsch, L.W. and Morgan, D.J., 1987, *Electromagnetic and Electromagnetic Machines*, J. Wiley & Sons.
13. King, S., 1986, *Modeller User's Manual*, School of Studies in Computing, Bradford University UK.
14. McMillan, *et al.*, 1993, *Intech*, pp.24–26.
15. Wu, Q.H. and Hogg, B.W., 1988, *IEEE Proc. Pt. D*, **135**(1), 35–41.
16. Kanniah, J., Malik, O.P. and Hope, G.S., 1984, *IEEE Trans. PAS*, **103**(5), 897–910.
17. Wu, Q.H. and Hogg, B.W., 1991, *Automatica*, **27**(5), 845–852.
18. Malik, O.P. *et al.*, 1992, **PES Winter Power Meeting**.
19. Isermann, R. and Lachmann, K.H., 1985, *Automatica*, **21**(6), 625–638.
20. Astrom, K.J. and Eykhoff, P., 1971, *Automatica*, **7**, 123–162.
21. Bayne, J.P., Kundur, P. and Watson, W., 1975, *PAS*, **94**, 1141–1146.
22. Ahson, S.I. and Hogg, B.W., 1979, *Int. J. Cont.*, **30**(4), 533–548.
23. Flynn, D., Hogg, B.W., Swidenbank, E. and Zachariah, K.J., *Control Engineering Practice*, **3**(11), 1571–1579.
24. Flynn, D., 1994, *Expert self-tuning control for turbogenerator systems*, PhD Thesis, The Queen's University of Belfast.
25. Cheng, S.J. *et al.*, 1986, *IEEE Trans. PWRS*, **1**(3), 101–107.
26. Astrom, K.J. and Wittenmark, B., 1980, *Proc. IEEE Pt. D*, **127**, 120–130.
27. Wittenmark, B. and Astrom, K.J., 1980, *Symp. on Methods and Applications in Adaptive Control*, Bochum, pp.21–30.
28. Brown, M.D., Irwin, G.W., Swidenbank, E. and Hogg, B.W., 1994, *Control Engineering Practice*, **2**(3), 405–414.
29. Ibrahim, A.S., Hogg, B.W. and Sharaf, M.M., 1989, *IEEE Proc. Pt. D*, **136**(5), 238–251.
30. Flynn, D., Hogg, B.W. and Swidenbank, E., 1994, *Trans. Inst. MC*, **16**(1), 40–47.
31. Parkum, J.E., Poulsen, N.K. and Holst, J., 1992, *Int. J. Control*, **55**(1), 109–128.
32. Astrom, K.J. and Wittenmark, B. 1989, *Adaptive control*, Addison-Wesley.
33. Wellstead, P.E. and Zarrop M.B., 1992, *Self-Tuning Systems — Control and Signal Processing*, Wiley.
34. Hirayama, K. *et al.*, 1993, *IEEE Trans EC*, **8**(4), 602–609.
35. Pullman, R.T., 1987, *Coordinated Excitation and Governor Control of Turbogenerators using State-Space Techniques*, PhD Thesis, University of Liverpool U.K.
36. Brown, M.D. and Swidenbank, E., 1995, *Int. Journal of Electric Power & Energy Systems*, **17**(1), 21–38.
37. Swidenbank, E., 1994, *Measurement of Speed and Rotor Angle*, The Queen's University of Belfast UK, Department of Electrical and Electronic Engineering.
38. White, B.J. and Zachariah, K.J., 1989, *24th UPEC*, Belfast U.K.
39. Brown, M.D. and Swidenbank, E., 1991, *1st IFAC Symp. on Parallel Algorithms and Architectures for Real-Time Control*, Univ. of Bangor, Wales.
40. Zachariah, K.J., Finch, J.W. and Farsi, M., 1990, *25th UPEC*, Aberdeen, pp.623–626.
41. *Using Professional OS-9 — Operating System Manuals & Language Manuals*, Microware Systems Corp, 1986.
42. *Turbines, Generators and Associated Plant*, Modern Power System Practice Series, Pergamon Press, 1992.

43. Green, N.J. and Hutchinson, A.M., 1994, *Control '94*, Univ. of Warwick, Coventry, 21–24 March, pp.385–389.
44. Flynn, D., Hogg, B.W., Swidenbank, E. and Zachariah, K.J., *IEEE PES Winter Power Meeting*, Baltimore, USA, 21–25 January 1996.
45. Chawdry, P.K. and Hogg, B.W., 1989, *IEEE Proc. C*, Gener. Transm. Distrib., **136**(5), 261–271.
46. Swidenbank, E. and Hogg, B.W., 1989, *IEEE Proc. D, Control Theory Appl.*, **136**(3), 113–121.
47. Takebe, T., Iwaasa, T., and Christie, D., 1993, *Advances in Instrumentation and Control, Int. Conf. and Exhibition*, **48**(part 1), 511–519.
48. Irwin, G., Warwick, K. and Hunt, K. (Eds), 1995, *Neural network applications in control and systems*, IEEE Control Engineering Series.
49. Lu, S., A thermal power plant model, *Year report 1992*, EDC the Queen's University of Belfast.
50. Forsythe, T., 1993, *Adaptive control of a power station boiler*, PhD thesis, the Queen's University of Belfast, U.K.
51. Irwin, G., Brown, M., Hogg, B.W. and Swidenbank, E., 1995, *IEEE Proc. D, Control Theory Appl.*, **142**(6), 529–536.

3 ARTIFICIAL NEURAL NETWORKS FOR PATTERN RECOGNITION AND DATA CLASSIFICATION

J. McDONALD, G. BURT, J.A. STEELE and A. UL-ASAR

University of Strathclyde, Royal College Building, 204 George Street, Glasgow, Scotland, G1 1XW, UK

The principal aim of this chapter is to investigate the application of a maturing technology, called Artificial Neural Networks (ANNs), to the short term load forecasting problem in power systems demonstrating, via examples, that ANNs can provide an effective forecasting system. Additional investigations were also carried out to exploit the promise of the ANN methodology in the analysis of electrical plant data for condition monitoring purposes.

3.1. INTRODUCTION

Over the past few years neural networks have received a great deal of attention and are being treated as one of the most significant computational tools ever developed. Much of the excitement is due to the apparent ability of neural networks to imitate the humans brains ability to make decisions and draw conclusions when presented with complex, noisy, irrelevant, and/or partial information.

Artificial neural networks go by many names such as connectionist models, parallel distributed processing models, and neuromorphic systems. Whatever the name, all these models attempt to achieve good performance via dense interconnection of simple computational elements. Instead of performing a program of instructions sequentially, as in conventional computation, neural network models explore many competing hypothesis simultaneously using potentially massively parallel nets composed of many computational elements connected by links with variable weights.

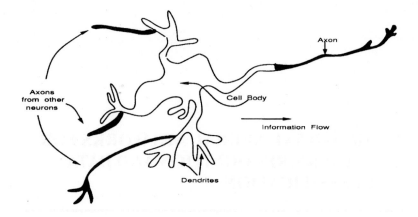

Figure 1. The biological model of a neuron.

3.1.1. Biological Neuron

Studies over the past few decades have shed some light on the construction and operation of the human brain and nervous system [1–3]. The basic building block in the nervous system is the neuron. The major components of a neuron include a central body, dendrites, and an axon. Figure 1, a conceptual diagram of a neuron, is a sketch of only one representation of a neuron. The signal flow goes from left to right, from the dendrites, through the cell body, and out through the axon. The signal from one neuron is passed on to another by means of a connection between the axon and of the first and a dendrite of the second. This connection is called the synapse. Axons often connect onto the trunk of a dendrite, but they can also connect onto the cell body. The human brain has a large number of neurons or processing elements. Typical estimates are in the order of 10–500 billion [1,3]. According to one estimate [2], neurons are arranged into about 1000 main modules, each with about 500 neural networks. Each network has in the order of 100,000 neurons. The axon of each neuron connects to about 100 (but sometimes several thousand) other neurons, and this value varies greatly from neuron to neuron. The artificial neural network is roughly based on the biological neural network. In biological neural networks, the memory is believed to be stored in the strength of the interconnection between layers of neurons. Using neural networks terminology, the strength or the influence of an interconnection is known as its weight. Artificial neural networks borrow from this theory and utilise variable interconnection weights between layers of simulated neurons.

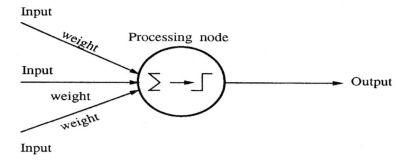

Figure 2. The outline of the basic ANN model.

3.1.2. The Artificial Neural Networks

An artificial neural network is a data-processing structure containing processing element, called neurons, which are fully interconnected with one-way
signal channels, called connections. Each neuron can take in data from many
sources but can send data out in one direction. This output connection can
branch into a number of other connections, but each connection then carries
the same signal. Each input is multiplied by a corresponding weight, which
is analogous to the brain's synaptic strength. All the weighted inputs are then
summed to determine the activation level of the neuron. Nearly all neural
network architecture's are based on this model. The diagram of the basic
neural networks is shown in Figure 2.

   Neural networks are not programmed; they learn by example. Typically,
a neural network is presented with a training set consisting of groups of
examples from which the network can lean. These examples, known as
training patterns, are represented as vectors, and can be taken from such
sources as images, speech signals, sensor data, robotic arm movements,
financial data, diagnostic information etc.

   Training may be supervised, in which case the network is presented with a
target answer for each pattern that is input. In some architecture's, training is
unsupervised — the network adjusts its weight in response to input patterns
without the benefit of a target answer. In unsupervised learning, the network
classifies the input pattern into similarity categories.

3.1.3. What you can and cannot do with a Neural Network?

In principle, ANNs can compute any computable function, i.e. they can do
everything a normal digital computer can do. Especially anything that can

be represented as a mapping between vector spaces can be approximated to arbitrary precision by feedforward ANNs.

In practice, ANNs are especially useful for mapping problems which are tolerant of some errors, have lots of example data available, but to which hard and fast rules can not easily be applied. ANNs are, at least today, difficult to apply successfully to problems that concern manipulation of symbols and memory.

3.1.4. Who is concerned with Neural Networks?

Neural Networks are interesting for quite a lot of very dissimilar people with diverse applications

- Computer scientists want to find out about the properties of non-symbolic information processing with neural nets and about learning systems in general;
- Engineers of many kinds want to exploit the capabilities of neural networks on many areas to solve their application problems;
- Cognitive scientists view neural networks as a possible apparatus to describe models of thinking and conscious;
- Neuro-physiologists use neural networks to describe and explore medium-level brain function;
- Physicists use neural networks to model phenomena in statistical mechanics and for a number of other tasks;
- Biologists use Neural Networks to interpret nucleotide sequences.

3.1.5. How many learning methods for ANNs exist? Which should be chosen?

There are now many learning methods for ANNs. Nobody knows exactly how many. New methods are invented every week. The main categorisation of these methods is the distinction of supervised from unsupervised learning.

In **supervised learning**, there is a "teacher" function which in the learning phase "tells" the net how well it performs or what the correct behaviour should have been.

In **unsupervised learning** the net is autonomous: it just processes at the data it is presented with, finds out about some of the properties of the data set and learns to reflect these properties in its output. What exactly these properties are, that the network can learn to recognise, depends on the particular network model and learning method.

Many of these learning methods are closely connected with certain network topologies. The following is a list of some networks:

1. Unsupervised Learning

Feedback Nets:	Feedforward-only Nets:
Additive Grossberg	Learning Matrix
Shunting Grossberg	Driver-Reinforcement Learning
Binary Adaptive Resonance Theory	Linear Associative Memory
Analog Adaptive Resonance Theory	Optimal Linear Associative Memory
Discrete Hopfield	Sparse Distributed Associative Memory
Continuous Hopfield	Fuzzy Associative Memory
Discrete Bidirectional Associative Memory	Counterprogation
Temporal Associative Memory	
Adaptive Bidirectional Associative Memory	
Kohonen Self-organising Map	
Competitive learning	

2. Supervised Learning

Feedback Nets:	Feedforward-only Nets:
Brain-State-in-a-Box	Adaline, Madaline
Fuzzy Congitive Map	Backpropagation
Boltzmann Machine	Cauchy Machine
Mean Field Annealing	Adaptive Heuristic Critic
Recurrent Cascade Correlation	Time Delay Neural Network
Learning Vector Quantization	Associative Reward Penalty
Backpropagation through time	Avalanche Matched Filter
Real-time recurrent learning	Backpercolation
Recurrent Extended Kalman Filter	Artmap
	Adaptive Logic Network
	Cascade Correlation
	Extended Kalman Filter

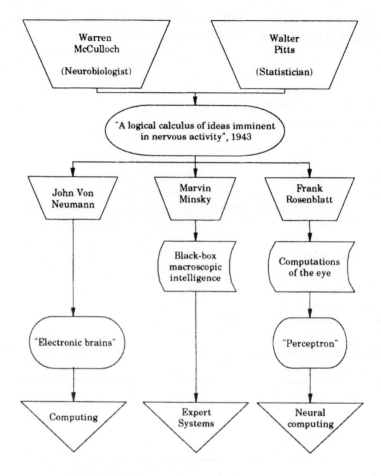

Figure 3. A flow chart showing the launch of three diverse fields.

3.1.6. The History of Neural Computing

During the Second World Walter Pitts, a Statistician, and Warren McCulloch, a Neurobiologist, published a watershed paper titled "A logical calculus of ideas imminent in nervous activity". This paper inspired numerous researchers including John Von Neumann, the father of computing, Marvin Minsky and Frank Rosenblatt to develop new fields of interest, as shown in Figure 3.

The first conference on Artificial Intelligence was held in 1956. It was organised by the pioneers of this new and exciting domain Marvin Minsky, John

McCarthy, Nathanial Rochester and Claude Shannon. Nathanial Rochester, of IBM research, presented results of the first known software simulation of a neural computing network. From this conference the field of Artificial Intelligence exploded, and a new area called Neural Computing was born.

The following year Neural Computing was strengthened by Frank Rosenblatt, at Cornell, who developed the "perceptron". This is a pattern classification system which can identify both abstract and geometric patterns. Although limited in use, it attracted a great amount of interest into the field for over a decade.

By the end of the 50's Neural Networks were being applied to major real-world problems. One of the first was adaptive filtering of echoes on the phone line. This was achieved by Bernard Widrow, at Stanford, who developed the "Adaline" and "Madaline" networks, which consist of simple neuron like elements.

During the 60's Neural Computing continued to grow, until the publication of **Perceptrons** by Marvin Minsky and Seymour Papert in 1969. This was a detailed mathematical analysis of an abstract version of Rosenblatt's perceptron, which basically concluded that Neural Computing was not "interesting", and caused the funds for research in this field to quickly dry up!

Some researchers continued work in this field despite the publication of **Perceptrons**. One such researcher was James Anderson, of Brown University, who developed the linear associator, which is based on the Hebbian principle of strengthening connections through their use. Anderson went on to develop the brain-state-in-a-box model, which is an extension of the earlier linear associator.

In the early 70's Teuvo Kohonen of Helsinki Technical University in Finland was working on adaptation rules and associative memories, which have the linear associator and brain-state-in-a-box models as special cases. Kohonen later developed the principle of Competitive learning, where processing elements compete to respond to an input stimulus and the winner adapts itself to react stronger to that stimulus.

Stephen Grossberg was another researcher who continued in the Neural Computing field despite **Perceptrons**. In the mid 70's he developed the sigmoid threshold function, and later studied Adaptive Resonance Theory models.

The return of respect for Neural Computing happened in 1982 when John Hopfield, of Caltech, presented the first paper in the field to the National Academy of Sciences since the 60's. He showed results of the Crossbar Associative network, which is a system consisting of interconnected processing elements that seek an energy minimum. Due to his reputation, charisma and clarity of explanation researchers became interested in Neural Computing again.

Over the past decade many groups have started to specialise in Neural Computing. David Rumelhart, who helped start the Parallel Distributed Processing Group, is credited with the development of one of the most popular models, the back-propagation network. Geoffrey Hinton, of the University of Toronto, and Terrance Sejnowski at the Salk Institute developed the Boltzmann machine, which is a modification of the Hopfield network. Bart Kosko at the University of Southern California has designed a Bi-directional Associative Memory model, which is Grossberg-like and recalls by oscillating between the two layers of the network until stability is achieved.

3.2. INTRODUCTION TO THE Anns USED IN THIS CHAPTER

In this chapter two different Artificial Neural Networks will be used. The first, and most widely used, is Back-Propagation, which is applied to the short term load forecasting sections. The second is Self-Organising Maps which are used in the data classification of transformer data section.

3.2.1. Back-Error Propagation

There are several types of neural architecture's. Although all consist of processing elements joined by the multiplicity of connections, they differ in the learning laws incorporated into their transfer functions, the topology of their connections, and the weight assigned to their connections. The real breakthrough in ANN research came with the discovery of the Back-Propagation training method based on the systems developed in the mid-1980s by David Rumelhart, a psychologist at Stanford (CA) University [4]. Back-Propagation algorithms tend to do well at function estimation and time series tasks. It is also good at representing complex, non-linear relationships in the form of a compact, efficient network [5]. Finally, it is easy to use. The power of back-propagation lies in its ability to train hidden layers and thereby escape the restricted capabilities of single-layer networks.

The back-propagation network has been applied successfully in test studies in a broad range of areas. Application studies have spanned tasks from military pattern recognition, medical diagnosis, and from speech recognition and synthesis to robot and autonomous vehicle control. Back-propagation has been applied to character recognition, image classification, signal encoding, and a variety of other pattern analysis problems [6,7]. Back-propagation can attack any problem that requires pattern mapping. Given an input pattern, the network produces an associated output pattern.

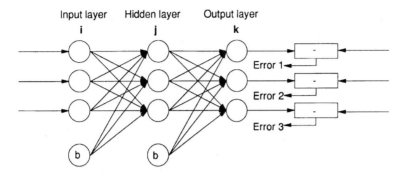

Figure 4. The Back-propagation model.

3.2.1.1. An overview of back-propagation ANNs

Typically back-propagation based Ann's employs three layers of processing units. The input layer receives external inputs while the output layer is responsible for generating output. The layer "sandwiched" between these two layers is called the hidden layer. The input layer is effectively a passive layer and only serves the purpose of accepting the inputs from the outside world without any change. Thus the output of each input layer node is exactly equal to the applied input. A weight is associated with each connection to nodes in the hidden layer. The layers in Figure 4 are fully interconnected i.e., each node of the input layer is connected to every node of the hidden layer. Likewise, each node of the hidden layer is connected to every node of the output layer. Units are not connected to others in the same layer. The number of hidden layers and hidden nodes may vary in number and are empirically chosen for a given problem. Figure 4 shows the topology for a typical three-layer back-propagation network.

A back-propagation neural network is trained by supervised learning. The network is presented with a training data set made up of pairs of patterns i.e., an input pattern paired with a target output. Upon each presentation, weights are adjusted to minimise the difference between the network's output and the target output. The back-propagation learning algorithm involves a forward propagation step followed by a backward propagation step. Both the forward-and the back-propagation steps are done for each pattern presentation during training.

The forward-propagation step begins with the presentation of an input pattern to the input layer of the network, and continues as activation level calculations propagate forward through the hidden layers. Figure 5 shows the neuron used as the fundamental building block for back-propagation

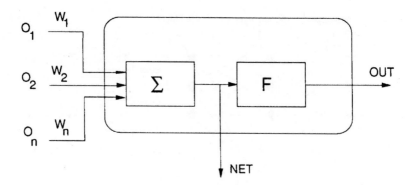

Figure 5. The fundamental building block for Back-propagation network.

networks. A set of inputs is applied, either from the outside or from previous layers. Each of these is multiplied by a weight, and their products are summed in the processing node. This summation of products is termed *net* and must be calculated for each neuron in the network. After each *net* is calculated, an activation function is applied to modify it, thereby producing the signal in a more general form. This is explained by the following two equations:

$$\text{net}_i = \sum_i \omega_{ji} \tag{2.1}$$

$$o_j = f(\text{net}_j) \tag{2.2}$$

where $w_{ij} = o_j w_j$.

Figure 6 below shows the activation function usually used for back propagation.

This function is called a sigmoid function or simply a squashing function and is given by the following two equations:

$$o_j = f(\text{net}) = \frac{1}{(1 + e^{-\text{net}_j})} \tag{2.3}$$

$$f' = (\text{net}_j) = \frac{\delta o_j}{\delta \text{net}_j} = o_j(1 - o_j) \tag{2.4}$$

The non-linear nature of the sigmoid transfer function plays an important role in the performance of ANNs. Multilayer networks have greater representational powers than a single layer network only if a non linearity is introduced [7]. The squashing function produces the needed non linearity.

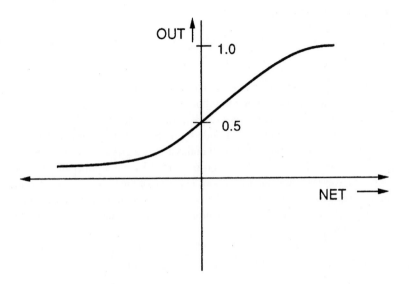

Figure 6. The sigmoid function.

In the analysis, all nodes used in the back-propagation ANNs are of the additive sigmoid variety. This is desirable in that it has a simple derivative. The back-propagation algorithm requires only that the function be every where differentiable. The sigmoid function satisfies this requirement. It compresses the range of *net* so that o_j lies between zero and one. It has the additional advantage of providing a form of automatic gain control. For small signal *net* near zero, the slope of the input/output curve is steep, producing high gain. As the magnitude of the signal becomes greater, the gain decreases. In this way large signals can be accommodated by the network without saturation, while small signals are allowed to pass through without excessive attenuation. To summarise, during the feed forward calculations, two mathematical operations are performed by each node, and the output state, or activation, is obtained as a result. The first is a summation of previous layer node outputs times the connecting weights, and the second is the squashing function. The squashing function serves to mitigate problems caused by possible dominating effects of large input signals.

The backward propagation step begins with the comparison of the network output pattern to the target vector, when the difference, or "error" is calculated. The error of any output node in layer k is

$$e_k = t_k - o_k \qquad (2.5)$$

and a total error function is written as

$$E_p = \frac{1}{p} \sum (t_k - o_k)^2 \tag{2.6}$$

Equation 2.5 defines an error term that depends on the difference between the output value an output node is supposed to have, called the target value t_k, and the value it actually has as a result of the forward pass calculation, o_k. This error term is calculated on a node by node basis over the entire set of patterns. The error is summed over all nodes giving a grand total for all nodes and all patterns. Then this grand total is divided by the number of patterns to give an averaged sum squared value. The goal of the training process is to minimise this averaged sum squared error over all training patterns. Learning comprises changing weights so as to minimise this error function in a gradient descent manner [4].

To obtain a rule for adjusting weights and thresholds, the gradient of E_p with respect to w_{ji} is used and is represented as follows:

$$-\partial \frac{E_p}{\partial w_{kj}} = d_{pk} o_k \tag{2.7}$$

where ∂_{pk} is defined in two ways. There is a requirement to estimate what ∂_{pk} should be for each node in the network.

For nodes in the output layer, it is given by:

$$\partial_{pk} = (t_{pk} - o_k)(f_k'(\text{net}_{pk})) \tag{2.8}$$

where f_k' is the derivative of f defined earlier by Equation 2.4. Thus this equation becomes

$$\partial_{pk} = (t_{pk} - o_k) o_{pk} (1 - o_{pk}) \tag{2.9}$$

For nodes in the hidden layer where the output is unknown, it is calculated in the following way:

$$\partial_{pj} \& = f_k'(\text{net}_{pj}) \sum \partial_{pk} \omega_{kj} \tag{2.10}$$

$$\partial_j \& = o_{pj}(1 - o_k) \sum \partial_{pk} w_{kj} \tag{2.11}$$

Using these δ's, the rule of adjusting weights can be derived using the following equation:

$$\omega_{kj}(t+1) = \omega_{kj}(t) + \eta \delta_k o_k \tag{2.12}$$

In this equation, $\omega_{kj}(t)$ is the old weight for the connection feeding the output layer from the hidden layer at time $\omega_{kj}(t+1)$ is the new value of this weight and δ_k is the error term for the node k. The error associated with each processing unit reflects the amount of error associated with that unit. This parameter is used during the weight correction procedure, while learning is taking place. A larger value for δ indicates that a larger correction should be made to the incoming weights, and its sign reflects the direction in which weights should be changed. The η is the learning coefficient. It serves to adjust the size of the average weight changes and can take values between zero and one.

Rumelhart [4] describes a method for improving the training time of the back-propagation algorithm. Called momentum α, the method involves adding a term to the weight adjustment that is proportional to the amount of previous weight change. The momentum factor α can take values between zero and one. Equation 2.12 with the momentum term added becomes Equation 2.13 which is as follows:

$$\omega_{kj}(t+1) = \omega_{kj}(t) + \eta\delta_k o_k + \alpha[\Delta\omega_{kj}(t-1)] \qquad (2.13)$$

The weight changes for the connectors feeding the hidden layer from the input layer are calculated in a manner analogous to those feeding the output layer. Thus

$$\omega_{ji}(t+1) = \omega_{ji}(t) + \eta\delta_j o_j + \alpha[\Delta\omega_{ji}(t-1)] \qquad (2.14)$$

In the above description, we have processing elements called bias nodes as indicated by the nodes with the letter b in Figure 4. These nodes always have an output of 1. They serve as a threshold unit for the layers to which they are connected and weights from the bias nodes to each of the nodes in the following layer are adjusted exactly like the other weights. In Equation 2.8 for each of the nodes in the output layer, the subscript k takes on the values from 0 to n_k, which is the number of output nodes. The n_kth value is associated with the bias nodes. In a similar way, Equation 2.9 applies to hidden nodes and subscript j takes on values from 0 to n_j where n_j implies the bias node. The training of a neural network may require many iterations until the values of the interconnected weights change and satisfy the prespecified error criterion. The use of trained networks requires only a forward pass in the testing phase.

3.2.1.2. The back-propagation algorithm

The objective of training the neural network is to adjust weights so that the application of the set of inputs produces the desired set of outputs. Training assumes that each input vector is paired with the target vector representing the desired output, together, these are called a training pair. Before the training process, all weights must be initialised to small random numbers. This ensures that the network is not saturated by large weight values.

Training the Back-Propagation network requires the following steps [8]:

1. Select the next training pair from the training set, apply the input vector to the network input;
2. Calculate the output of the network;
3. Calculate the error between the network output and the desired output (the target vector from the training pair);
4. Adjust the weights of the network in a way that minimises the error;
5. Repeat step 1 through 4 for each vector in the training set until the error for the entire set is acceptably low.

The operations required in steps 1 and 2 above are similar to the way in which the trained network will ultimately be used for testing; that is an input vector is applied and the resulting output is calculated. Calculations are performed on a layer by layer basis.

In step 3 each of the network outputs is subtracted from its corresponding component of the target vector to produce an error. This error is used in step 4 to adjust the weights of the network. After enough repetitions of these four steps, the error between actual outputs and target outputs should be reduced to an acceptable value, and the network is said to be trained. At this point, the network can be used for recognition and weights are not changed. It may be seen that steps 1 and 2 constitute a forward pass in that the signal propagates from the network input to its output. Steps 3 and 4 are a reverse pass; here the calculated error signal propagates backward through the network where it is used to adjust weights. Figure 7 shows the flow-chart of the Back-Propagation ANN

3.2.1.3. *Development steps of ANN models*

The development of neural network architecture's for the given problem can be characterised by a group of discrete activities on the basis of which ANN models are built. They are briefly described below:

3.2.1.3.1. *The data for training and testing*

The first and usually longest step in development, and generally the most critical to eventual success, is the creation of data sets. Tasks here include

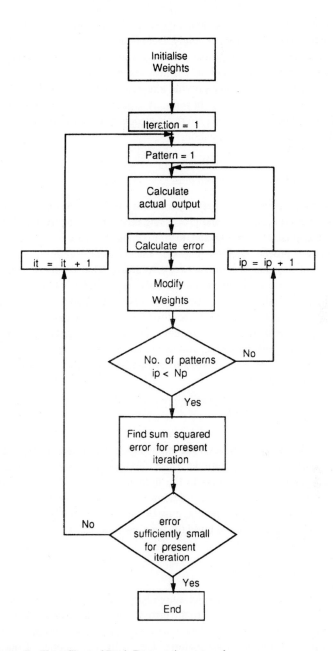

Figure 7. Flow Chart of Back-Propagation network.

gathering raw data, analysing it, selecting variables, and processing the data so that the network can learn efficiently. Data is, in any case, paramount for a neural network; it is an empirical system. It can recognise new examples of the patterns used to train, but only if they resemble the training patterns. In general, the training set must provide a representative sample of the data which the network will process in the ultimate application. The amount of data required for a generation of test patterns is closely tied to user requirements and to the specifics of the application.

3.2.1.3.2. *Pre-processing of given data*

Generally, the majority of the effort in neural-network development goes into collecting data examples and processing them appropriately. Once that task is accomplished, training and testing of the network is a relatively straightforward procedure. Pre-processing means transforming the data so that it becomes easier for the network to learn. The standard form is normalisation. For back-propagation the only requirement is that the input to each input node should be in the interval between 0 and 1. Normalising input patterns can actually provide a tool for pre-processing data in different ways. The data can be normalised by considering all the inputs together, or normalising each input channel separately or normalising groups of channels. In some cases, the way one chooses to normalise the input can affect the performance of the ANN.

3.2.1.3.3. *Selection of inputs*

Data enters the network through the input layer. The nodes in the input layer are passive, not computational; each simply broadcasts a single data value over weighted connections to the hidden nodes. Building the input pattern generally requires choosing among many measurable values. Data analysis helps "weed out" the potential input variables so that only the most "explanatory" ones are used to build the training patterns.

3.2.1.3.4. *Selection of outputs*

The key to the back-propagation learning algorithm is its ability to change the values of weights in response to the errors. For this to be possible i.e. to calculate the errors, the training data must contain a series of input patterns labelled with their target output patterns. Generally, the number of output nodes is fairly straightforward to determine as compared with the number of input nodes. If the data is to be classified into several classes, one output node needs to be assigned for each. In general, the output nodes cost relatively little computationally.

3.2.1.3.5. Building the ANN architecture

The next step is to configure the ANN by setting the number of input and output nodes to agree with the number of inputs and outputs in the data. The architecture is a specification of neural network topology to build, train and test it. The back-propagation training requires a preliminary value for the number of hidden nodes; the final value depends on experiments during training. It should be kept in mind that the number of hidden layers and number of nodes in the hidden layer are problem dependent and are empirically selected. Also, the values of the learning rate and momentum coefficient are chosen to give fast convergence. Trying different configurations is a matter of changing parameters that control network structure and behaviour. The parameter values assigned can have a large impact on the performance of the system. As there are seldom any way of deducing the best values, training requires experimentation. The nomenclature used to represent the ANN structure in the analysis is I-H1-H2-O where I represents the input layer nodes, H1 and H2 being the first and second hidden layer nodes, and O is the output layer nodes.

One of the issues in the application of a network to a problem domain is the size of the network as measured by the number of free parameters of the network. Although there is no general method to determine the optimal size of the network for a parameter task, there are statistical arguments which suggest that the number of training patterns required to fully determine the weights in a network is approximately proportional to the number of weights in the network. A rule of thumb often cited is that the number of weights should be less than one tenth of the number of training patterns [11]. However, the question of theoretical upper and lower bounds on the sample size vs ANN size for MLP (multiple perceptron i.e., Back propagation) with real valued functions such as sigmoid functions is still largely an open problem [9,10]. The ANNs reported here were created with a relatively limited data set.

3.2.1.3.6. Training and testing of ANNs

The final development step is training and testing the network. During training, the network cycles through the data repeatedly, changing the values of its weights to improve performance. Each pass through the training data is called an epoch, and the neural network learns through the overall change in weights accumulating over many epochs. Training continues until the values of the weights cause the network to map input patterns to appropriate results. Training is therefore an interactive process; the trainer tries a configuration, evaluates a result, makes a change, tries it again, and so on until satisfied.

During testing, the network passes the testing patterns forward and calculates the output without testing the weights. The testing patterns are not labelled with target outputs.

3.2.2. Self-Organising Maps

Classifying similar objects is a difficult yet fundamental and frequent activity. The self-organising map neural network tackles this by creating a two-dimensional feature map of input data in such a way that order is preserved, i.e. if input vectors are close to each other they will be mapped to the same node.

These networks were originally developed to view topologies and hierarchical structures of higher dimensional input spaces. At the heart of a SOM is the Kohonen layer which can transform any n-dimensional space into an ordered z-dimensional map.

3.2.2.1. Network Architecture

The SOM typically has two layers. The input layer is fully connected to the two-dimensional Kohonen layer. The nodes of this layer each measure the Euclidean distance of its weights to the arriving input values. During recall, the Kohonen node with the minimum distance is called the winner and has an output of 1.0, while the other Kohonen nodes have an output of 0.0. Thus the winning node is, in a measurable way, the closest to the input value and thus represents the input value.

During the training stage the neighbours of the winning nodes adjust their weights in order to be closer to the same input values. This causes the order of the input space to be preserved.

3.2.2.2. Finding the winning nodes

The winning node is determined by the distances between each weight and the input value, or vector.

If the input data has M values and is denoted by:

$$X = (x_1, x_2, \dots, x_m) \tag{2.14}$$

Thus each Kohonen node 'i' will have M weights and can be denoted as:

$$W = (w_{il}, w_{i2}, \dots, w_{iM}) \tag{2.15}$$

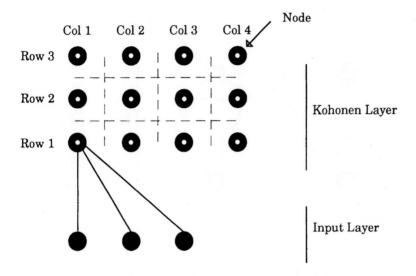

Figure 8. Input layer fully connected to a 3x4 PE, 2D Kohonen layer.

The Euclidean distance Di between each of the N Kohonen nodes can be computed as follows:

$$D_i = \|X - W_i\| \\ = \sqrt{(x_1 - w_{il})^2 + (x_2 - w_{i2})^2 + \cdots + (x_M - w_{iM})^2}$$ (2.16)

The node closest to the input vector will be the winner, however during the training phase a conscious mechanism is employed which requires extra steps.

To develop a uniform distribution in the SOM layer the conscious mechanism encourages nodes that are losing below an average frequency, and discourages nodes that are winning above an average frequency. This is done by appropriately adjusting the weights.

The adjusted distance D_i' is simply the original value minus a bias B_i. The bias is computed using the formula below:

$$B_i = \gamma(NF_i - 1)$$ (2.17)

where F_i is the frequency with which the node i has historically won. Initially the bias is set to zero. Now the adjusted distance D_i' can be worked out as follows:

$$D_i' = D_i + B_i$$ (2.18)

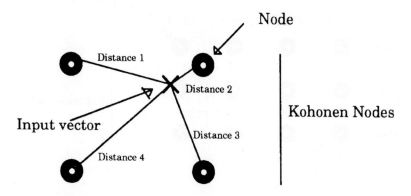

Figure 9. The distances of the input vectors to the nodes.

3.2.2.3. Adjusting the weights of Neighbouring Nodes

Once the closest node to the input vector has been found, the neighbouring nodes can be updated using the following equation:

$$W_{ij(\text{new})} = W_{ij(\text{old})} + \alpha(X_j - W_{ij(\text{old})}) \qquad (2.19)$$

The learning coefficient α decreases with time and may start with a value such as 0.4 lowering to 0.1 over the training phase.

3.2.2.4. Estimating the winning frequency of nodes

The mechanism of 'consciousness' was introduced to stop any one node representing too much of the input data. This is done by keeping a record of each nodes winnings and adjusting the distance correspondingly. If a node has won more than $1/N$ times the distance is increased, otherwise it is decreased.

The frequency for the winning nodes are adjusted by as follows:

$$F_{i(\text{new})} = F_{i(\text{old})} + \beta(1.0 - F_{i(\text{old})}), \qquad (2.20)$$

and for the other nodes;

$$F_{i(\text{new})} = F_{i(\text{old})} + \beta(0.0 - F_{i(\text{old})}), \qquad (2.21)$$

3.3. SHORT TERM LOAD FORECASTING

3.3.1. Introduction

Load forecasting occupies a central role in the operation and planning of electric power systems. The lead times range from a few minutes ahead for the economic loading of power plant to over forty years for the economic planning of new generating capacity and transmission networks.

The techniques of load forecasting may be broadly divided into three areas of application, these are:

(i) Long-term econometric forecasts for system planning;
(ii) Medium-term forecasts for the scheduling of fuel supplies and mainte-nance programmes;
(iii) Short-term predictions for the day to day operation and scheduling of the power system.

The problem of short term load forecasting (one to twenty-four hour) in the electric utility industry has received extensive attention in the last twenty years. A number of algorithms have been suggested for the load forecasting problem. They include statistically based techniques [12–14], the expert system approach [15,16] and an artificial neural network approach [17–22].

3.3.2. Artificial Neural Network Approach

The advantage of ANNs over statistical models lies in their ability to model a multivariate problem without making complex dependency assumptions among input variables. Furthermore, the ANN extracts the implicit non-linear relationship among input variables by learning from training data. They do not rely on human experience, as in the expert system approach, but attempt to draw links between sets of input data and observed outputs. Neural networks offer the potential to overcome the reliance on a functional form of the forecasting model and large historical databases. These neural networks represent a pattern or load shape, and in reality, perform a pattern recognition function. Since this pattern is based on training cases provided to the network, the problem of the application of a neural network reduces to two fundamental parts. Appropriate training cases must be selected and the structure of the network must accommodate the size and complexity of patterns.

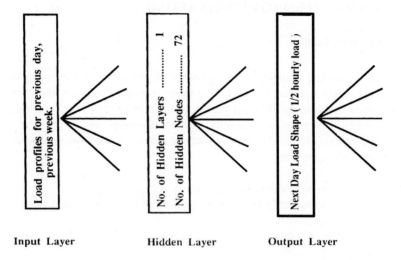

Input Layer   Hidden Layer   Output Layer

Figure 10. The ANN architecture used for predicting load shape using actual utility data (NGC,UK).

3.3.3. Application of ANNs to short term load profile forecasting using Actual Electric Utility data (National Grid Company, UK)

3.3.3.1. Introduction

This section demonstrates the ANN models used to forecast load shape a day ahead. The analysis utilised actual utility data (NGC,UK) comprised of half hourly recorded load values spreading over a period between 1991 to 1993. No weather information was used in this case. The analysis was carried out for weekdays in the winter season only.

3.3.3.2. Application

In this case, a single ANN model was built to forecast the half hourly load sequentially over the next twenty four hour period in the weekdays only. The inputs used for such an ANN were previous day and week load shape from the date of prediction thus giving a total of 96 nodes used in the input layer. The output layer is comprised of 48 nodes and thus represents the load shape for the forecasting day. The previous day was taken as Friday in the case of the Monday load forecast i.e. the previous week day was considered as the same day type as that of the forecast day. The training cases were spread over the period from 1991 to 1993. The months considered for the winter season include November, December, January and February. After experimenting with different ANN architectures, the final ANN selected for testing was 96-72-48. Figure 10 shows the outline of such an ANN model.

3.3.3.3. Results and Conclusion

The testing phase for the ANN involves four representative case sets selected from November, December, January and February in the year 1993. Each set is comprised of 5 test cases representing weekdays from Monday to Friday and thus giving a total of 20 test cases all together. The test results are graphically shown in Figure 11 through Figure 14 for November, December, January, and February respectively.

The results for the all of the test cases are summarised in Table 1 in terms of overall average forecasting error and show that the ANN methodology has been able to predict the load shape within a reasonable accuracy.

Table 1. Average Percentage Forecasting Error for Four Months in the Winter Season.

Months	Mon	Tue	Wed	Thu	Fri
November	6.52	2.19	1.82	3.05	2.79
December	3.61	2.53	3.68	1.70	3.32
January	1.40	1.22	2.24	1.79	1.28
February	2.97	2.62	1.61	1.12	1.07

3.3.4. Application of ANNs to short term peak, profile and half hourly load forecasting using Actual Utility Data (ScottishPower, UK)

3.3.4.1. Introduction

This section explores the application of artificial neural networks to the short term load forecasting problem using ScottishPower Data. Separate cases have been considered to exploit the learning ability of neural networks using the back propagation algorithm. The cases include: prediction of peak load using load data for the same "day type" over a historical "window" or time frame; forecasting the load shape a day ahead using load data alone as well as a combination of load and temperature data; half hour ahead load forecasting; and prediction of load shape a day ahead using half hour ahead forecast values. The networks were trained and tested on actual power utility load data (half hourly recorded load) and weather data obtained from ScottishPower.

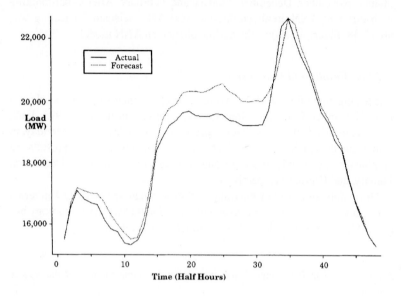

Figure 11. Load shape forecast for a weekday in November 1992.

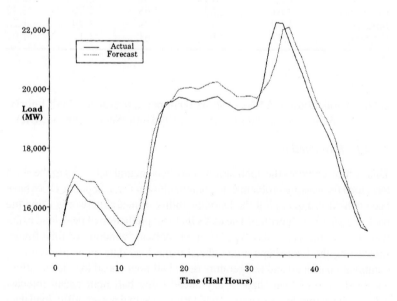

Figure 12. Load shape forecast for a weekday in December 1992.

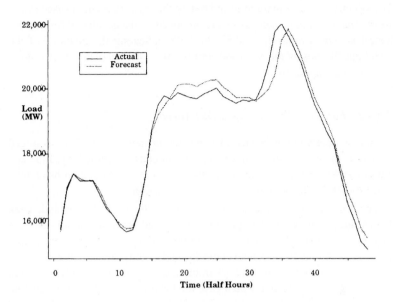

Figure 13. Load shape forecast for a weekday in January 1992.

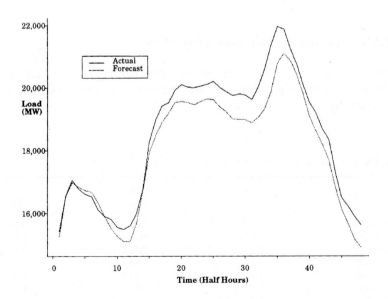

Figure 14. Load shape forecast for weekday in February 1992.

The selective features in training and testing the network is emphasised in this work. The absolute percentage errors produced by the developed forecasting neural networks range from 0.5% to 2.5% indicating the potential of the technique for most of the cases and confirms the potential of the methodology for economic applications.

3.3.4.2. *Forecast of Peak Load a Day Ahead*

In this case, the training data used for generating pattern examples was spread over four months of data and includes the months of April, May, June and July in 1990. The testing data constitutes test cases for the months of June and July 1991. Thursday was chosen for prediction.

The 3 inputs used in the neural network for this prediction include peak load values of previous day, previous week, and previous month (of same day type, i.e., Thursday) from the day of prediction. There is only one output node representing the peak load of the day under study. A number of networks have been investigated for network training error performance (average sum squared error) and the one which gave the lowest error had a 3-3-1 structure and was thus selected for testing. Table 2 illustrates the actual and forecast values of peak loads for different dates in June and July; the overall average error comes to be 1.72%.

3.3.5. Prediction of Load Shape a Day Ahead

This section details some of the possible correlations explored between the lagged load values alone and also between lagged load values and temperature.

Table 2. Forecast results for peak demand.

Test Date	Actual Load	Forecast Value	Error (%)
06/06/91	2643	2623	0.76
13/06/91	2904	2822.9	2.79
20/06/91	2655	2645.8	0.35
27/06/91	2687	2589.4	3.63
04/07/91	2584	2582.9	0.04
11/07/91	2507	2473.1	1.35
08/07/91	2408	2350.3	2.4
25/07/91	2296	2352.2	2.45

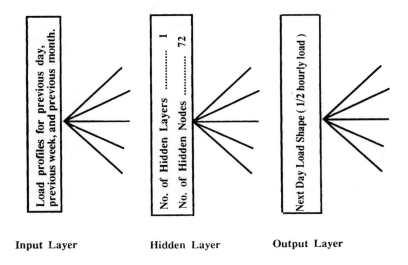

Input Layer   **Hidden Layer**   **Output Layer**

Figure 15. The ANN architecture used for predicting load shape using load data only.

3.3.5.1. Neural Network Using Load Data Only

In this case, the previous day, previous week and previous month load shape for the day of prediction is used as inputs. Each load shape constitutes 48 half-hourly load data points starting from 0030 hours to 2400 hours. Thus, a total of 144 inputs were used in the neural network. The training data covers the same period as mentioned in the previous section. The output of the neural network is a single 48 node layer representing the load shape in the time frame between 0030 to 2400 hours. A number of different neural network architectures were used in terms of the number of hidden layers and hidden nodes. The one giving the lowest network training error was chosen for testing. The best performance was obtained from a network having a 144-72-48 structure. The outline of the ANN structure is shown in Figure 15.

Some of the test results in forecasting the load shape for Thursday in the months of June and July were obtained and are illustrated in Figures 16 and 17.

The summary of these results in terms of overall average error over a 24 hour period is given in Table 3.

It is evident from Table 3 that the error is quite small over the 24 hour period in all the cases and shows the promise of the neural network methodology. In order to compare these forecast results with the other days in one of the weeks of July 1991, the same trained network was tested for Monday through

Table 3. Summary of forecast results using load data only.

Test Date	Average prediction Error%
06/06/91	1.69
13/06/91	1.25
20/06/91	2.27
27/06/91	0.97
04/07/91	2.15
11/07/91	2.01
08/07/91	2.24
25/07/91	2.75

Sunday and their summary in terms of overall average error is included in Figure 18.

It shows that the average forecasting error is relatively high for Monday, Friday, Saturday and Sunday. Monday as the beginning of a week and Friday as the last working day of a week show a slightly modified load shape compared with other days because of "start-up" activities in the former case and "shut-downs" in the latter case. For days like Tuesday and Wednesday,

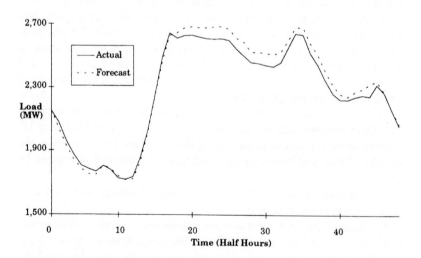

Figure 16. Forecast results for load shape for 6.6.91.

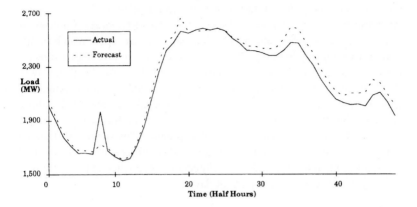

Figure 17. Forecast results for load shape for 4.7.91.

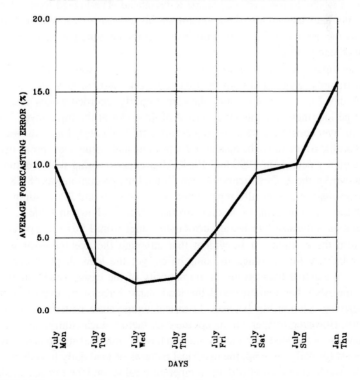

LOAD SHAPE COMPARISON FOR DIFFERENT DAYS

Figure 18. Comparative summary of overall average error.

the test results indicate that the error is not significantly high which shows that these days have matching load shapes with that of Thursday. Similarly, as one can expect, Saturday and Sunday load shapes were a poor match for the normal working day load shapes and therefore liable to give higher forecast errors. Using the same trained network, the load shape for the Thursday in the month of January 1991 was also tested and it shows again a completely different picture from summer as can be expected. In view of the above analysis, the results indicate that the given ANN performs very well for the type of days for which it is trained to predict. This suggest that neural networks are able to predict more accurately if carefully selected data is used for training and that specific ANN architectures are applied to particular data sets, e.g. in this case days or parts of days.

3.3.5.2. Neural Network Using Temperature Data Only

In this case, the weather information was utilised in terms of temperature only to train ANNs to forecast load shape a day ahead. The lagged load values have not been used at all. The main aim of this test was to see how an ANN can extract a relationship among the historical temperature patterns alone to predict load profile.

The input layer was comprised of previous day, week and month's temperatures recorded at both the east and west coasts of Scotland for the given day. Since the temperature data are 3 hourly recorded values over 24 hour period, therefore this gives a total of 48 nodes at the input layer. The output layer constitutes 48 nodes representing the half hourly load values as a load profile over a day. The data used for training and testing was spread over the summer period for 1990 and 1991. After experimenting with a number of networks, the ANN structure chosen for the final prediction was 48-48-48 The results are graphically represented in Figures 19 and 20. A summary of these results in terms of absolute average errors is shown in Table 4. The results show the absolute average error in the range from 1.96 to 5.46 percent.

From these results, it is evident that although the average error is in the relatively higher range in some cases, yet the load shapes forecasts mirror the actual load profile in the majority of the cases which suggests that the ANN has demonstrated the load shape forecast by extracting an underlying relationship between historical temperature patterns and load values. However, temperature is not the only factor which contributes to the behaviour of the load demand. In order to have a more realistic model of the forecast for daily demand, the historical patterns of load daily load demand were used in conjunction with the temperature values and the forecast made. This is discussed in the following section.

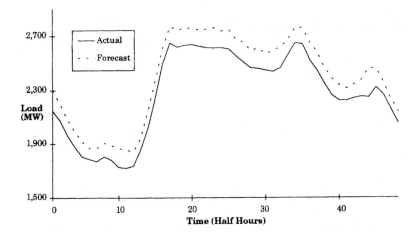

Figure 19. Load shape forecast results for 6.6.91 using temperature data only.

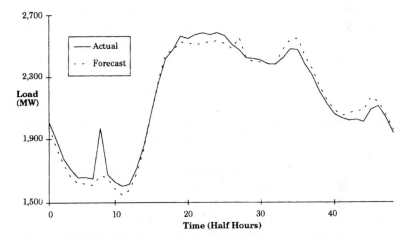

Figure 20. Load shape forecast results for 4.7.91 using temperature data only.

3.3.5.3. Neural Network Using Load Data as well as Temperature Data

In this case, a neural network was constructed which takes both temperature values and load values as inputs to the network. Thus in addition to the load inputs outlined as in previous sections, the inputs comprising temperature data include 3 hourly recorded values over 24 hour periods for both east and west coast (of Scotland) for previous day, week and month. In this way,

Table 4. A summary of forecast results using temperature data only.

Test Date	Average prediction Error%
6/6/91	5.46
13/6/91	3.77
20/6/91	2.17
27/6/91	4.78
4/7/91	1.67
11/7/91	1.97
8/7/91	4.40
25/7/91	4.24

a total of 192 inputs are used for this neural network. The output is a 48 node layer representing load shape over a day. Among several different trained networks, the one giving minimum network training error was selected for testing and was found to have a 192-96-48 structure. The outline of the ANN model is shown in Figure 21. Some of the results from June and July 1991 are included in Figures 22 and 23 and the summary in terms of overall average error is given in Table 5. These results indicate that the average forecasting error is improved in general in most of the cases. The results confirm the promise of the methodology and indicate that a strong correlation exists between temperature and load demand.

It is worth mentioning here that in addition to temperature variables, wind speed and wind direction was also utilised in the ANN-based forecast model which could not, however improve the accuracy of the forecasting model at all. However, further investigations are necessary to see the effect of other weather parameters on load demand. These may include cloud cover, illumination factor, precipitation etc.

Table 5. Summary of Forecasting Errors%.

Test Date	Average prediction Error (%)
06/06/91	0.96
13/06/91	3.88
20/06/91	2.43
27/06/91	1.52
04/07/91	1.76
11/07/91	2.47
18/07/91	1.54
25/07/91	1.94

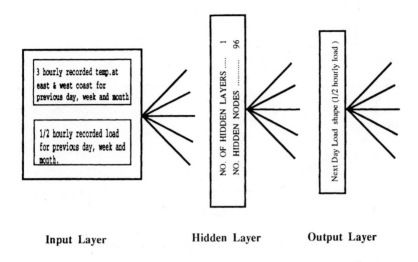

Figure 21. The ANN architecture for predicting load shape using load as well as temperature data.

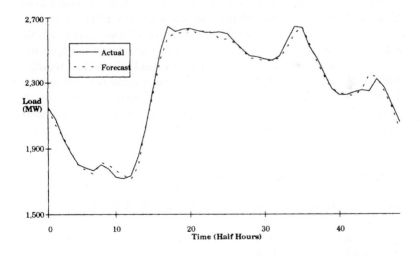

Figure 22. Load shape forecast for 6.6.91 using load and temperature data.

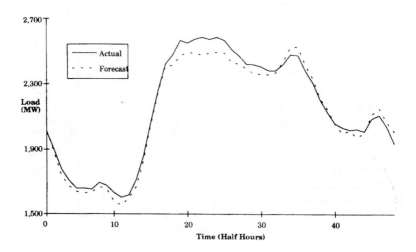

Figure 23. Load shape forecast for 4.7.91 using load and temperature data.

2.3.5.4. A Comparative Summary of Forecasting Results

Comparing Tables 3 through 5, it is evident that all the three ANN-based models performed well within a reasonable forecasting accuracy. The main reason behind forming a temperature based forecasting model was to see how well an ANN can respond to just temperature values as inputs. It is clear from Table 4 that such a model did not prove to be a failure but rather, the average error was under 6%. This shows that the inclusion of temperature variable in the forecasting model can perhaps play a significant role in improving the accuracy of the forecast. Using temperature and load variables together in the input pattern of an ANN model shows that average forecasting error (Table 5) is improved in over 50% of test cases compared with Table 3. These investigations show that, in general, there is a strong correlation exists between temperature and load demand.

3.3.6. Half Hour Ahead Prediction

For this case, four inputs were used comprising of the following:

(a) Load at $(t - 1/2)$ hour
(b) Preceding day load at t hours
(c) Preceding week load at t hours
(d) Preceding month load at t hours,

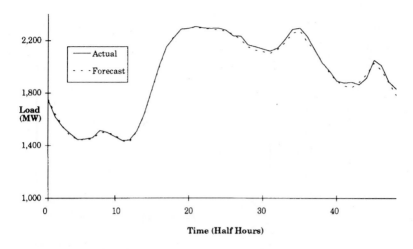

Figure 24. Half hour ahead forecast from 0300 to 2400 hours for 25.7.91.

where t is the time of prediction and varies from 0030 hours to 2400 hours. The output is comprised of one node representing the load to be predicted at t hours ranging from 0300 hours to 2400 hours. There were a total of 48 networks trained using load data over April, May, June, and July 1990. Figure 24 shows results for 25.7.91. in terms of actual and forecasted load. The average forecasting error over 48 half hours is 0.76% which shows that very short term load forecast gives more accurate results than relatively long range forecasts.

3.3.6.1. Prediction of load shape on the basis of half hourly predicted values

The 48 trained networks described above were also used for predicting the load shape for a particular day. In this case, the arrangement used during the testing phase is that in addition to regular fixed inputs described in training phase, a flexible input is used which takes the previous half hour forecast and includes it in the set of inputs to forecast the next half hour ahead. In this way, the complete load shape is predicted. Figure 25 shows the topology used for obtaining the load forecasts over 48 data points.

The test results for dates in June and July are graphically represented in Figures 26 and 27. Table 6 gives the summary of the results in terms of over all average errors. These error range from 1.51

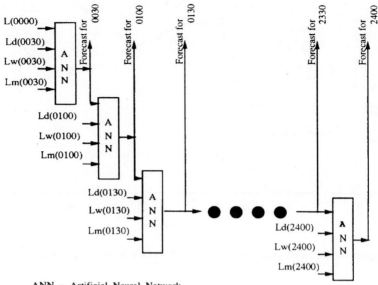

ANN – Artificial Neural Network

L – Load at the time of prediction
Ld – Previous day load
Lw – Previous week load
Lm – Previous month load

Figure 25. ANN model for predicting load shape using forecast values for next half hour.

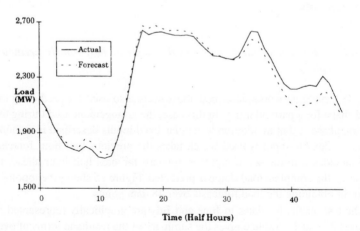

Figure 26. Load shape forecast for 6.6.91 using ANN dynamic model.

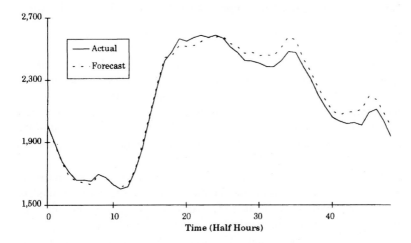

Figure 27. Load shape forecast for 4.7.91 using ANN dynamic model.

3.3.6.2. Conclusions

A number of cases studies have been presented to demonstrate the application
of ANNs in forecasting various targets and using a variety of input pattern
combinations. In the analysis, different case studies show that ANNs
performed well when trained and tested over carefully selected data. The
tests indicate that ANN models have also been able to respond to temperature
variations in addition to load values with reasonable forecasting results. The
promise of the ANN methodology is also demonstrated using smaller ANNs
for predicting half hour ahead load and load shape a day ahead.

Table 6. Summary of Forecasting errors.

Test Date	Average Error (%)
6/6/91	1.97
13/6/91	2.05
20/6/91	2.11
27/6/91	1.51
04/7/91	1.76
11/7/91	2.32
18/7/91	2.73
25/7/91	1.53

3.3.7. A Modularised ANN Approach to Short Term Load Forecasting

3.3.7.1. General

This section discusses a novel approach to the Short Term Load Forecasting problem from a modularised ANN perspective. The issues of concern within this approach are threefold: (i) load forecasting sensitivity to the nature of the selected load data, (ii) avoiding undue complexity in the learning process with an increase in the forecast lead time, (iii) increasing the speed of the learning process and to make the resultant ANN potentially viable for full implementation in a control room environment.

Items (i) and (ii) are addressed through a step-wise refinement consisting of repeatedly decomposing the problem into smaller forecast models. Eventually, one has a collection of small tasks/models each of which are trained using data carefully selected for hourly periods, daily, weekly and seasonal time varying windows. This section presents a number of results to illustrate the importance of these issues. Sixteen months of actual electric power utility load data and regional temperature data have been utilised to train and test a family of ANNs. The structures thus obtained can be used in line with a supervisory expert system to form an "expert network" which could potentially offer a solution for full implementation in the control room environment. The results obtained are highly encouraging and show great potential for the improvement of the forecasting function by using a modularised ANN-based approach.

3.3.7.2. Introduction to the modularised approach

The most unique feature of the ANN performance is its ability to form its own working model for a given problem through learning by training examples based on domain data. Thus, the network can generalise more readily than articulated rules from humans as in expert systems or through fetch-execute store cycles of an ordinary sequential computer programme. However, the notion that a single ANN model can perform the generalisation tasks for all purposes is not supported through experience [23]. Modelling all features through a single ANN and fitting the model by "brute force" to the available data leads to more complex systems which are difficult to train and hard to understand. The forecasting results using ANN techniques as discussed in the previous sections indicate that ANNs are sensitive to domain training sets upon which they are trained; e.g., an ANN trained on a summer training set performed poorly when applied to a January day. Similarly, a weekday trained ANN cannot be expected to perform well over weekends. Even, to be more selective, an ANN trained over Thursday data produced relatively high errors when tested for Mondays and Fridays. These observations therefore suggest the decomposition of the load forecasting problem into a variety of scenarios which can be represented with carefully selected training data. Furthermore, the learning process can be improved and network complexity can be reduced with decomposition of the application tasks into subtasks that are trained separately. One reason for care in selecting input and output variables is to keep the networks small, so that less data is needed to train it, and to avoid the problem of "over-parameterisation."

One of the major benefits of the decomposition approach is the possibility of evaluating the success of each sub-task independently. For each sub-task, a partial success criterion and verification procedure can be defined. If the entire application fails to operate correctly, the search for the cause of the failure can, in turn be decomposed with respect to the partial tasks [24]. This provided the motivation to reconfigure the whole forecasting problem into a decomposed structure.

3.3.7.3. The outline of the approach

The key concept in arriving at the modular ANN-based solution for STLF is a step-wise refinement consisting of repeatedly decomposing the problem into smaller forecast models. Eventually, one has a collection of small tasks/models each of which are trained using data selected for hourly periods, daily, weekly and seasonal time varying windows. There are many factors that affect the STLF; the proposed model is roughly based on these factors. Some of the most important ones are as follows:

(a) The load shows daily, weekly and seasonal behavioural patterns: The seasonal variation dictates four broad boundaries namely spring, summer, autumn and winter. This provides the basis to form four sets of ANNs each set having subsets which can have training attributes covering a range of scenarios for that season.
(b) Weekly variations: The load variation pattern may change from week to week. A lower demand for electricity during week-ends will distinguish days in a week-end from week days. Similarly holidays consecutive with a week-end may appear as a long week-end in a week. Such features have to be carefully taken into consideration while selecting the training data for the neural networks to recognise these patterns. Thus separate ANNs have to be formed for weekdays, weekends and "special days".
(c) Daily variations: Daily load shapes may differ from day to day in a week. For instance, in weekdays, Monday may prove to be different from typical Thursday or Wednesday because of start-up activities in the former case. The same may be true for Friday, but in this case it

will be due to early finishing in commerce and industry. It is therefore essential for ANN models to recognise periodic variations due to social patterns. This brings a need to allocate one set of neural networks to each day.

(d) Holidays: The load shape of weekdays are usually different from weekends. To be more selective, the Saturday load pattern can be captured by separate ANNs from Sunday. Similarly all established conventional holidays such as Christmas day, New Year etc. can be modelled by customised ANNs. This, however, brings a requirement for a large amount of data so that classified holiday patterns can be used to train ANNs.

(e) The effects of weather can have a strong impact on load and is influenced by the seasonal and climatic fluctuations. The seasonal variations in the load have already been taken into consideration through trained ANNs working under 4 different seasonal boundaries as mentioned in (a). The fluctuations in load due to prevailing climatic conditions can be considered by including the geographic area temperatures as inputs in addition to historical load inputs.

A general outline of the decomposed structure of the STLF model is shown in Figure 28. A family of ANNs have been formed/ trained and tested and results are included to demonstrate this feature.

3.3.7.4. Application

Based on the factors outlined above, the load shape for each day was forecast over one year. The ANNs developed in the present work are for weekdays and weekends only. However, the analysis can be readily extended to model holidays, special events etc. provided adequate data is available over several years so that appropriate training examples can be formed. The database used for the present case consists of load and weather information. For each day, 48 ANNs have been formed and trained separately. Each ANN represents a model for half hourly load consisting of a 3-node input layer and 1-node output layer. The input layer nodes represent previous day load (one value), previous week load of the same day type (one value) for the targeted time and temperature data (one value) for the time of prediction. For training, actual temperature values have been used whereas, for testing, the forecast values were taken. Since the previous day load for Monday happens to be Sunday, therefore the previous Friday load curve was used instead. Similarly, for Saturday, the previous day is replaced by the Saturday of the same day during the previous week load. Thus in this case, the previous two consecutive Saturdays' load

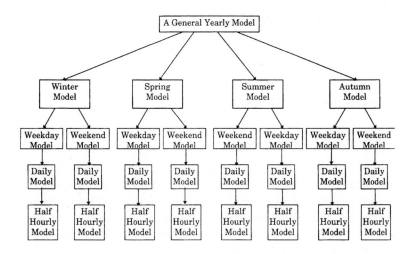

Figure 28. A modularised ANN model.

curves were used for preparing the training data. In the preparation of training sets for different ANNs for weekdays, the data was carefully selected so that any holiday period can be avoided in the context of previous day or previous week load. If such a situation appears in any particular case, these days were either replaced by an alternative date or the whole pattern was excluded from the training set. The temperature data available was comprised of 3 hourly recorded readings over a 24 hour period. This gives a total of 8 data points for each day. Therefore, the scheme used for preparing temperature data for 48 ANNs was such that each temperature measurement in a 24 hour period was used for 6 ANNs. The scheme of selection is as follows:

Recording time(hrs) of temperature	Time band for ANNs (half hourly)
0300	0030 to 0300 (6 ANNs)
0600	0330 to 6000 (6 ANNs)
0900	0630 to 0900 (6 ANNs)
1200	0930 to 1200 (6 ANNs)
1500	1230 to 1500 (6 ANNs)
1800	1530 to 1800 (6 ANNs)
2100	1830 to 2100 (6 ANNs)
2400	2130 to 2400 (6 ANNs)

The seasonal boundaries identified by the co-operating utility company define various months in each season. This is as follows:

Winter: November, December, January, February.
Spring: March, April, May.
Summer: June, July, August.
Autumn: September, October.

3.3.7.5. Results and discussion

From 16 months of available data spreading over a period between April 1990 to July 1991, four sets of test cases have been prepared to represent load for each season. Each test case is comprised of days from Monday through Sunday. The absolute relative error (in percent) with respect to the actual value is calculated for each half hourly forecast. The forecasting results are evaluated in terms of overall average error over a given day for the respective month in the season. These results are tabulated in Table 7. The overall summary of results is given in Table 8.

Table 7. Average forecasting error% for twelve months.

Months	Mon	Tue	Wed	Thu	Fri	Sat	Sun
Jan	3.97	4.9	1.96	1.36	4.88	5.16	2.15
Feb	2.69	1.28	1.22	1.81	3.69	8.08	3.4
Mar	7.15	3.04	4.5	1.19	1.85	3.26	2.73
Apr	7.61	4.51	2.47	2.23	3.38	5.87	2.43
May	3.54	3.89	2.81	1.55	5.54	2.21	3.92
June	1.53	1.84	1.12	1.65	2.1	2.68	1.21
July	1.64	1.49	1.89	1.82	3.41	4.01	4.73
Aug	4.67	4.55	2.52	3.59	2.88	3.71	3.65
Sep	4.34	4.31	1.84	3.46	5.86	5.6	5.09
Oct	6.77	4.37	2.69	2.98	4.61	2.19	4.54
Nov	2.88	2.19	4.65	1.19	2.33	3.5	2.47
Dec	3.29	1.68	4.64	1.4	1.23	7.63	3.03

Table 8. Summary of forecasting error% for four seasons.

Season	Mon	Tue	Wed	Thu	Fri	Sat	Sun
Winter	3.2	2.51	3.11	1.44	3.03	6.09	2.76
Spring	6.6	3.81	3.26	1.65	3.59	3.78	3.02
Summer	2.61	2.62	1.84	2.35	2.79	3.46	3.19
Autumn	5.55	4.34	2.65	3.22	5.23	3.89	4.81

The advantages obtained in using a modularised ANN approach includes: (i) a reduction in training time to a calculated level of one fortieth of that required for the non-decomposed case, (ii) better insight to the ANN structure, and (iii) the forecast can be generated for any length of time between half-an-hour to a 24 hour period. The results of Table 7 show that the forecasting error is within reasonable accuracy in the majority of the cases. However, there are a few occasions where there is some inconsistency in the behaviour of the error. These are analysed in the following two observations:

Daily variations : The forecast error is within reasonable accuracy for all days in a week except for Monday and Saturday where for a few occasions, the error is out of line with the others. The main reason behind this odd behaviour is likely to be the absence of an input node related to the previous day load which may effect the following day load. However, in both these cases, the previous day load has not been used due to the reasons given in the previous section. Further investigations are needed to examine this effect so that these days can be modelled in a more efficient way.

Seasonal Variations: From Table 8. the forecast results show consistency for summer compared with the other seasons. In a country like Britain, the heating load is a significant factor in raising or lowering the demand in the winter season and is therefore more susceptible to changes in the weather. Due to the reduced involvement of heating load in summer, the demand is more uniform and hence the forecast could be more consistent. Autumn and Spring are the two most unpredictable seasons in Britain and represent the border blocks of pre- and post-Winter seasons respectively. With weather such as in Britain, these two seasons are more inclined to Winter season patterns and thus make load demand more volatile. However, with the inclusion of other weather parameters, like wind speed and direction, chill factor etc., the accuracy of the forecast can possibly be improved.

3.3.7.6. Conclusion

The ANN methodology has been applied to short term load forecasting from another perspective which is based on repeatedly decomposing the short term load forecasting problem into a number of smaller forecasting models each of which are trained using data carefully selected over hourly periods, daily, weekly and seasonal time varying windows. A number of results have been included to illustrate the applicability of this approach. Sixteen months of actual power utility load data and regional temperature data have been utilised to train and test a family of ANNs. The results are promising and indicate that modular approach may offer advantages in terms of reducing the complexity of the forecasting methodology and increasing the efficiency of ANNs.

The modularised approach has provided the basis to introduce the novel concept of the 'expert network' which provides a future direction in related research. The emerging trend of the use of a hybrid approach [25] in the context of combining the learning capabilities of an ANN and the functionality of expert systems is potentially suited to the problem of STLF. This idea can help avoid the pitfall of the scepticism of human users for ANN-based forecasts and will thus include the same kind of heuristics which are implicit in the forecasting strategy accepted by the control engineers.

3.4. VIBRATION IN POWER TRANSFORMERS

Power transformers are an essential part of an ac power system. Their prime function is to transform the voltage at which power is transmitted power from one level to another, for example from the transmission level to distribution level. An important subsidiary function of large transformers is often the control of system voltage stability through the process of tap changing, where changes can be made to the overall transformer voltage ratio by tapping one (or more) of the windings at different points. Thus, the importance of these items of plant cannot be over emphasised, and when one considers the extremely high capital costs involved, it is clear that the "health" of a power transformer is of prime concern to the owning company, for both economic and operational reasons [26].

3.4.1. Causes of Vibration in Power Transformers

Power transformers have an inherent level of vibration which cannot be eliminated and does not represent a problem under normal operational conditions. However, this level of vibration, either in the core or in the windings, can be increased due to a number of reasons [27,28]:

- Loose core clamping bolts and bolts tying together the core structure;
- Repeated switching of the transformer into circuits on no-load, particularly for transformers located close to a generating source;
- Heavy external short circuit faults subjecting the transformer to short term high mechanical stresses as a result of internal unbalanced electromagnetic conditions;
- Rapidly fluctuating loads causing high levels of mechanical stress.

3.4.2. Effects of Abnormally High Vibration in Power Transformers

The principal effect of vibration in a transformer's windings is damage to the winding insulation [27]. This is because vibration can cause adjacent turns of the winding to abrade, leading to a progressive degradation in the dielectric strength of the winding insulation. This winding insulation degradation could ultimately lead to a short-circuit fault between winding turns, which may remain undetected for some time. The culmination of such faults could be catastrophic, for example a cascade series of breakdowns in the winding insulation could occur, causing irreversible damage to the windings or even the complete destruction of the transformer.

3.4.2.1. The Transformer Monitoring Data

The equipment from which the train/test data was derived comprises a small power transformer which was fitted with an extensive set of measuring instruments and associated data retrieval and archiving hardware. The measurements were taken at the Massachusetts Institute of Technology for Electromagnetic and Electronic Systems. The parameters which were measured from the transformer are listed below:

- Electrical terminal data, i.e. H.V and L.V current and voltage measurements;
- Thermal data, which consists of 27 individually monitored points distributed around and within the transformer;
- Winding current data, i.e. the current present in the windings (directly related to the primary current being supplied through the transformer);
- Vibrational data, consisting of vibration measurements taken from both the core and windings of the transformer via accelerometers. The output signal from the accelerometers has been processed (using a fourier transform technique) in order to obtain magnitude and phase representations of the harmonic components of the core and winding vibration. For the purpose of this chapter, the first, second and third harmonic components are of interest.

The total extent of the data is as shown below:

- 9 weeks of "healthy" (or undistorted winding) transformer data.
- 6 weeks of "unhealthy" (or distorted winding) transformer data. This distortion was achieved via a hydraulic ram connected to the windings of the transformer.

3.4.2.2. ANN Train/Test Strategy & results

3.4.2.2.1. Overall train/test strategy

The available train/test data was divided into three major train/test files each representing 4 weeks worth of measured transformer data:

- week 1–4 (Files containing 4 weeks of "healthy" transformer data).
- week 5–8 (Files containing 4 weeks of "healthy" transformer data).
- week 10–13 (Files containing 4 weeks of "unhealthy" transformer data).

The major strategy is to train and test ANNs on various combinations of "healthy" and "unhealthy" data in order to achieve the following objectives:

- Individual estimations of the harmonic components of winding vibration from complete sets of available input data;
- Simultaneous estimations of all six harmonic components from complete sets of available input data;
- Evaluation of network performance under different conditions. For example, experimentation with different network topologies/paradigms, changes to the employed learning rules, varying the number of training iterations, presenting incomplete and subsetted input data etc.;
- The use of ANNs to perform classification type analysis on the available input data, given our prior knowledge that certain train/test data relates to the transformer under various "unhealthy conditions".

Packages, such as NeuralWorks Professional II/ PLUS, used for constructing, training and testing the ANNs allow the training files to be presented in a random fashion. This helps to optimise the learning/training process and avoids the network becoming "trapped" in local minima.

3.4.2.2.2. Explanation of testing procedures and results: Winding vibration estimation.

ANNs of the back propagation type were trained on week 1–4 data and tested on the files from week 5–8 and week 10–13 data.

Figure 29 shows the results for a back propagation network (16 node input, 10 node hidden layer and 6 node output), which was trained on week 1–4, using inputs of: electrical data (4 fields); winding current data (6 fields); thermal data (6 fields). The network has 6 outputs which are related to the six harmonic components of winding vibration which are being estimated. Core vibration measurements were omitted due to them having little effect on the accuracy of the ANN. Emphasis should be placed on the fact that the network is being tested on totally new or "unseen" data.

The following figure shows how the same network used to produce the results shown in Figure 29, but tested on the week 10–13 ("unhealthy" transformer data) file performs. These results are particularly encouraging in that they illustrate how a network, trained on "healthy" transformer data,

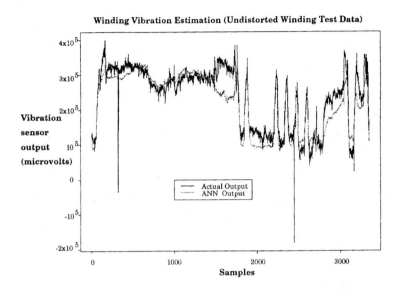

Figure 29. The results for a back propagation network.

estimates a level of winding vibration which is significantly different from the actual vibration as measured by the winding accelerometer. This could be useful in the field of condition monitoring as potential transformer problems could be identified and alarmed through a comparison of actual and estimated parameters.

3.4.2.2.3. Explanation of testing procedures and results: Classification of transformer data

An ANN of the self organising map (SOM) variety was used in order to attempt classification, or categorisation of the overall transformer train/test data set. The results from this exercise can then be compared with our prior knowledge of the data relating to the transformer in the "healthy" and "unhealthy" states.

The particular network used in this application consists of an input layer of 22 neurons (4 electrical terminal measurements, 6 thermal measurements, 6 winding current and 6 core vibration harmonic components). It is important to note that winding vibration measurements are not included as inputs to the network, thus reducing the input data requirements.

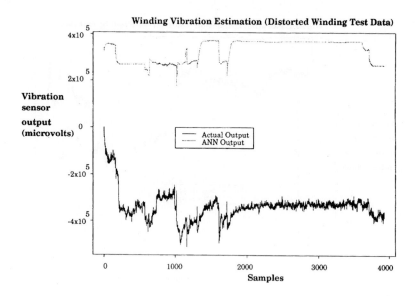

Figure 30. The same network performance tested on "unhealthy" transformer data.

The 22-node input layer is fully connected to the 4x4, 16 node Kohonen layer. The training/testing procedure for this network is fundamentally different from the earlier estimation work in that the training process is "unsupervised". In this particular experiment, the entire data set ("healthy" and "unhealthy") is presented to the network and it then attempts to characterise the data set in the manner described earlier. In order to test the network, the individual subsets of data representing the transformer under "healthy" and "unhealthy" conditions respectively are presented to the trained network and the response is then analysed in terms of the distribution of winning nodes.

Figure 31 shows the distribution of winning nodes for the trained network after presentation of the data subset relating to the "healthy" transformer. A grey-scale image plot representation has been used, i.e. the darker the colour relating to a node, the greater the relative "winning" frequency of that node. For this example, it is clear that the winning nodes are predominantly placed in the area beneath the leading diagonal of the Kohonen map, with nodes (3,2) and (4,1) particularly active.

In Figure 32 the distribution of winning nodes for the same network after presentation of the data subset relating to the "unhealthy" transformer are shown. In this case, the winning neurons' physical distribution is in almost

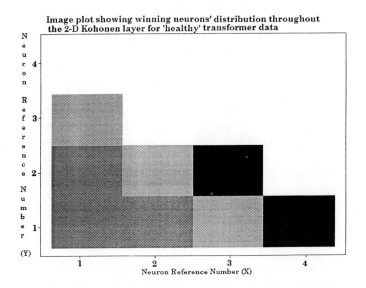

Figure 31. The distribution of winning nodes relating to the "healthy" transformer data.

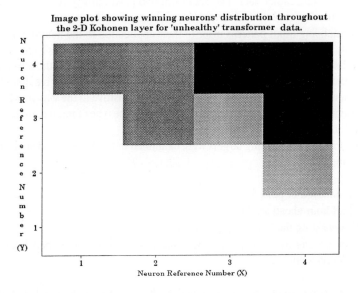

Figure 32. The distribution of winning nodes for "unhealthy" transformer data.

complete contrast to that for the "healthy" case, with the winning nodes predominantly placed in the area above the leading diagonal of the Kohonen map, with nodes (4,4), (3,4) and (4,3) particularly active.

3.4.3. Conclusions

This section has shown that there is potential for an accurate estimation of transformer winding vibration from other available sensor data through the use of back propagation ANNs. The use of ANNs to classify/categorise the train/test data sets has also been investigated. Networks of the self-organising map (SOM) type (e.g. Kohonen networks) have been shown to be suitable for this type of function

3.5. OVERALL CONCLUSION

The aim of this chapter was to investigate the applicability of the novel methodology of artificial neural networks to short term load forecasting problem and for the purposes of electrical plant condition monitoring. The objective of load forecasting was pursued by carrying out detailed studies in terms of training and testing the neural nets with historical electrical utility load data and weather data. The data from two power utility companies had been utilised to exploit the learning capabilities of ANNs.

Normalising input patterns prior to their application to the neural net can actually provide for pre-processing data in different ways. This aspect has been discussed by normalising the given data in various ways for the different input patterns applied to the networks and through comparison of results. The experience shows that the forecasting error was sensitive to the way in which the training data was handled and it depends upon the type of input parameters used for ANNs.

A variety of forecasting targets were examined in the ANN-based analysis, these include:

 (i) prediction of peak load;
 (ii) half hour ahead load forecasting;
 (iii) forecasting the load shape a day ahead;
 (iv) forecasting over a flexible time window varying from one half hour to
       24 hours ahead.

Forecasting targets such as in (i) and (ii) have proved to be handled easily and the forecast accuracies were also better. The reason behind this behaviour involves the use of smaller ANNs and the reduced network complexity involved. Moreover, they required less time both in terms of training and testing phases. The same argument applies to the target mentioned in (iv) , as in this case also, the processing nodes for the given ANN are few in number and hence there is less complexity involved.

The prediction target given in (iii) has been investigated through the use of two approaches namely the lumped and the distributed (discrete) approaches. Both approaches produced results with reasonable accuracies. However, the latter one has merit in terms of its smaller network structure, better insight into the model, less complexity and speed of operation.

A number of input pattern combinations were used to model ANNs for predicting various targets. The input parameters include lagged load values including the previous half hour's, day's, week's month's load; and past and estimated weather information. The historical load values used alone in the input layer have successfully trained ANNs to predict various targets. However, the immediate past hour loads have contributed only in the prediction targets of very short term ahead forecasts for obvious reasons. Among weather parameters, the data for temperature, wind speed and directions was available from the utility. The inclusion of wind speed and its direction along with load values in the input patterns were also used which did not improve the forecasting results at all and was therefore discarded from the analysis. The temperature data proved to be the most effective weather parameter in ANN analysis and results show interesting indications towards the forecasting of load. The ANN model using temperature values alone in the input pattern combination has given this indication that ANNs can model the forecast process on the basis of temperature variation. However, load demand is not dependent on weather parameters alone but also depends on several other factors which can be captured through incorporating historical load values which have significant influence on the given load pattern. In this context, therefore, a number of ANN forecasting models have been developed using load data alone and the results show that the ANN has successfully captured the features which are implicit in the load pattern.

The combination of temperature and load values have also been used in the input pattern to observe the intuitive correlation between these two input categories for different forecasting targets. The results indicate that both strong and weak correlations apparently exist between historical load data and temperature values under different scenarios. It is also observed that forecasting accuracies become generally improved in those cases where more regional weather data was used.

The selective feature in training and testing the ANN is emphasised by illustrating a number of examples. The indication is that the given ANNs can perform well for the types of day for which they are trained. Thus, it suggests that ANNs are able to predict more accurately if carefully selected data is used for training.

This experience has given the motivation to apply a discrete approach to STLF in an ANN perspective. Thus the application of the ANN methodology in short term load forecasting has been taken to another level of abstraction which is based on repeatedly decomposing the short term load forecasting problem into a number of smaller forecasting models each of which are trained using data carefully selected over hourly periods, daily, weekly and seasonal time varying windows. A number of results have been included to illustrate the applicability of this approach. The results are promising and indicate that the modular approach may offer more advantage in terms of reducing the complexity of the forecasting methodology and increasing the efficiency of ANNs.

The promise of the ANN methodology has also been extended to an other problem namely the analysis of electrical plant performance in the form of test transformer data. This problem is classified in the general category of pattern recognition. In all the problems studied it was found that the ANN algorithm converged to an acceptable solution.

In short, the ANN methodology is emerging as a new tool in the field of artificial intelligence. In this context, an exposition of the practical applications of neural networks is appropriate and the authors believe that this chapter demonstrates a number of cases to show that the ANN approach can be successfully used for the purpose of short term load forecasting modelling and condition monitoring applications. These results are viewed as being rather encouraging for the future of ANNs as they can potentially be a significant tool in power system and more general applications.

REFERENCES

1. House, E.L. and Pansky,B., 1967, *A fundamental approach to neural anatomy*, McGraw Hill, New York.
2. Stubbs, D., 1988, *Neurocomputers*. M.D. Computing.
3. Rumelhart, D.E., Hinton, G.E. and Williams, R.J., 1986, *Parallel distributed processing, explorations in the microstructure of cognition, Vol.2: Psychological and biological models*, MIT Press.
4. Rumelhart, D.E., Hinton, G.E. and Williams, R.J., 1986, *Learning internal representations for error propagation in Parallel distributed processing*, Vol. 1, MIT Press, pp.318–362.
5. Hammerstorm, D., 1993, "Neural networks at work", *IEEE Spectrum*, pp.26–32.
6. Lippmann, R.P., 1987, "An introduction to computing with neural nets", *IEEE ASSP Magazine*.
7. Eberhart, R.C. and Dobbins, R.W., 1990, "Neural network PC tools", Academic Press, Inc..
8. Pao, Y-H., 1989, "Adaptive Pattern Recognition and Neural Networks", Addison-Wesley Publishing Company Inc.
9. Wasserman, P.D., 1989, "Neural computing, Theory and Practice", Van Nostrand Reinhold.

10. Weigend, A.S., Huberman, B.A. and Rumelhart, D.E., 1990, "Predicting the future: A connectionist approach", *International Journal of Neural Systems*, World Scientific Publishing Company, Vol. 1. No. 3.
11. Baum, E.B. and Haussler, D., 1989, "What size net gives valid generalization?", *Advances in neural information processing system I*, Morgan Kaufmann Publishers Inc.
12. Chauvin, Y., 1989, "Dynamic behaviour of constrained back-propagation networks" *Advances in neural information processing system I*, Morgan Kaufmann Publishers Inc.
13. Christiaanse, W., 1971, "Short term load forecasting using general exponential smoothing", *IEEE Trans. on Power App. and Sys.*, Vol. PAS-90, pp.900–910.
14. Box, G.E. and Jenkins, G.M., 1976, "Time series analysis — forecasting and control", Holde n-Day, San Francisco.
15. Rahman, S. and Bhatnagar, B., 1988, "An expert system based algorithm for short term load forecast", *IEEE Trans. on Power Systems*, Vol. 3, No. 2.
16. Ho, K-L., Hsu, Y-Y., Chen, C-F., Lee, T-E., 1990, "Short term load forecasting of Taiwan Power system using a knowledge-based expert system", *IEEE Power Engineering Review*.
17. Park, D.C., El-Sharkawai, M.A., MarksII, R.J., Atlas, L.E. and Damborg, M.J., 1991, "Electric load forecasting using an artificial neural network", *IEEE Trans. on Power Systems*.
18. Lee, K.Y. and Park, J.H., 1992, "Short term load forecasting using an artificial neural network", *IEEE Trans. on Power Systems*, Vol. 7, No. 1.
19. Kun-Long Ho, Yuan-Yih Hsu, Chien-Cluen Yang, 1992, "Short term load forecasting using a multilayer neural network with an adaptive learning algorithm", *IEEE Trans. on Power Systems*, Vol. 7, No. 1.
20. Peng, T.M., Hubele, N.F. and Karady, G.G., 1992, "Advancement in the application of neural networks for short term load forecasting", *IEEE Trans. on Power Systems*, Vol. 7, No. 1.
21. Lu, C.N., Wu, T.H. and Vemuri, S., 1993, "Neural network based short term load forecasting", *IEEE Trans. on Power Systems*, Vol. 8, No. 1.
22. Azzam-ul-Asar and McDonald, J.R., 1994, "A specification of neural network applications in the load forecasting problem", *IEEE Transactions on Control Systems Technology*.
23. Azzam-ul-Asar, and McDonald, J.R., "A modularised ANN approach to short term load forecasting", American Power Conference, April, 1994, Illinois Institute of Technology, Chicago.
24. Hrycej, T., 1992, "Modular Learning in Neural Networks", John Wiley & Sons, Inc.
25. Kandel, A. and Langholz, G., 1992, "Hybrid architectures for intelligent systems", CRC Press Inc.
26. Hiirononniemi, E. *et al.*, 1992, "Experiences of On- and Off-line Condition Monitoring of Power Transformers in Service", paper 12–102 Cigr,, August-September, 1992
27. Franklin, A.C. and Franklin, D.P., 1983, "The J&P Transformer Book", 11th Edition, Butterworths.
28. McDowell, G.W.A. and Lockwood, M.L., 1994, "Real Time Monitoring of Movement of Transformer Windings", *IEE Colloquium on Condition Monitoring and Remnant Life Assessment in Power Transformers*.

4 MECHATRONIC TECHNIQUES IN INTELLIGENT HEATING, VENTILATING AND AIR CONDITIONING SYSTEMS

HONG ZHOU, MING RAO and KARL T. CHUANG

Department of Chemical and Materials Engineering, University of Alberta, Edmonton, Alberta, Canada T6G 2G6

Heating, ventilating, and air conditioning (HVAC) processes provide a comfort environment, but consume a great deal of energy. Many efforts have been put into building energy conservation since the energy crisis of 1973. On other hand, the energy conservation efforts led to tight building envelopes and low ventilation rates, which causes poor indoor air quality (IAQ), the so called "Sick Building Syndrome". Conflict exists between energy saving and indoor air quality improvements. In this paper, an intelligent system approach is proposed to support HVAC operations, aiming at improving energy conservation and IAQ control [25].

This chapter describes the construction of an Intelligent Operation Support System (IOSS) for heating, ventilating, and air conditioning (HVAC) processes. The system contains important expertise, qualitative reasoning and quantitative computation. It consists of expert system for operation planning or operation mode consulting , comfort setting, conflict reasoning and some other functions. The expert system for operation planning provides recommendation of energy saving operation modes for HVAC processes. The comfort indoor setting system sets indoor temperature by comfort-stat strategy instead of setting the thermostat, which has advantages of energy saving, thermal comfort and indoor air quality. The conflict reasoning system achieves conflict resolution for energy saving and IAQ control. An integrated distributed intelligent system framework is introduced to integrate these systems, including qualitative reasoning and quantitative computation. IOSS provides a real time integrated operation planning method in HVAC processes. It can be used to assist or train operators to achieve better operation in HVAC systems. The mechatronics methodology paves the way for us to upgrade and improve this industrial capability.

4.1. INTRODUCTION

Heating, ventilating and air conditioning (HVAC) processes provide a comfortable environment inside buildings, but consumes a great deal of energy. Roughly speaking, buildings account for one-third of total energy consumption in the world, most of which is used in HVAC processes.

Since the oil embargo of 1973, extensive research and development on energy conservation in buildings has been pursued. Computer technology has played an important role in energy management and control systems (EMCS) during the past 10 to 15 years.

In summary, the HVAC energy saving control strategies may be classified as follows:

(1) *Time scheduling:* This is the most effective energy-saving control strategy in the early use of EMCS in HVAC. It turns equipment ON or OFF at preset times.

(2) *Advanced HVAC process control:* Optimization, adaptive control and predictive control have been applied to HVAC processes for better control and energy conservation [1,13,15,17].

(3) *Economic operations:* There are several alternative air handling processes which can achieve desired indoor setpoints. The following control methods [5] are energy conservation operations:

   (a) *Economy-cycle outside-air control:* This strategy is to use outside air for cooling whenever possible, thus minimizing energy consumption in the refrigeration cycle.

   (b) *Enthalpy control:* This strategy is achieved by calculating the enthalpy of air being processed and choosing the minimum enthalpy cost for air handling operation. The calculation of the enthalpy is not an easy task. Alternatively, we can compare the enthalpy cost of air processing operations on the psychrometric chart in order to select the minimum-enthalpy cost operation.

   (c) *Conversion of dual-duct to variable-air-volume (VAV):* Replacing a dual-duct system by a VAV system is a very effective energy conservation technique. Mixing loss and fan horse power savings can be made by adding fan volume control [12,22].

   (d) *Energy-source shutdown:* Depending on the season, shutting down heating or cooling sources also provides energy conservation. However, the shutdown should not cause the loss of environmental control.

(4) *Comfort technology:* Comfort research has provided a number of alternatives for energy conservation [3,18]. We believe that "comfort" and "energy" could be simultaneously optimized through the use of operation strategies which consider the dynamics of comfort and control system.

Comfort technology has not yet been fully utilized for energy conservation. Many users and designers are still unaware of the opportunities available for energy conservation. Comfort technology education has been suggested for transferring this knowledge to the public [9].

(5) *New energy efficient equipment:* Heat-pump air conditioners, solar energy heaters, and inverter control systems are the recently developed energy-efficient HVAC equipment [16]. Most of these items use microprocessor-based control and significantly reduce energy consumption [2].

From the system science viewpoint, the current energy management and control strategies used in HVAC processes are a collection of remedies or recipes, with neither coordination nor integration. The reduction in energy consumption was easy to achieve in earlier times with only a small capital investment. But these simple remedies have now been exhausted. It is very difficult to model the HVAC process. Most problem solutions are based on heuristics or experience [7,11].

On the other hand, energy conservation efforts of the last decade have led to reduced infiltration and ventilation in occupied space, and increased use of unvented appliances, as well as new materials and new equipment introduced into buildings. The concentrations of internally generated contaminants increase as a result. Air sampling in many buildings has indicated that indoor concentrations of known pollutants often exceed standards set for outdoor and industrial exposures. Complaints by occupants have also drawn attention to indoor pollutant levels, and raise questions as to the adequacy of indoor air quality (IAQ) to protect the health of the building occupants [14].

Poor indoor air quality causes health problems, such as sensory irritation in the eye, nose, or throat, skin irritation, neurotoxic symptoms, unspecific hyperreactivity reactions, and odors, the so called "sick building" syndrome. As people spend up to 90 percent of their time in a residential or commercial environment, the sick building problem has been characterized as "one of the most serious public health challenges in the next decade.

The sources of pollutants of indoor air are classified into three groups [10]: outdoor air pollution, emission products of building materials, and indoor human activities. Human activities, such as smoking, cooking, perspiration, and particles emitted by skin, are major parts of indoor air pollution. Indoor air pollution caused by occupants can be removed by proper ventilation, but the increase of the air exchange rate costs more energy.

Currently indoor air quality control has not been integrated into HVAC operation and control systems, so IAQ and EMCS become two conflicting goals. It is realized that we need to develop new control strategies to seek IAQ answers and avoid extensive energy consumption. This chapter focuses on the research of new methods for HVAC control, which aims at reducing energy consumption and improving indoor air quality in buildings.

This chapter outline is as follows: first we give a brief introduction about the background of heating, ventilating, and air conditioning, as well as conventional control strategies, and address the problems related to the control strategies. An intelligent system methodology is proposed to solve the problems. An expert system for HVAC operation planning is presented. The models for indoor air quality and ventilation are identified. Then conflict resolution for energy conservation and indoor air quality control is discussed. Finally a new control strategy, integrated intelligent control framework for HVAC processes is proposed.

4.2. BACKGROUND OF HVAC SYSTEMS AND OPERATIONS

It is important for readers to understand the HVAC process before discussing the research project. Therefore, a description concerning HVAC system structures and air handling processes as well as control schemes will be presented. In addition, the problems with the HVAC conventional control strategies will be discussed.

4.2.1. System Structure and Classification

There exist many types of HVAC systems and various air handling equipment to meet the requirements for all kinds of buildings and weather. The system entity structure for HVAC systems is demonstrated in Figure 1, which contains decomposition, coupling and taxonomy information. Most residential or office buildings only require temperature conditioning, whereas some industrial environment need to condition both temperature and humidity. A single zone system sets all conditioned rooms at the same temperature and humidity, but a multizone system sets different indoor parameters for the conditioned rooms. Air washer and cooler both are used to cool and dehumidify air, which are chosen by the HVAC system designer depending on the weather and other factors. For a large commercial building, terminal controller or a distributed control system may be required. In this research project, we are only concerned about central HVAC systems.

There exists a total of sixteen types of central HVAC systems (Figure 1), which may be classified as follows:

- **TSC:** conditioning temperature, single zone and constant air volume system. Outside air is mixed with return air, then filtered, and heated or cooled to meet supply air requirement. A constant air volume fan supplies processed air to rooms.

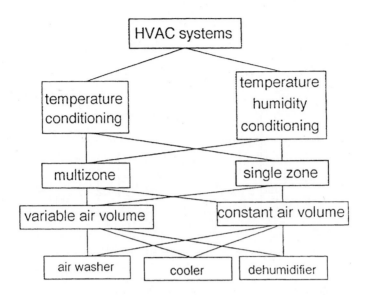

Figure 1. Structure classification for HVAC systems.

- **TSV:** the same system as above but with variable air volume.
- **TMC, TMV:** the same systems as TSC, TSV, but with multizone. In multizone system. Air amount which enters each zone is distributed by terminal thermostat, which can also add heat to supply air if it is needed.
- **THSCD:** shown in Figure 2, conditioning both temperature and humidity, single zone, constant air volume, and dehumidifier are used for the air dehumidifying process. Fresh air is mixed with recirculated air, which passes humidifier or dehumidifier, heating or cooling coils. After processing, air is supplied to rooms by fan. The differential pressure is usually kept positive by adjusting exhaust air damper.
- **THSCC:** the same system as THSCD but use cooling coil for the air dehumidifying process.
- **THSCW:** the same system as THSCD but using air washer for the dehumidifying process. Fresh air is mixed with return air, then partially goes through an air washer, and mixes with bypass air. This process is used for both cooling and dehumidifying.
- **THSVD, THSVC, THSVW:** the same systems as THSCD, THSCC, THSCW but with variable air volume (VAV). Being processed, air is supplied to rooms by a VAV fan. The supply air amount changes responding to heat and/or wet load.

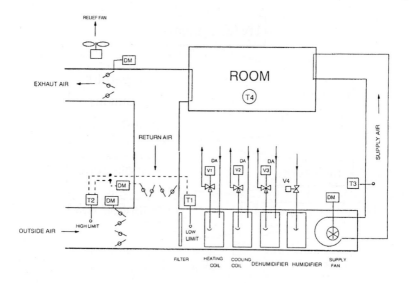

Figure 2. A scheme of temperature and humidity conditioning, single zone, and constant volume (THSCD) system.

- **THMCD, THMCC, THMCW:** conditioning both temperature and humidity, multizone, constant air volume, and dehumidifier, cooling coil, air washer used for the dehumidifying process, respectively. A variable-volume damper to supply each zone will be operated by the static pressure controller.

- **THMVD, THMVC, THMVW:** the same systems as above but with variable air volume.

4.2.2. Air Handling Process

The air processes, such as air washing, heating, humidifying, cooling, dehumidifying and mixing can be shown in an air property chart (Figure 3), where RH stands for air relative humidity. It should be pointed out that the enthalpy-wet coordinate air property chart is just another representation of Psychrometrics, which is convenient for further analysis.

- **Mixing:** Assuming a is fresh air, N is return air, then K stands for the mixed air status, $\overline{akN}$ shows a mixing process.

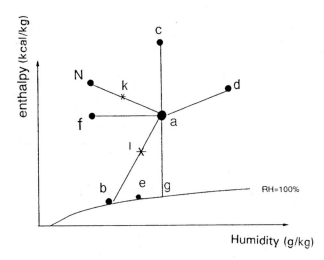

Figure 3. Air handling process.

- **Washing:** Assuming b is air washer water, air partially goes through the air washer and mixes with the air from by pass. $\overline{aLb}$ shows an air washing process. L stands for the processed air status. The process can be used to cool and dehumidify air.
- **Heating:** Heating process increases air temperature or enthalpy, but does not affect absolute humidity. $\overline{ac}$ demonstrates an air heating process.
- **Humidifying:** Ideally, a humidifying process should change only air dampness, and not affect air enthalpy. In fact, if the temperature of water or steam sprayed is higher or lower than that of air processed, a humidifying process causes enthalpy to go up or down. $\overline{ad}$ demonstrates a humidifying process (Figure 3).
- **Cooling:** It decreases air temperature or enthalpy. As air temperature goes down below saturated water vapor temperature, water is condensed out from the air. The process is shown as $\overline{a}g\overline{e}$ in Figure 3, and used to decrease temperature and/or humidity.
- **Dehumidifying:** As demonstrated by $\overline{af}$ in Figure 3, when air passes through a chemical dehumidifier, the water in the air is absorbed by the dehumidifier.

4.2.3. Control

Currently there are several control schemes for HVAC systems, 100% outdoor air, 10% outdoor air, economy cycle, and enthalpy control [6].

- **100% outdoor air:** this type of operation is for some special area such as chemistry laboratories and special manufacturing. 100% fresh air passes through an filter, then enters an air handling unit (AHU), in which the heating/cooling coil is controlled by a room thermostat. This control scheme is simple, and the indoor air quality is high, but it consumes a great deal of energy.
- **10% outdoor air:** By far the simplest method of outdoor air control is to open a "minimum outside air" damper whenever the supply fan is running. 10% outdoor air is mixed with 90% return air, then enters AHU to be processed and supplied to rooms. Compared to 100% outdoor air, it provides energy saving, but sacrifices indoor air quality. This was one of the first ideals developed and is still used extensively.
- **Economy cycle:** It is found that with minimum 10% outdoor air control, we need to operate the cooling coil even when the out door air temperature is near or below the freezing mark. This gives rise to the so called "economy cycle", with outside, return and relief dampers controlled by temperature.

SINGLE-ZONE SYSTEM: Figure 4 shows a single-zone air handling unit (AHU); economy cycle outside air control system. Outside air and relief dampers are in the minimum open position at the winter design temperature, and the return air damper is correspondingly in its maximum open position. As outside air temperature increases, the mixed air thermostat (T1) gradually opens the outside air damper to maintain a low-limit mixed air temperature. Return and relief dampers modulate correspondingly. As the outside air temperature continues to increase, an outdoor air high-limit thermostat (T2) is used to cut the system back to minimum outside air, thus decreasing the cooling load. The room thermostat can be used to reset the low-limit mixed air controller. This will provide greater energy conservation than with a fixed low-limit set point. The room thermostat resets the supply air control point. The supply air thermostat controls the hot and chilled water valves in sequence. This is the system most commonly used today.

MULTIZONE SYSTEM: Figure 5 demonstrates a typical variable air volume system with discriminator control. It provides multizone control with only a single duct. The supply air is maintained at a constant temperature which varies with seasons. The individual zone thermostat varies the air supply

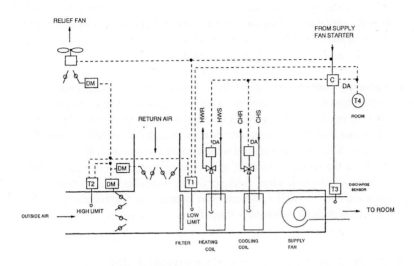

Figure 4. Control system for single zone, constant volume and temperature conditioning (TSC) AHU.

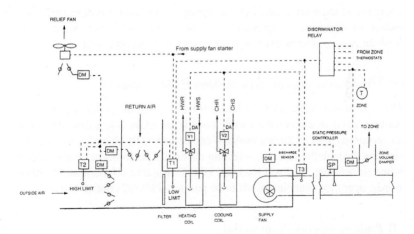

Figure 5. Control system for mulitzone, VAV, and temperature conditioning (TMV) AHU.

quantity to the zone to maintain the desired temperature condition. Minimum supply air quantity is usually not less than 40% of design airflow to provide sufficient ventilation. Supplemental heating is used in all exterior zones. The zone thermostat controls the damper down to its minimum setting then starts to open the heating valve if heating is required.

- **Enthalpy control:** It can be shown that outside air "economy cycle" control based on dry bulb temperatures is not always the most economical. That is, in very humid climates the total enthalpy of the outside air may be greater than that of the return air even though the dry bulb temperature is lower. Since the cooling coil must remove the total heat from the air to maintain the desired condition, it is more economical in this case to hold outside air to a minimum. While enthalpy control is ideal in theory, it is questionable in practice, because the accuracy of commercial humidity sensors is difficult maintain without frequent calibration.

4.2.4. Problems with Conventional Control Strategies

There are several problems with current existing control strategies, which are summarized as follows:

(1) Lack of indoor air quality monitoring and control

There is no indoor air quality control strategy in the control system, therefore indoor air quality is not guaranteed. If the number of people increases, or fresh air mixing decreases, indoor air quality may become poor.

(2) Fixed control strategy versus variable HVAC operations

The control system has a fixed strategy for all seasons. it is often seen that the cooling coil is in use in winter while outside air intake is less than 100% for an "economy cycle". Such a problem can be solved by changing the control strategy through the seasons, selecting different operations or combinations based on weather, heat/wet load and HVAC structure information. In other words, operation planning needs to be put into a HVAC control system.

(3) Problem with supply air control

In a multizone VAV system, supply air temperature is fixed, it is not coordinated with the supply air volume control, which possibly gives rise to both energy waste and poor indoor air quality.

(4) Disadvantages of thermostat indoor setting

The control system sets indoor temperature at a constant value for thermostat strategy. Except temperature, the other six variables: mean radiant temperature, relative humidity, air velocity, clothing, activity level (metabolic rate) and exposure time, also have significant affect on comfort. The thermostat does not maintain the optimal thermal comfort inside building. It possibly results in energy waste because the indoor temperature could be set unnecessarily high or low. Comfort technology has not been fully utilized in HVAC control to achieve energy conservation and comfort. It is suggested that setting indoor temperature by comfort-stat, may give better comfort and energy saving than thermostat setting.

(5) Difficult to accumulate and utilize private knowledge

The building operator usually gains some valuable expertise for economic operations from his experiences. Current existing control strategies have no facilities to update and utilize the operator's valuable knowledge.

4.3. INTELLIGENT SYSTEM APPROACH

In order to solve the problems encountered, an intelligent system approach is developed. First, we introduce an operation planning concept and the problem definition. Then an intelligent operation support system (IOSS) is presented, which aims at coordinating the different operations in an HVAC system in order to achieve better energy saving, comfort and IAQ than that can be obtained from the conventional control strategy. The development of the intelligent system, including knowledge acquisition, representation, and organization is discussed. The system configuration is based on the concept of integrated distributed intelligent systems.

4.3.1. Operation Planning

Operation planning is the phase of the process that is concerned with the selection and sequencing of different operations and combinations to transfer an initial state to a goal state [23].

As weather or load changes, an air conditioning system must change air handling equipment. For example, in winter, the heating coil is used to condition temperature, but in summer, instead of using heating, a cooling coil is used to control temperature. In some cases, alternative air processing is able to meet the conditioning requirements. The problem of operation

Table 1. Energy saving operation modes for THSCD systems.

actuator mode	return air damper U1	outside air damper U2	cooling coil U3	dehumidifier U4	humidifier U5	heater U6
1	90%	10%	off	off	Ⓓ	Ⓣ
2	Ⓣ	1-U1	off	off	Ⓓ	off
3	0%	100%	Ⓣ	off	Ⓓ	off
4	90%	10%	Ⓣ	off	Ⓓ	off
5	90%	10%	Ⓣ	Ⓓ	off	off
6	0%	100%	Ⓣ	Ⓓ	off	off
7	Ⓣ	1-U1	off	Ⓓ	off	off
8	90%	10%	off	Ⓓ	off	Ⓣ
9	Ⓓ	1-U1	off	off	off	Ⓣ
10	Ⓓ	1-U1	Ⓣ	off	off	off
11	0%	100%	off	Ⓓ	off	Ⓓ
12	0%	100%	off	off	Ⓓ	Ⓓ

planning is to choose an optimal operation mode for air processing according to the information about the weather, HVAC system structure, air handling equipment, heat and wet load, indoor setting, and indoor air quality. Table 1 shows some typical operation modes for the air conditioning system in Figure 2. HVAC operation planning is a bounded multi-objective optimization problem with non-algorithm and ill-structure. It seeks the operation mode for various HVAC systems which optimize energy consumption, thermal comfort, and indoor air quality.

4.3.2. Problem Solving Strategy

In order to solve the problems encountered above, our strategy is to develop an intelligent system that suggests the optimal operation mode for air processing based on the information concerning weather, HVAC system structure, air handling equipment, heat and wet load, indoor setting, and indoor air quality.

An intelligent system is one of the important engineering applications of artificial intelligence research, which acquires and codes the knowledge from human experts to solve ill-structured problems. The expertise about the operation planning is gained over a relatively long period of time from our research on energy saving control for HVAC systems, and verified by the operators. Heuristics and common sense reasoning play an important role in the problem solving.

The intelligent system provides assistance to operators to optimize the operation environment. It transfers and accumulates the expert knowledge

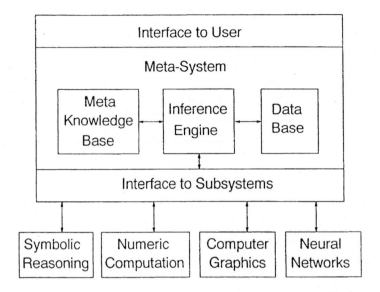

Figure 6. Integrated distributed intelligent system.

into computer programs so that those who do not have well-trained operation experiences can also control HVAC processes at the level of an expert.

4.3.3. System Configuration

The knowledge of IOSS covers HVAC control systems, energy conservation management, and comfort technology as well as indoor air quality control. Qualitative and quantitative information processing have to be coordinated to reach these goals. Here, we face the problem of knowledge integration and management. The implementation of IOSS requires an advanced intelligent system architecture and software environment.

An integrated distributed intelligent system structure was first proposed in 1987 [19]. It is a large knowledge integration environment, which consists of several symbolic reasoning systems and numerical computation packages. The architecture of an integrated distributed intelligent system is illustrated by Figure 6.

In IOSS, expert system for operation planning, indoor comfort setting system, and conflict reasoning system are integrated and coordinated by a meta-system. Symbolic reasoning, and numerical computation are integrated

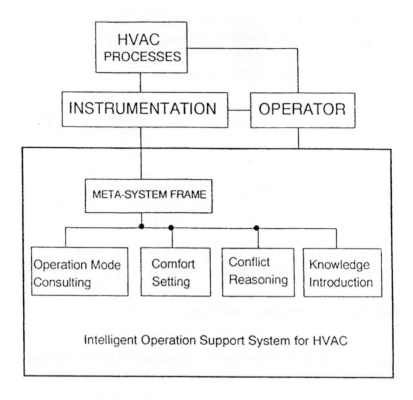

Figure 7. Integrated architecture of IOSS.

to facilitate the functionalities of IOSS. The integrated architecture of IOSS is demonstrated in Figure 7. Several systems, such as operation mode consulting, comfort setting, conflict reasoning and knowledge introduction, including a commercial package, such as DBASE III, Personal Consultant Plus (PC-PLUS, an expert development tool), are integrated by the meta-system. The software menu screen is shown in Figure 8. The knowledge organization and functions of each system will be described in the following sections.

4.4. EXPERT SYSTEM: OPERATION PLANNING

The operation planning recommends an energy saving operation mode for air handling based on the input information about HVAC structure, weather,

Figure 8. Software menu screen.

load, and indoor setting. It is an expert system for operation planning (ESOP), which is implemented under PC-PLUS. PC-PLUS provides external access interface, which allows the execution of the external program and retrieving of an external database.

4.4.1. Knowledge Acquisition

The key issue in the operation mode consulting is to identify the optimal operational mode subject to the conditions of system structure, weather, indoor setting, air handling equipment, and changing of heat/wet load.

Expertise about optimal operation mode identification

The expertise for the operation mode identification can be demonstrated in an air property chart [20,24]. The following example gives a description about the expertise.

For a constant air volume, single zone, temperature and humidity conditioning (THSCD) system in Figure 2, the operation modes can be divided in an enthalpy-wet coordinates air property chart. The operational mode identification expertise is demonstrated in Figure 9.

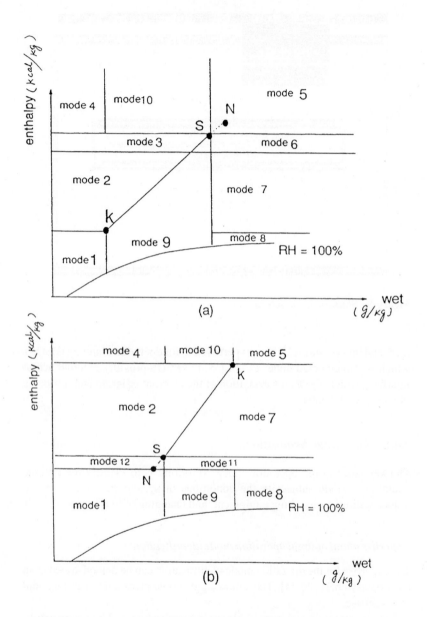

Figure 9. Operation mode identification for THSCD system.

N is indoor air setting, which is determined by thermal comfort requirement.

S is supply air, which is decided by the heat and wet load of an HVAC system.

K is reached by mixing minimum allowed fresh air (usually 10%) with return air, $\overline{SN}/\overline{KN} = 10\%$.

t_w is washer water temperature.

In Figure 9, if fresh air is in area mode 1, the optimal operation should choose mode 1, i.e. return air is 90%, fresh air 10%; cooling coil and dehumidifier OFF; heating coil and humidifier ON. If fresh air is in area mode 2, the air conditioning system should be operated under mode 2, and so on (Table 1).

Air handling processes for various operation modes are described below:

Mode 1: 10% Outside air (1a) is mixed with 90% return air, and heated to supply air temperature, then water or steam is sprayed to humidify air in order to meet the supply air requirement. The processes $(1a - 1b - 1c - S)$ are shown in Figure 10.

Mode 2: Outside air (2a) is properly mixed with return air to reach supply air temperature, then water or steam is sprayed to humidify the air to supply air moisture. The processes $(2a - 2b - S)$ are shown in Figure 10.

Mode 3: 100% outside air intake and is cooled to supply air temperature, then water or steam is added to reach the supply moisture requirement. The processes $(3a - 3b - S)$ are demonstrated in Figure 10.

Mode 4: 10% outside air is mixed with 90% return air, and then cooled to supply air temperature, then water or steam is added to reach the supply moisture requirement. The processes $(4a - 4b - 4c - S)$ are shown in Figure 10.

Mode 5: Outside air mixed with return air at ratio of 1:9, then the mixed air is cooled to supply air temperature. Afterwards, the air passes through dehumidifier to meet supply air requirements. The processes are $5a - 5b - 5c - S$ in Figure 10.

Mode 6: 100% fresh air is cooled and then passed through a dehumidifier to satisfy the supply requirement. $6a - 6b - S$ in Figure 10 shows the air handling processes.

Mode 7: Outside air is mixed with return air and controlled to supply air temperature, then dehumidify it to supply it to the system. The processes are shown as $7a - 7b - S$.

Mode 8: 10% outside air is mixed with 90% return air, then heating and dehumidifying process are applied, shown as $8a - 8b - 8c - S$ in Figure 10.

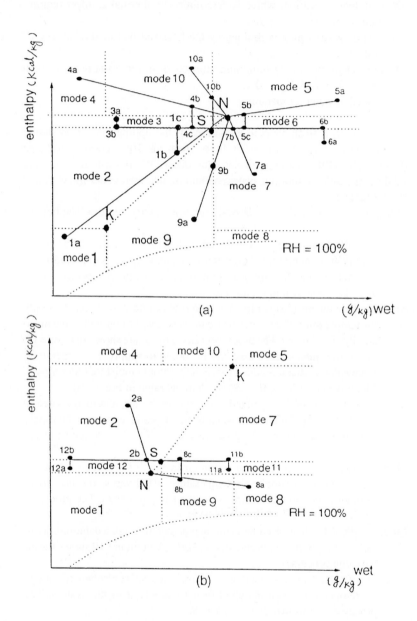

Figure 10. Air handling processes of energy saving operation modes for THSCD system.

Mode 9:   Mixing outside air with return air to control supply air moisture, then heating the mixed air to supply air temperature. The processes are shown as $9a - 9b - S$ in Figure 10.

Mode 10:   Outside air is mixed with return air to control moisture, then the mixed air is cooled to supply air requirement. The processes are demonstrated as $10a - 10b - S$ in Figure 10.

Mode 11:   100% outside air is taken in, dehumidified and heated to supply requirements. $11a - 11b - S$ in Figure 10 shows the processes.

Mode 12:   100% fresh air is humidified and heated to reach supply requirements as shown in $12a - 12b - S$ of Figure 10.

According to Figure 10, the operation modes are identified as:

MODE 1: $i \leq iik$ AND $w \leq wk$ AND $iin \geq iis$ AND $wn \geq ws$

MODE 2: $i \leq iis$ AND $i \geq iik$ AND $\frac{iin-i}{wn-w} \leq \frac{iin-iis}{wn-ws}$ AND $iin \geq iis$ AND $wn \geq ws$

......

Where i, iis, iin, iik are the enthalpy of fresh air, supply air, desired indoor air, and minimum mixed fresh air respectively; w, ws, wn, wk are the wet value of fresh air, supply air, desired indoor air, and minimum mixed fresh air, respectively.

It should be pointed out that S is a dynamic point, and the relative position of S and N reflects the system heat and wet load. Considering various load cases, S moves around N, so the integrated conditions to determine the operation modes become that in mode 1:

$$i \leq iik \; \text{AND} \; w \leq wk \; \text{AND} \; iin \geq iis \; \text{AND} \; wn \geq ws$$

OR

$$i \leq iin \; \text{AND} \; w \leq ws \; \text{AND} \; iin < iis$$

OR

$$i \leq iik \; \text{AND} \; w \leq ws \; \text{AND} \; iin \geq iis \; \text{AND} \; wn < ws$$

in mode 2:

$$i \leq iis \; \text{AND} \; i \geq iik \; \text{AND} \; \frac{iin - i}{wn - w} \leq \frac{iin - iis}{wn - ws}$$
$$\text{AND} \; iin \geq iis \; \text{AND} \; wn \geq ws$$

OR

$$i \leq iis \; \text{AND} \; i \geq iik \; \text{AND} \; \frac{iin - i}{wn - w} \geq \frac{iin - iis}{wn - ws}$$
$$\text{AND} \; iin \leq iis \; \text{AND} \; wn \geq ws$$

OR

$$i \leq iis \text{ AND } i \geq iik \text{ AND } \frac{iin - i}{wn - w} < \frac{iin - iis}{wn - ws}$$
$$\text{AND } iin \leq iis \text{ AND } wn \leq ws$$

OR

$$i \leq iis \text{ AND } i \geq iik \text{ AND } \frac{iin - i}{wn - w} > \frac{iin - iis}{wn - ws}$$
$$\text{AND } iin \geq iis \text{ AND } wn \geq ws$$

....

These conditions are used to issue the rule base to identify the operation modes. There exists a total of ten operation modes that respond to various load cases and weather in this type of air conditioning system. Each rule identifies one optimal operation mode corresponding to the input information.

For various HVAC structures (Figure 1), a total of sixteen types and over one hundred rules have to be be issued for the operation mode consulting knowledge base.

4.4.2. Knowledge Representation

Once the process operation and the related problems are studied, all the system inputs and outputs can be investigated.

Operation Planning Inputs:

HVAC SYSTEM STRUCTURE shown in Figure 1: Conditioning-type (temperature-humidity, temperature), zone-type (single zone, multizone), air volume (constant, variable), dehumidifying unit (air washer, cooler, or dehumidifier).

WEATHER: Outdoor air temperature, relative humidity.

INDOOR SETTING: Indoor air temperature, relative humidity.

SUPPLY AIR PARAMETERS: Supply air temperature, humidity, and air volume.

Operation Planning Output:

RECOMMENDED OPERATION MODE: Indication of all air handling units status and combinations.

Qualitative Knowledge Representation

Qualitative knowledge representation for HVAC system structure and symbolic information process determine the classification of the HVAC system.
A typical rule used to classify HVAC systems may be described as:

IF CONDITIONING-TYPE = TEMPERATURE-HUMIDITY
    AND ZONE-TYPE = SINGLE
    AND AIR-VOLUME-TYPE = CONSTANT
    AND DEHUMIDIFY-AIR-HANDLING-UNIT = AIR-WASHER
THEN HVAC-SYSTEM = THSCW

Quantitative Computation

The weather information, indoor setting and supply air parameters are quantitatively represented. Numerical information processing are applied for operation mode identification, calculation of supply air parameters and indoor setting. The processed quantitative information is coupled into the symbolic reasoning.

EVALUATION OF AIR ENTHALPY AND WET VALUE:

To identify the operation mode, as shown in the above expertise, we need to evaluate air enthalpy and wet value by the input of air temperature and humidity. Air enthalpy and wet value are evaluated as follows [6]: in English units:

$$i = 0.24t + w(1061.2 + 0.444t)$$
$$w = \frac{0.6129 Ps\phi}{(14.696 - Ps)}$$
$$\phi = \frac{1}{P_s}\left[P_{ws} - \frac{(14.696 - P_{ws})(t - W_t)}{2831 - 1.43}\right] \tag{1}$$

while in SI units:

$$i = t + w(2501.3 + 1.86t)$$
$$w = \frac{0.6129 Ps\phi}{1.0132 \times 10^{-5} - Ps}$$
$$\phi = [P_{ws} - 6.748 \times 10^{-9}(t - W_t)]/Ps \tag{2}$$

Where t is dry bulb temperature of air; ϕ is relative humidity of air; w is absolute humidity of air; W_t is wet bulb temperature of air; P_s and P_{ws} are saturated water vapor pressure with respect to the dry bulb temperature and the wet bulb temperature, respectively.

SUPPLY AIR PARAMETERS ESTIMATION:

Supply air parameters response to HVAC system heat and wet load, which behave uncertainty in most situations. Some air conditioning systems have no measurements for supply air. The supply air parameters are relatively stable or slow time changing parameters after an air conditioning system reaches stable working condition, and can be measured on line in most cases. For the optimal operation mode identification, the data need to be processed as follows:

CASE 1. On-line data for supply air parameters available and off-line operation mode consulting needed: The mean of the last five sampling data are used as the steady state supply parameters.

CASE 2. On-line data for supply air parameters available and on-line operation mode consulting needed: modeling the output and input relationship for indoor air parameters and supply air parameters by self-tuning algorithm or other adaptive control techniches [21]. The steady state supply air parameters can be predicted by the models.

CASE 3. On-line data for supply air parameters unavailable and off-line operation mode consulting needed: the steady state supply air parameters are estimated as:

$$I_{ss} = [(V_i + V_{ss})ii_n + Qt - i \cdot V_i]/V_{ss} \qquad (3)$$

$$W_{ss} = [(V_i + V_{ss})W_n + Q_w - W \cdot V_i]/V_{ss} \qquad (4)$$

For a variable air volume system,

$$\frac{V_{ss}}{V_{\max}} = \frac{1}{2K} \qquad (5)$$

Where I_{ss}, W_{ss}, V_{ss} is the enthalpy, wet, air volume of steady state supply air respectively; V_i is infiltration air volume; Q_t, Q_w is heat and wet load of the HVAC system. K is the mixed air percentage of fresh air and return air, i.e. $K = \frac{\text{Fresh Air Volume}}{\text{Recycle Air Volume}}$. As mentioned previously, supply air parameters are estimated based on steady state, the HVAC system design data can be directly applied to Q_t, Q_w, V_i and $V_{\max}$, and these data are usually available for operators.

4.4.3. Knowledge Organization and Implementation

PC-PLUS, an expert system development tool, is chosen to develop this expert system.

Knowledge base structure: In PC-PLUS, a knowledge representation mechanism consists of three parts: frames, parameters, and rules. Each frame, parameter, or rule has its properties that define its characteristics. Every frame holds a set of parameters and a rule base.

Frames can be hierarchically structured in PC-PLUS. A parent frame can invoke the rules in a child frame, and a child frame can access the parameters in its parent frame.

Hierarchy of the Frames: The hierarchical structure of the frame in operation mode consulting is based on the classification of HVAC systems, which is demonstrated in Figure 11. It consists of two levels of frames. The first level frame classifies HVAC systems, which decides the subframe to activate according to the structure of the HVAC systems. In the second level, consisting of sixteen frames, each frame contains a set of rules to choose energy saving operation modes with external access rules to complete numerical computation.

Root Frame: In this frame, the user inputs or confirms the information about the HVAC system structure. Symbolic reasoning decides the subframe to be activated in responding to the HVAC system structure information.

Subframe: According to the classification of the root frame, there are sixteen types of HVAC systems, and each corresponds to a subframe. As air enthalpy and wet value are used in the rules to identify the operation mode, numerical calculation programs for evaluating air enthalpy and wet are written in BASIC and executed by external access rules. Each frame contains a rule base to choose an energy saving operation mode for the corresponding structure of the HVAC system.

A typical rule for system classification and activating a subframe in root frame is shown as:

IF CONDITIONING-TYPE = TEMPERATURE-HUMIDITY
    AND ZONE-TYPE = SINGLE
    AND AIR-VOLUME-TYPE = CONSTANT
    AND DEHUMIDIFYING-UNIT-TYPE = DEHUMIDIFIER
THEN HVAC-SYSTEM = THSCD AND CONSIDERFRAME THSCD

In external access, input data about temperature and humidity are sent to external data files, executing the external computation program, then the enthalpy and wet data are retrieved. Except for air enthalpy and wet

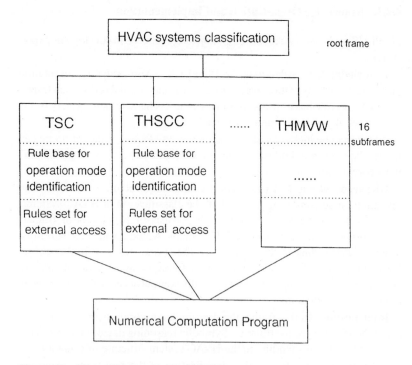

Figure 11. Hierarchical structure of ESOP frames.

evaluation, the external numeric computation programs also handle some data processing, i.e., the estimation of supply air parameters.

For a constant air volume, single zone or central air conditioning system with a dehumidifier, a typical rule in the subframe THSCD is:

IF $i \leq iik$ AND $w \leq wk$ AND $iin \geq iis$ AND $wn \geq ws$

OR

$\qquad i \leq iin$ AND $w \leq ws$ AND $iin < iis$

OR

$\qquad i \leq iik$ AND $w \leq ws$ AND $iin \geq iis$ AND $wn < ws$

THEN OPERATION-MODE = MODE-1

AND

PRINT "The recommended operation mode is:
OUTSIDE-AIR-DAMPER at 10%; RETURN-AIR-DAMPER

HVAC operation planning

The recommended operation mode is: RETURN-AIR DOOR open 90%; FRESH-AIR DOOR open 10%; COOLER off; HEATER used to control TEMPERATURE.

┌─ Review: ───

Yes

↔	•	air conditioning type:	: :	TEMPERATURE
↔	•	The zone type matched your system.	: :	SINGLE-ZONE
↔	•	The type of air volume matched your . . .	: :	CONSTANT
↔	•	unit system: English units and SI units	: :	SI-UNITS
↔	•	outdoor fresh air temperature	: :	2
↔	•	expected or desired temperature in room	: :	20
↔	•	two types of measurement for humidit . . .	: :	NO
↔	•	*IMPORT* : : TK1 (7.677787) TK (1.448332E-3) TS2 (23.676. . .

1. Use arrow key or first letter of item to position the cursor.
2. Select all applicable responses.
3. After making selections, press ENTER to continue.

** End – press ENTER to continue.

Figure 12. An consulting example.

at 90%; DEHUMIDIFIER off; HUMIDIFIER **ON**; DEHUMIDIFIER **OFF**; HEATING COIL **ON**; COOLING COIL **OFF**."

There are twelve operation modes for this type of air conditioning system, so twelve such kinds of rules are contained in the rule base. For all various sixteen types of HVAC systems, the knowledge base is made of over a hundred such rules to identify energy conservation operation modes.

An illustration of a consulting example (Figure 12) by using ESOP is given below:

Input: HVAC system structure information:
      TEMPERATURE
      SINGLE-ZONE
      CONSTANT AIR VOLUME
Air parameters:
      UNIT-SYSTEM = SI-UNIT
      WET-MEASUREMENT = RELATIVE-HUMIDITY
      T1 = 5 (Outside air temperature is 5°C.)
      TN = 24 (Desired indoor temperature is °C.)

TS = 30 (Supply air temperature is 30°C.)

Output: OPERATION-MODE = MODE-1

"The recommended operation mode is: OUTSIDE-AIR-DAMPER at 10%; RETURN-AIR-DAMPER at 90%; HEATING COIL ON; COOLING COIL OFF."

4.5. COMFORT INDOOR SETTING

According to ASHRAE 55-1981 standard, a desired indoor temperature is 65–80°F or 18–27°C, and the relative humidity is 35–62%. Comfort research has provided a number of alternatives for energy conservation [3,18]. So far, comfort technology has not been fully utilized for energy conservation. In this section, a brief introduction to current existing indoor setting strategy is given. Then we discuss the problems with this strategy, and propose comfort indoor setting to solve the problems. The system for comfort indoor setting is presented.

4.5.1. Thermostat Strategy

Currently, most HVAC systems are controlled by thermostat strategy, which maintains indoor temperature constantly. The disadvantages of this strategy are as follows:

1. It does not maintain the optimal thermal comfort inside buildings, because, except for temperature, the other six variables: mean radiant temperature, relative humidity, air velocity, clothing, activity level (metabolic rate) and exposure time, also have significant affect on comfort.
2. It possibly results in energy waste because the indoor temperature could be set unnecessary high or low.
3. It hinders the air exchange in the areas where air is not well distributed, thereby decreasing the indoor air quality.

4.5.2. Comfort-stat Strategy

To overcome these drawbacks, we present a comfort-stat control strategy, which sets indoor air temperature by the thermal comfort equation [4]. So indoor temperature setting changes responding to the coordinations among the other variables to maintain constant comfort are implemented. This strategy provides thermal comfort-stat rather than keeping temperature stable. It optimizes comfort and energy conservation simultaneously. It is also beneficial to the indoor air quality because temperature changes increase the air exchange where air is not well distributed.

The comfort temperature indoor setting can be determined by the comfort equation, which was developed by Fanger in 1970:

$$t_{c1} = 35.7 - 0.032 \frac{M}{A_{Du}}(1 - \eta) - 0.18I_{cl}\{\frac{M}{A_{Du}}(1 - \eta)$$

$$- 0.35[43 - 0.061\frac{M}{A_{Du}}(1 - \eta) - P_a] - 0.42[\frac{M}{A_{Du}}(1 - \eta) - 50]$$

$$- 0.0023\frac{M}{A_{Du}}(44 - P_a) - 0.0014\frac{M}{A_{Du}}(34 - t_a)] \tag{6}$$

$$\frac{M}{A_{Du}}(1 - \eta) - 0.35[43 - 0.061\frac{M}{A_{Du}}(1 - \eta) - P_a]$$

$$- 0.42[\frac{M}{A_{Du}}(1 - \eta) - 50] - 0.0023\frac{M}{A_{Du}}(44 - P_a)$$

$$- 0.0014\frac{M}{A_{Du}}(34 - t_a)$$

$$= 3.4 \times 10^{-8} fc1[(t_{cl} + 273)^4 - (t_{mrt} + 273)^4] + f_{cl}h_c(t_{cl} - t_a) \tag{7}$$

$$h_c = \begin{cases} 2.05(t_{cl} - t_a)^{0.25} & \text{for } 2.05(t_{cl} - t_a)^{0.25} > 10.4\sqrt{V} \\ 10.4\sqrt{V} & \text{for } 2.05(t_{cl} - t_a)^{0.25} < 10.4\sqrt{V} \end{cases} \tag{8}$$

Where the parameters are:

t_{cl}: Clothes temperature (°C).
$\frac{M}{A_{Du}}$ Human metabolic rate (kcal/hrm^2).
η: Mechanical Efficiency.
v: Relative velocity in still air (m/s).
$I_{cl} = R_{cl}/0.18(clo)$, where R_{cl} is the total heat transfer resistance from skin to outer surface of the clothed body (m^{2}hr°C/kcal).
f_{cl}: the ratio of the surface area of the clothed body to the surface area of the nude body.
p_a: the partial pressure of water vapor in inspired air (ambient air) (mmHg).
t_a: Comfort temperature (°C), i.e. the desired indoor temperature setting.
t_{mrt}:Mean radio temperature (°C).

The input information for solving the equation is $\frac{M}{A_{Du}}$, η, v, I_{cl}, f_{cl}, P_a, and t_{mrt}. Data for $\frac{M}{A_{Du}}$, η, v, I_{cl} and f_{cl} are shown in Table 17–18 of [4]. P_a and t_{mrt} can be measured.

The output information is comfort temperature t_a, which should be the indoor setting temperature for an HVAC control system. It is also one of the input information items for the operation mode consulting expert system.

To simplify the computation of solving the comfort Equations (6)–(8), the result for engineering application was provided as Tables 19–20 [4], which is used in this system.

An example of indoor temperature setting is shown below: Input information:

ACTIVITY: Resting

Seated

CLOTHING: Business suit

Output information:

COMFORT TEMPERATURE: 23°C

PREDICTED MEAN VOTE: − 0.02

4.6. MODELING AND CONFLICT REASONING FOR ENERGY CONSERVATION AND IAQ CONTROL

Indoor air pollution sources can be classified into three groups, outdoor air, modern building materials, and human activities [10]. The main pollutions from the ambient environment are streets with heavy traffic density, industries, and house heating equipment. The processes of furnishing a room, ozone from a photocopier, and emission products of modern building materials, are the second group of sources. The third source of indoor air pollution is human activity, such as respiration, particles emitted from the skin, smoking, and cooking. Here we are mainly concerned about the pollution source from human activities, which is a major indoor pollution source. The number of people in a given room or the space per person, coupled with the supply of fresh air determines the accumulation rate of the contamination.

Most indoor air pollution can be removed by proper ventilation, but increasing ventilation rate consumes more energy. Conflict exists between energy saving and indoor air quality (IAQ) improvement. In this section, we first give a conflict analysis, then conduct the field test for IAQ and ventilation. Based on the analysis and test, conflict reasoning rules for a compromised resolution between energy saving and IAQ are determined.

Table 2. Conflict Analysis.

HVAC Operation	Cooling/Heating Status	Energy Consumption	Indoor Air Quality
Outside Air Intake DECREASE	ON OFF	DECREASE Unchange	DECREASE
Outside Air Intake INCREASE	ON OFF	INCREASE Unchange	INCREASE
Ventilation Rate DECREASE		DECREASE	DECREASE
Ventilation Rate INCREASE		INCREASE	INCREASE

4.6.1. Conflict Analysis

In HVAC operations, IAQ is mainly affected by ventilation, outside air intake, and the effect of filters, and the number of the occupants. Conflict analysis for operations are described in Table 2.

When OUTSIDE AIR intake **increases**, IAQ **increases**; energy consumption **increases** if heating or cooling is in use; but energy consumption remains **unchanged** if it is in the free cooling or heating operation mode.

When the VENTILATION RATE **increases**, IAQ **increases**; energy consumption **increases**. Whereas,

If OUTSIDE AIR intake **decreases**, IAQ **decreases**; energy consumption **decreases** if heating or cooling is in use; but energy consumption remains **unchanged** if it is in the free cooling or heating operation mode.

If VENTILATION RATE **decreases**, IAQ **decreases**; energy consumption **decreases**.

4.6.2. Field Test and Model Identification

Indoor air pollution caused by the occupants can best be assessed through odors and objectively by measuring the carbon dioxide in the room. Concentration of carbon dioxide in ambient air lies between 0.03% and 0.04% by volume; this concentration can be twice as high in cities and in industrial areas. The recommended maximum concentration of CO_2 in indoor air of living places is 0.1–0.15%, and a minimum fresh air supply of 12–15 m^3/person/h is required.

The air pollutants caused by people were investigated with the help of carbon dioxide and odor measurements [10]. The comparison of the curve for odor intensity with that for carbon dioxide concentration shows correlations between these two parameters. Such a relationship would be of a high practical value because the relatively easy carbon dioxide measurement can be related to a momentary odor situation in a room.

To obtain the dynamics among IAQ, ventilation and outside air intake, a field test was conducted in the Administration Building, University of Alberta. The carbon dioxide of return air, supply air, and fresh air were measured at the square wave change of the supply fan speed and mixing damper opening respectively. Figure 13(a–d) shows the test results.

Based on the data from the field test, we identify the following model by using the least square method:

$$CO_2(k + 1) = 1.8835\, CO_2(k) - 0.8819\, CO_2(k - 1) + 0.0184\, U(k)$$
$$- 0.0446\, U(k - 1) + 0.0216\, V(k) - 0.0901\, V(k - 1) \quad (9)$$

$$Var = 3.6076$$

Where U: Speed (%) of supply fan.

V: Opening (%) of mixing damper, i.e. outside air damper opening is at $V\%$, and return air damper opening is at $(1 - V\%)$.

It should be pointed out that the above model is time varying, because the indoor CO_2 level is strongly related to the number of occupants inside the building, and we treat it as a disturbance of the control model.

4.6.3. Conflict Reasoning

The field test and model identification provide us the knowledge to investigate conflict reasoning for a compromise solution between energy conservation and IAQ control.

Before we investigate the reasoning rules, definitions for the operation mode are given as follows:

IAQ Operation Mode: Mixing-damper is used to control CO_2 concentration; Heating or cooling is applied to meet supply temperature requirements; Humidifier or air washer/ cooler/ dehumidifier is used to meet supply humidity requirements.

Energy Conservation Mode: The operation mode recommended by the expert system for operation planning.

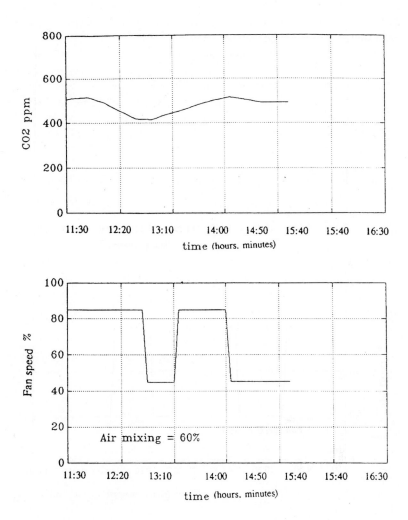

Figure 13a. Field test 1.

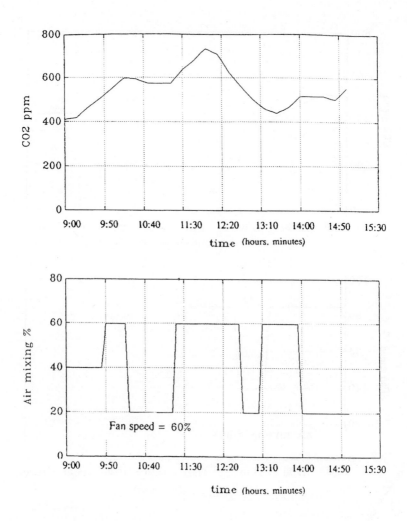

Figure 13b. Field test 2.

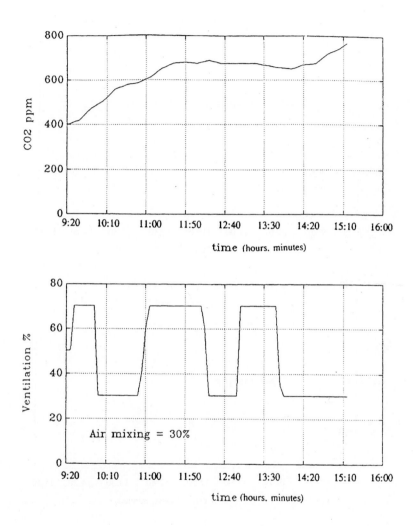

Figure 13c. Field test 3.

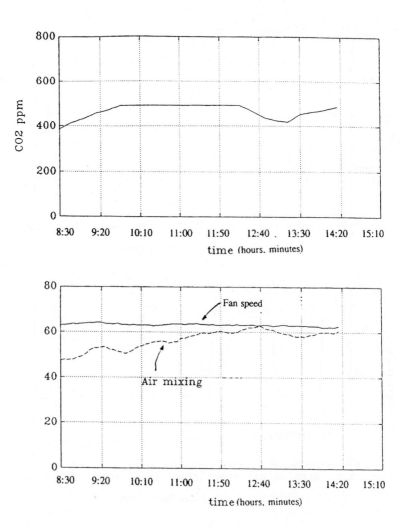

Figure 13d. Field test 4.

Typical conflict reasoning rules are described below:

Rule-1: IF CO_2 concentration $\geq$ **Limit**
AND
HVAC processes operated under *Energy Conservation Mode*
THEN Changing to *IAQ Operation Mode*

Rule-2: IF CO_2 concentration $\geq$ **Limit**
AND
Outdoor-Air intaken $\leq$ 10% for the last period of 30 minutes
AND
HVAC processes operated under *IAQ Operation Mode*
THEN Changing to *Energy Conservation Mode*

Rule-3: IF CO_2 concentration $\leq$ **Limit**
AND
HVAC processes operated under *Energy Conservation Mode*
THEN Remain the operation mode

Rule-4: IF CO_2 concentration $\leq$ **Limit**
AND
Outdoor-Air intaken $\geq$ 10%
AND
HVAC processes operated under *IAQ Operation Mode*
THEN Remain the operation mode

The conflict reasoning system is implemented in C language. The conflict reasoning does not directly exchange data with the other subsystems. The data exchange between the subsystems goes through the meta-system. The system input is indoor CO_2 concentration data and the recommended maximum concentration of CO_2 in indoor air of living places. Output information is the operation mode change recommendation.

4.7. SUMMARY

A mechatronic system approach to an intelligent operation support system for HVAC process is presented. This original research will pave the way for new research directions for future process operation, energy management and control in the HVAC industry. The following conclusions about IOSS are summarized:

(1) IOSS codifies the important expertise about operation planning, which for the first time considers supply air parameters with complete dynamics and provides a real time and integrated operation planning method in HVAC processes. It overcomes the disadvantages with some currently used methods, and offers energy conservation, better comfort and indoor air quality. As the expertise rules are expressed as air enthalpy and wet conditions, they are easy to apply to real time control systems in the HVAC processes.

(2) The objective of this intelligent system is to provide assistance for human operators to achieve better operation in the HVAC processes or to train HVAC operators to identify energy conservation operation modes. It increases the interactions between the process and the operators.

(3) The development of an IOSS for a real industrial application requires the knowledge from different disciplines and representations. The integrated distributed intelligent system, meta-system provides a framework to integrate and manage the knowledge. In IOSS, four subsystems such as operation mode consulting, indoor comfort setting, conflict reasoning and knowledge introduction are integrated by the meta-system.

(4) An indoor comfort setting system provides comfort-stat instead of temperature-stat, which offers comfort and energy saving simultaneously.

(5) Conflict reasoning gives the conflict resolution for energy conservation and indoor air quality control in HVAC processes. IAQ operation mode is recommended when indoor CO_2 concentration is over the limit, which increases outdoor air intake. If indoor CO_2 concentration is under the limit, the energy saving operation mode is in use, which utilizes the energy of fresh air as much as possible.

(6) Integrating the IOSS into a real time control HVAC system is suggested for further research.

REFERENCES

1. Athienitis, A.K., 1988, A Predictive Control Algorithm for Massive Buildings. *ASHRAE Transactions*, **94**(2), 1050–1067.
2. Cooper, K.W., 1983, Microcomputers in an Energy Saving Residential Heat Pump. *IEEE Transactions on Industry Applications*, **IA–19**(4), 486–490.
3. Doherty, T.J. and Arens, E., 1988, Evaluation of the Physiological Bases of Thermal Comfort Models. *ASHRAE Transactions*, **94**(1), 1371–1378.
4. Fanger, P.O., 1970, Thermal Comfort Analysis and Application in Environmental Engineering, McGraw-Hill, New York.
5. Haines, R.W., 1984, Retrofitting Reheat-type HVAC System for Energy Conservation. *ASHRAE Transactions*, **90**(2B), 185–191.
6. Haines, R.W., 1987, Control Systems for Heating Ventilating, and Air Conditioning, Van Nostrand Reinhold, New York.

7. Hartman, T.B., 1989, An Operator's Control Language to Improve EMS Successes. *Energy Engineering*, **86**(2), 6–11.

8. Hartman, T.B., 1988, Dynamic Control: Fundamentals and Considerations. *ASHRAE Transactions*, **94**(1), 599–609.

9. Hayter, R.B., 1987, Comfort Education for Energy Conservation. *ASHRAE Transactions*, **93**(1), 1080–1083.

10. Hubber, G. and Wanner, H.U., 1983, Indoor Air Quality and Minimum Ventilation rate. *Environment International*, **9**, 153–156.

11. Jedlicka, A.D., 1985, Improving Conservation Behavior. *Energy Engineering*, **82**(5) 29–34.

12. Johnson, G.A., 1984, Retrofit of a Constant Volume Air System for Variable Speed Fan Control. *ASHRAE Transactions*, **90**(2B), 201–211.

13. Maxwell, G.M., Shapiro, H.N. and Westra, D.G., 1989, Dynamics and Control of a Chilled Water Coil. *ASHRAE Transactions*, **95**(1), 1243–1255.

14. McNall, P.E., 1986, Control of Indoor Air Quality By Means of HVAC Systems. Proceedings of The ASHRAE Conference IAQ '86: Managing Indoor Air for Health and Energy Conservation, pp.541–547.

15. Mehta, D.P., 1983, Effects of Control Dynamics on Energy Consumption in Residential Heating Systems. *Energy Engineering*, **80**(3), 7–24.

16. Nahar, N.M. and Gupta, J.P., 1989, Energy Conservation and Pay Back Periods of Collector-Cum-Storage Type Solar Water-Heaters. *Applied Energy*, **34**(2), 155–163.

17. Nelser, C.G., 1986, Adaptive Control of Thermal Processes in Buildings. *IEEE Control System Magazine*, **6**(4), 9–13.

18. Nelson, T.M., Nilsson, T.H. and Hopkin, G.W., 1987, Thermal Comfort: Advantages and Deviations. *ASHRAE Transactions*, **93**(1), 1039–1047.

19. Rao, M., Jiang, T.S. and Tsai, J.P., 1987, A Frame Work of Integrated Intelligent System. Proc. 1987 IEEE International Conference on System, Man, and Cybernetics, pp.1133–1137.

20. Shi, J., 1981, Identification of Optimal Energy Saving Operating Condition in an Air Conditioning System. *Heating & Air Conditioning*, **17**(4), in Chinese.

21. Shi, J. and Zhou, H., 1985, Air Conditioning Control System With Microcomputer Using an Adaptive Regulator. Proceeding of CLIMA 2000 World Congress on Heating, Ventilating, and Air Conditioning. Copenhagen.

22. Teji, D.S., 1987, Controlling Air Supply For Energy Conservation. *Energy Engineering*, **84**(3), 4–14.

23. Tsatsoulis, C. and Kashyap, R.L., 1988, A Case-Based System for Process Planning. *Robotics & Computer-integrated Manufacturing*, **4**(3), 557–570.

24. Wang, Y. and Zhou, H., 1987, Identification of Optimal Energy Saving Operation Conditions in a Cooling Type Air Conditioning System. *Heating & Air Conditioning*, **23**(4), in Chinese.

25. Zhou, H., Rao, M. and Chuang, K.T., 1993, Knowledge-Based Automation For Energy Conservation and Indoor Air Quality Control in HVAC Processes. *Engineering Application of Artificial Intelligence*, **6**(2), 131–144.

5 MECHATRONIC SYSTEMS TECHNIQUES IN STEEL REHEAT FURNACES — INTEGRATION OF CONTROL, INSTRUMENTATION AND OPTIMIZATION

Y.Y. YANG and D. A. LINKENS

Department of Automatic Control and Systems Engineering, University of Sheffield, Sheffield S1 3JD, UK

In this research an integrated control, instrumentation, and optimization scheme for reheat furnace operations has been developed by using mechatronic system techniques. The concept of integration advocated in mechatronics has been extensively exploited in the development of advanced reheat furnace operation control described in this chapter, from instrumentation, data collection to control and optimization. As a result of this integration, different control functions involved in the reheat furnace process can be carried out on a wider information basis which leads to more efficient coordination, hence better overall operation performance. General requirements for the integration scheme are given, along with the detailed techniques of algorithms, strategies, and implementation. Essentially, the integrated scheme consists of the on-line reheat furnace model, slab tracking system, dynamic setpoint control, and DDC level control. The reheat furnace model provides a real time estimation of the slab temperature distribution during heating in the reheat furnace, which provides critical information for control and optimization strategies. The slab tracking system is developed to trace the slab movement within the furnace area as well as along the whole production line, which provides the necessary data for setpoint control and on-line slab temperature estimation. The dynamic setpoint control comprises steady-state optimization and dynamic compensation, which provide optimized furnace zone temperature setpoints based on the characteristics of the heated slabs, the current mill rolling rate, and other dynamically changing reheat furnace process variables. The reheat furnace DDC level control aims at the best regulation for the zone temperature, furnace pressure, fuel flow, combustion air flow, etc. according to their given setpoints. It also provides functions such as process variable monitoring and recording. The hardware system and software structure for the implementation of the proposed integrated scheme is based on a Micro Vax II computer and a WDPF system. The integrated system strategy has been successfully applied in production scale reheat furnaces involved in a hot rolling steel plate mill complex (The 5th Plate Mill of Chongqing Iron and Steel Corporation, Shichuan, China), and the resulting control performance is quite satisfactory with significant energy saving and product quality improvement.

5.1. INTRODUCTION

As is well known, a large amount of energy in the steel industry is consumed in reheat furnaces for heating steel stock to achieve the required rolling temperature. The reheating process is still an important part in the iron and steel industry, although nowadays many modern steel processes are trying to employ slab (ingot) hot direct rolling which (partially) avoids the reheat furnaces. There are still a considerable number of steel mills where reheat furnaces are essentially used to produce hot slabs for the subsequent rolling mills. Energy consumption by the reheat furnaces is huge and it absorbs a significant part of the total operational costs. Furthermore, the quality of reheating operation has a significant effect on the downstream mill operation as well as on the quality of the final product. Due to the importance of this process, much research has been carried out in modelling, optimization, and control of the reheat furnaces [1–8]. Some of the research has concentrated on the model development of the heating process, while other has focused on the control strategies aimed at heating quality and energy optimization. However, research on integrated control, instrumentation and optimization of the reheating process, together with the consideration of the rolling mill operation dynamics has not been sufficient, despite the fact that the potential benefit of doing so is apparent.

The increasing impact of information and computer technology on mechanical engineering has changed substantially the approach to solving technical problems in the last decade. As a result, a multi-disciplinary technology called "Mechatronics" has emerged [9,10]. Although much time has been spent in trying to define it, no universally accepted definition has yet been reached. One definition given by Bradley *et al.* [11] states that mechatronics is the synergetic combination of mechanical engineering, electronic control, and system thinking in the design of products and processes. Another definition given by Schweitzer [12] states that mechatronics is an interdisciplinary field of engineering science, combining the classical areas of mechanical engineering, electrical engineering, and computer science. Many other definitions exist in the literature, however, most definitions generally agree that mechatronics differs from "traditional" engineering in integrating the previously separately considered disciplines of mechanical, electrical, electronic, control, computer and software engineering. Typically, a mechatronic system picks up signals, processes them, as provides an output of forces and motions. The critical concept of integration in mechatronics has found many applications in the modern engineering world, ranging from design, management, optimization and control. In this chapter, we will illustrate how this principle of integration can be applied for the design of reheat furnace control and optimization, along with the extra instrumentation and communication requirements to achieve such integration.

The aim of this chapter is to develop an integrated control, instrumentation, and optimization strategy for reheat furnaces involved in a hot rolling steel plate mill, with the application of a mechatronic approach in mind. It is believed that by proper integration of control, instrumentation, and optimization, the reheat furnace performance will be significantly improved in terms of both the heating quality and the operation cost. The chapter is organized as follows. In Section 5.2 a brief description of the reheat furnace process together with its original control scheme based on a practical steel mill complex is given. General concepts for the integrated control, instrumentation, and optimization strategies for reheat furnaces, together with the information flow relations within the different components inside the integrated scheme are also outlined in this section. Section 5.3 describes the development of the reheat furnace model which gives an on-line estimation of the slab temperature distribution during the heating period. This on-line information of slab temperature is of vital importance for reheat furnace control and optimization, as will be clear later. The setpoint optimization strategy for reheat furnace operation is developed in Section 5.4, using heuristic search techniques for steady-state optimization and dynamic compensation for frequent process disturbances. These setpoints are then sent to the corresponding DDC control loops of the reheat furnace process, such as the furnace zone temperature control, the fuel and air flowrate control, the furnace pressure control, etc. which are outlined briefly in Section 5.5. Section 5.6 describes the implementation of the proposed integrated control scheme, with focuses on its hardware and software structure using WDPF and Micro VAX II computer systems. Finally, concluding remarks are given in Section 5.7.

5.2. INTEGRATION OF CONTROL, INSTRUMENTATION AND OPTIMIZATION FOR REHEAT FURNACES

The function of a reheat furnace in a steel mill is to heat the slabs up to an appropriate temperature suitable for rolling when they reach the furnace exit, and to supply the slabs to the subsequent rolling mill at the right pace demanded by the rolling operation. As slabs may vary in size and grade, and the mill throughput may change unpredictably, it is not easy to control the reheat furnaces to meet the above requirements while maintaining the minimum energy consumption in a frequently changing environment. In this chapter we consider a five zone pusher type continuous reheat furnace involved in a practical hot rolling steel plate mill (The 5th Plate Mill of

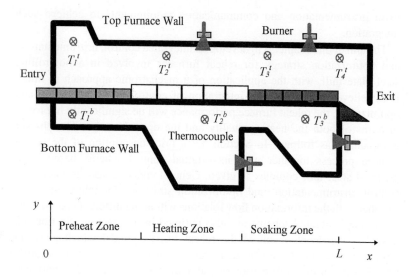

Figure 1. Sketch of a pusher type five zone continuous reheat furnace.

Chongqing Iron and Steel Corporation, Shichuan, China). Figure 1 shows
the sketch of the reheat furnace, while its specification is given in Table 1.

Cold slabs are charged into the reheat furnace via the furnace entry, usually
one at a time, by the slab pusher (refer to Figure 2). Meanwhile, a hot slab
is discharged (pushed out) at the furnace exit for subsequent rolling. All the
slabs heated in the furnace, which form a continuous heating stock, move
towards the furnace exit with the distance equal to the width of the charged
slab whenever a new cold slab is charged into the furnace by the slab pusher.
The slab pusher usually starts to charge a cold slab when a request for a
hot slab by the roughing mill is received. The hot rolling steel plate mill is
equipped with three identical reheat furnaces, a two high reversible roughing
mill and a four high reversible finishing mill, as shown in Figure 2.

Table 1. Specification of the reheat furnace.

Item	Description
Type	5-zone pusher type
Effective length × width	26.448 × 3.712 (m^2)
Capacity	40 (Tonnes/Hour)
Fuel	Natural gas

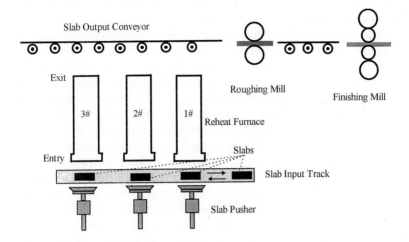

Figure 2. Flowchart of the hot rolling steel plate mill.

Before the integrated scheme was developed, the original control strategies were as follows. The gas (fuel) flow rates supplied to different heating zones were controlled by PID controllers, with their setpoints mandated by the furnace operators based on their operation experience, the reference heating table for different kind of slabs, and real time observation of the slabs in the reheat furnaces via the observation hole. The combustion air flow rate was also controlled by PID controllers with their setpoint being proportional to that of the gas controller (with a fixed constant air/gas ratio). The furnace zone temperature is treated as a reference variable, without direct control loop for it. The furnace pressure was controlled by a PID controller with a fixed setpoint around 5 mmH$_2$O. Due to the frequently changing conditions of the production line, such as disturbances of the rolling rate of the subsequent mills, changing the size and the steel grade of the heating slabs, varying heating time of the slabs in the furnaces, etc., it was very difficult for the furnace operators to select the setpoints properly. As a result, performance deterioration occurred frequently due to either the over-heating or insufficient heating of the slabs. This caused extra energy consumption and severe slab oxidization when overheating, or reduction in the mill production rate with insufficient heating. To improve the reheat furnace operation, the mechatronic approach to integration can be applied. Since the final goal of the reheat furnace operation is to provide satisfactorily heated slabs at the right time interval (determined by the mill production) with minimum energy consumption, it is obvious that the slab temperature in the furnace during the heating process is vital. Due to the fact that no reliable measurement technique is available for detecting the slab temperature distribution in the furnace, a mathematical model to estimate the slab temperature distribution based on the measurable process variables and other known physical and thermal parameters becomes necessary. For the purpose of on-line slab temperature estimation, a slab tracking system is needed to identify the actual distribution of slabs in the reheat furnaces, as well as to provide the expected heating time for the slabs by considering the mill rolling characteristics. The furnace temperature setpoint should then be determined by the current temperature distribution of the slabs in the furnaces, the expected heating time of those slabs, and the desired slab temperature suitable for rolling, by using optimization techniques. Also, the control of gas flowrate, the air flowrate, the furnace pressure, etc. should be carried out in an integrated mode rather than separately, to achieve the full benefit of the new control strategies. Extra instrumentation is required to provide the necessary measurements demanded by the integrated strategies. Figure 3 shows the information flow within the integrated control and optimization system.

In the following sections, aspects related to this integrated control and optimization system will discussed, including the on-line estimation of the slab temperature, the slab tracking system, the setpoint optimization strategy, the DDC level control scheme, and the control system implementation, etc.

5.3. ON-LINE SLAB TEMPERATURE ESTIMATION

Much effort has been devoted to the development of reheat furnace models for slab temperature estimation [13]. In the following, the reheat furnace model developed by Yang and Lu [6,7] will be outlined. A reheat furnace is treated as a typical distributed parameter system where both the furnace temperature and the slab temperature are functions not only of time, but also of spatial positions. Because of the complicated nature of the reheat furnace, the following simplification assumptions are made for the model development (refer to Figure 1):

(1) The furnace temperature is considered to be one dimensional along the furnace length (x), with linear distribution between the adjacent temperature measurements.

(2) The slabs in a furnace are considered as a continuous steel stock with two-dimensional temperature distribution along the directions of x and y.

(3) The thermal and physical properties of the slabs, such as the specific heat, density, and heat conductivity are functions of the temperature as well as the steel grade of the slabs.

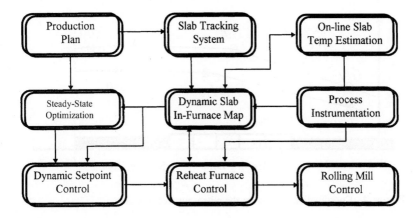

Figure 3. Information flow of the integrated control and optimization system.

(4) The heat transfer coefficients, such as the view factors, radiation emissivity, heat convection coefficients etc., are assumed to be constant within the same furnace zone.

We will use the coordinator system as shown in Figure 4 to derive the reheat furnace model. In a pusher type continuous reheat furnace, the slabs are moved toward the furnace exit during the heating process when a charge (or discharge) operation occurs. Due to the existence of the slab temperature gradient along the furnace length, this slab movement will cause certain heat transition under the fixed coordinator system. The unsteady-state heat conduction in the reheat furnace is given by the following two-dimensional partial differential equation, considering the heat transition caused by the slab movement:

$$\frac{\partial T(x,y,t)}{\partial t} = \frac{I}{c\rho} \left[\frac{\partial}{\partial x} \left(K_x \frac{\partial T(x,y,t)}{\partial x} \right) + \frac{\partial}{\partial y} \left(K \frac{\partial T(x,y,t)}{\partial y} \right) \right]$$
$$- v(t) \frac{\partial T(x,y,t)}{\partial x} \quad 0 < x < L, \quad 0 < y < d(x), \quad t > 0 \tag{1}$$

where K and K_x are the steel heat conductivity and the equivalent heat conductivity in the x direction, respectively, $T(x,y,t)$ is the temperature distribution of the slabs, c and ρ are the specific heat and density of the slabs located at x respectively, L is the effective length of the furnace, $d(x)$ is the slab thickness distribution along the furnace, and $v(t)$ is the equivalent slab moving rate along the furnace.

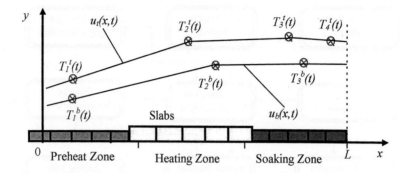

Figure 4. Coordination system for the reheat furnace model.

It should be pointed out that the equivalent heat conductivity K_x is smaller than the steel heat conductivity K due to the contact heat resistance between adjacent slabs. According to the assumptions made above, K_s and K should be functions of both time and slab position, i.e., $K = f_K(x, t)$ and $K_x = f_{Kx}(x, t)$, with their actual values determined by the average temperature and steel grade of the slab located at x at time t, which are obtained via the slab tracking system.

It is assumed that the top furnace temperature distribution $u_t(x, t)$ and the bottom furnace temperature $u_b(x, t)$ can be determined from the x position and the measurable zone temperatures (as indicated in Figure 4). The heat flux from the furnace to the slab surfaces can be calculated by using the concept of "seen temperature" to simplify the computation, i.e. the heat flux to the slab located at x can be determined (approximately) by the furnace temperature of the location above x. Based on this approximation, the heat flux to the slab top surface can be calculated by:

$$q_t(x, t) = e_t(x)\sigma[(u_t(x, t) + 273)^4 - (T(x, d(x), t) + 273)^4]$$
$$+ h_{ct}[(u_t(x, t) - T(x, d(x), t))]$$
$$e_t(x) = \varphi_{swt}\varepsilon_s(x) + (\varepsilon_w(x) + e_g(x))/2 \tag{2}$$

where e_t is the equivalent radiation heat transfer coefficient between the reheat furnace and the top surface of the slabs, φ_{swt} is the view factor between the furnace wall and the top surface of the slabs, ε_s, ε_g, and ε_w are the emissivity of the slabs, the furnace gas, and the furnace wall, respectively, σ is the Stefan-Boltzmann constant, and h_{ct} is the convective heat transfer coefficient between the furnace gas and the top surface of the slabs.

Due to the influence of the water cooled skid pipes (which are used to support the slabs in a top-bottom fired furnace), the heat flux from furnace to the bottom surface of the slabs is smaller than to the top one. Here an effective covering factor is employed to deal with the effect of the skid pipes, and the heat flux of the bottom surface of the slabs is calculated by:

$$q_b(x, t) = e_b(x)\sigma[(u_b(x, t) + 273)^4 - (T(x, 0, t) + 273)^4]$$
$$+ h_{cb}[(u_b(x, t) - T(x, 0, t)]$$
$$e_b(x) = [\varphi_{swb}\varepsilon_s(x) + (\varepsilon_w(x) + e_g(x))/2](1 - \eta_C) \qquad (3)$$

where e_b is the equivalent radiation heat transfer coefficient between the reheat furnace and the bottom surface of the slabs, φ_{swb} is the view factor between the furnace wall and the bottom surface of the slabs, h_{cb} is the convective heat transfer coefficient between the furnace gas and the bottom surface of the slabs, and η is the effective covering factor, representing the percentage of the bottom surface which is covered by the skid pipes. Using assumption we have made above, the view factors and the different emissivities are piecewise constants, which have the same value if they belong to the same furnace zone.

The boundary conditions of Equation 1 can be determined by:

$$\frac{\partial T(x, y, t)}{\partial t}\bigg|_{y=0} = -q_b(x, t)$$
$$\frac{\partial T(x, y, t)}{\partial t}\bigg|_{y=d(x)} = q_t(x, t) \qquad (4)$$

and the initial conditions and the entry condition takes the form:

$$T(x, y, 0) = T_0(x, y)$$
$$T(0, y, t) = T_{\text{Entry}}(y) \qquad (5)$$

where T_{Entry} is the slab temperature when charged into the furnace, and T_0 is the initial slab temperature distribution along the furnace.

Equations 1–5 constitute the mathematical model of the reheat furnace which involves a partial differential equation and the corresponding non-linear boundary conditions. For simplicity of simulation and on-line application, this distributed parameter model is converted into a discrete time state space model via finite difference approximation. The following discretization and approximation are used in Equations 1–5, along with the boundary energy conservative properties to derive the discrete time state space model.

$$t = k\Delta t, \quad k = 0, 1, 2, \ldots;$$

$$x = i\Delta x; \quad i = 1, 2, \ldots, N_x; \quad \Delta x = L/N_x;$$

$$y = j\Delta y; \quad j = 1, 2, \ldots, N_y; \quad \Delta y = d(x)/N_y;$$

$$\frac{\partial T(x, y, t)}{\partial y} = \frac{T(i\Delta x, j\Delta y, (k+1)\Delta t) - T(i\Delta x, j\Delta y, k\Delta t)}{\Delta t};$$

$$\frac{\partial T(x, y, t)}{\partial y} = \frac{T(i\Delta x, (j+1)\Delta y, k\Delta t) - T(i\Delta x, j\Delta y, k\Delta t)}{\Delta y};$$

$$\frac{\partial^2 T(x, y, t)}{\partial y^2} = \frac{\begin{array}{c} T(i\Delta x, (j+1)\Delta y, k\Delta t) + T(i\Delta x, (j+1)\Delta y, k\Delta t) \\ -2T(i\Delta x, j\Delta y, k\Delta t) + T(i\Delta x, (j-1)\Delta y, k\Delta t) \end{array}}{(\Delta y)^2} \tag{6}$$

where Δx, Δy are the discrete spatial step length along x and y directions respectively, and Δt is the discrete time step length. The resulting discrete equation can be summarized in the form of Equation 7.

$$T(i, j, k+1) = a_{j,j-1}(i, k)T(i, j-1, k) + a_{j,j}(i, k)T(i, j, k)$$
$$\quad + a_{j,j+1}(i, k)T(i, j+1, k) + c_j(i, k)T(i-1, j, k)$$
$$\quad + d_j(i, k)T(i+1, j, k) + b_{j1}(i, k)q_b(i, k) + b_{j2}(i, k)q_t(i, k);$$
$$T(0, j, k+1) = T(0, j, k)$$
$$i = 1, 2, \ldots, N_x; \quad j = 1, 2, \ldots, N_y; \quad k = 0, 1, 2, \ldots.$$
$$\{a_{j,m}(i, k) = 0 | m > N_y\}; \quad \{d_j(i, k) = 0 | i = N_x\};$$
$$\{b_{j1} = 0 | j \neq 1\}; \quad \{b_{j2} = 0 | j \neq N_y\}; \tag{7}$$

where $f(i, j, k)$ is the abbreviation for $f(i\Delta x, j\Delta y, k\Delta t)$.

Equation 7 can be arranged in matrix form by dividing the slabs into N_x subsystems, and selecting the state vector and input vector as:

$$\mathbf{X}_i(k) = [T(i, 1, k), T(i, 2, k), \cdots, T(i, N_y, k)]^t$$
$$\mathbf{U}_i(k) = [q_b(i, k), q_t(i, k)]^t, \quad i = 1, 2, \ldots, N_x \tag{8}$$

where the superscript t means the transpose of the corresponding component (vector or matrix). The final form of the discrete time state space model is given by:

$$\mathbf{X}_0(k+1) = \mathbf{X}_0(k)$$
$$\mathbf{X}_i(k+1) = \mathbf{A}_i(k)\mathbf{X}_i(k) + \mathbf{C}_i(k)\mathbf{X}_{i-1}(k) + \mathbf{D}_i(k)\mathbf{X}_{i+1}(k) + \mathbf{B}_1(k)\mathbf{U}_i(k);$$
$$\mathbf{X}_i(0) = \mathbf{X}_{i0}; \quad i = 1, 2, \ldots, N_x \tag{9}$$

with the coefficient matrices being given by:

$$\mathbf{A}_i(k) = \begin{bmatrix} a_{11}(i,k) & a_{12}(i,k) & 0 & \cdots & & & 0 \\ a_{21}(i,k) & a_{22}(i,k) & a_{23}(i,k) & 0 & \cdots & & 0 \\ 0 & \cdots & \cdots & \cdots & & & 0 \\ 0 & \cdots & & & 0 & a_{N_y,N_y-1}(i,k) & a_{N_y,N_y}(i,k) \end{bmatrix}$$

$$\mathbf{B}_i(k) = \begin{bmatrix} b_{11}(i,k) & 0 & \cdots & & 0 \\ 0 & 0 & \cdots & 0 & b_{N_y,2}(i,k) \end{bmatrix}^t$$

$$\mathbf{C}_i(k) = \mathrm{diag}[\, c_i(i,k) \quad c_2(i,k) \quad \cdots \quad c_{N_y}(i,k)\,]$$

$$\mathbf{D}_i(k) = \mathrm{diag}[\, d_i(i,k) \quad d_2(i,k) \quad \cdots \quad d_{N_y}(i,k)\,]$$

$$(10)$$

Using Equations 9–10, we can estimate the real time slab temperature distribution from its initial condition if the input vectors $(\mathbf{U}_i(k), i = 1, 2, \ldots, N_x)$ and other information about the slabs heated in the furnace, such as the steel grade and slab size, are available. Dynamic information on the slabs heated in the furnace is provided by the slab tracking system, which will be described in Section 5.4.4. The input vectors $\mathbf{U}_i(k), (i = 1, 2, \ldots, N_x)$ are obtained from on-line measurement of the furnace zone temperature and the assumption of linear distribution of the furnace temperature between two adjacent temperature measurements. This model can provide real-time estimation of the slab temperature distribution, which is vital to improve the reheat furnace operation and to provide guidance for the control design.

5.4. FURNACE TEMPERATURE SETPOINT OPTIMIZATION

The aim of reheat furnace temperature setpoint optimization is to determine the various zone temperature setpoints such that the slabs reach the desired temperature profile for rolling when they are at the furnace exit and are requested to be discharged, while minimising the overall energy consumption. It is apparent that for such optimization, we need information on the slab temperature distribution during the whole heating period. Although the reheat furnace model developed in the previous section gives the temperature distribution of all the slabs heated in the furnace, it is too complicated (mainly due to its high dimensionality) for the purpose of optimization. Here we will introduce a more simple model, referred to as a slab-following model, which focuses on the temperature profile of a single slab heated in the furnace. We assume that the temperature distribution of a slab is one dimensional along the y direction. Other assumptions made in the previous section will also be applied when appropriate.

5.4.1. Slab-Following Model

For any particular slab in the furnace, say slab s, the position of that slab changes with time during the heating period via the slab pusher movement. Applying the heat conduction principle and the heat transfer between the slab and the furnace environment, we can establish the following one-dimensional unsteady state heat conduction equation [6]:

$$\frac{\partial T^s(y, t)}{\partial t} = \frac{I}{c\rho} \frac{\partial}{\partial y} \left(K \frac{\partial T^s(y, t)}{\partial y} \right), \quad 0 < y < d$$

$$x^s(t) = \int_0^t v(\tau)d\tau, \quad 0 < t < t_f, \quad x^s(0) = 0, \quad x^s(t_f) = \int_0^{t_f} v(\tau)d\tau = L$$

$$\frac{\partial T^s(y, 0)}{\partial t} = -q_b(x^s, t), \quad \frac{\partial T^s(x, y, d)}{\partial t} = q_t(x^s, t)$$

$$T^s(y, 0) = T_{\text{Entry}}$$

$$\tag{11}$$

where $T^s(y, t)$ and d are the temperature distribution and the thickness of slab s respectively, $x^s(t)$ is the in-furnace position of the slab s at time t, and t_f is the total heating time for slab s. Supposing the heating begins at $t = 0$, then t_f equals the time when slab s is discharged. The calculations of heat flux from furnace to surfaces of the slab s are similar to those described in the reheat furnace model except that now we use the furnace temperature and slab temperature pertaining to slab s; refer to Equations 2–3 for details. Under steady-state heating conditions, i.e. slab thickness, steel grade, slab moving rate, furnace temperature distribution and other operation conditions do not change with time, the following relationship exists between the slab-following model and the reheat furnace model developed in the previous section:

$$T^s(y, t - t_0) = T(x^s, y, t) \tag{12}$$

where t_0 is the initial time when slab s is pushed into the furnace and begins heating.

For convenience of computation, the equations are converted to a discrete time state space model using similar techniques as mentioned in the previous section. The state vector (note here that the slab position is treated as an element of the state vector) and the input vector (note here that average slab moving rate is treated as a component of the input vector) are defined by:

$$\mathbf{X}^s(k) = [T^s(1, k), T^s(2, k), \dots, T^s(N_y, k), x^s(k)]^t$$

$$\mathbf{U}^s(k) = [q_b(x^s, k), q_t(x^s, k), v(k)]^t$$

$$\tag{13}$$

The discrete time state space model, using finite difference approximation similar to that described for the reheat furnace model, is given below:

$$X^s(k+1) = A^s(k)X^s(k) + B^s(k)U^s(k)$$

$$X^s(0) = T_{\text{Entry}}$$

$$A^s(k) = \begin{bmatrix} a_{11}^s(k) & a_{12}^s(k) & & & 0 & 0 \\ a_{21}^s(k) & a_{22}^s(k) & a_{23}^s(k) & & 0 & 0 \\ & \cdots & \cdots & \cdots & \cdots & \cdots \\ 0 & & & a_{N_y,N_y-1}^s(k) & a_{N_y,N_y}^s(k) & 0 \\ 0 & & & & 0 & 1 \end{bmatrix}$$

$$B^s(k) = \begin{bmatrix} b_{11}^s(k) & 0 & \cdots & 0 & 0 \\ 0 & 0 & \cdots & b_{N_y,2}^s(k) & 0 \\ 0 & 0 & \cdots & 0 & \Delta t \end{bmatrix}^t$$

This model gives an estimation of the temperature distribution of a specific slab during its whole heating period, along with the slab's position in the furnace. This information is sufficient for the purpose of steady-state optimization, which will be described in the next section.

5.4.2. Steady-state Optimization

In this section we shall consider the steady-state optimization problem for the reheat furnace zone temperature setpoints. Here steady-state assumes that the operation conditions, such as the slab size and grade, the slab moving rate, and other operation variables are constant against time for the time zone concerned. Under such steady conditions, the optimized zone temperature should also be constant, determined by the steady process conditions and the heating requirement of the slabs derived from the associated production plan, and the slab-following model becomes sufficient for optimization studies. For a specific batch of slabs (all with the same physical and chemical properties), the zone temperature setpoint optimization problem can be formulated, by considering the dynamics of slab heating, the furnace process constraints, the total heating time which is determined by the mill rolling rate, and the heating requirement determined by the steel grade and the associated production plan, resulting in the following equations:

$$J^* = \min_{T_F \in [T_{F\min}T_{F\max}]} \frac{1}{2} R_F T_F^2 \tag{15}$$

subject to:
  Equation 14, and

$$u_b(x,k) = f_{ub}(T_F, x, k), \quad u_t(x,k) = f_{ut}(T_F, x, k)$$

$$\overline{X}^2(k_f) = T^*, \quad \Delta X^s(k_f) \le \Delta T^*, \quad k_f = \text{int}(t_f/\Delta t) \tag{16}$$

where $\mathbf{T}_F = [T_1^t, T_2^t, T_3^t, T_4^t, T_1^b, T_2^b, T_3^b,]^t$ is the zone temperature setpoints vector, $\mathbf{T}_{F\,\mathrm{min}}$ and $\mathbf{T}_{F\,\mathrm{max}}$ are the minimal and maximal zone temperature vectors respectively, $\mathbf{R}_F$ is the cost weighting vector for the zone temperature setpoints, f_{ub} and f_{ut} are functions to determine the bottom and top furnace temperature distributions based on the zone temperature setpoints respectively, T^* and ΔT^* are the desired average slab temperature and the maximal slab temperature difference allowed when discharging, $\overline{X}^s(k_f) \triangleq \frac{1}{N_y} \sum_{i=1}^{N_y} T^s(i, k_f)$ is the estimated average temperature of slab s when discharging, and $\Delta X^s(k_f) \triangleq \max_{1 \le i \le N_y}(T^s(i, k_f)) - \min_{1 \le i \le N_y}(T^s(i, k_f))$ is the temperature difference of slab s when discharging.

The furnace zone temperature is usually controlled by a PID-based controller, with its setpoints coming either from a higher level supervisory control or from an operator. When dealing with optimization we assume that the PID controller is perfect in the sense that the furnace temperature is controlled precisely to its setpoint. Under this assumption there is no difference between the reheat furnace zone temperature and its corresponding setpoint value. For the reheat furnaces considered in this chapter, only four zone temperature setpoints, i.e. the top heating zone temperature, the top soaking zone temperature, the bottom heating zone temperature, and the bottom soaking zone temperature, are controllable. The optimization problem of equations 15–16 can be further reduced to selecting the controllable furnace zone temperature setpoints, $T_2^t, T_3^t, T_2^b, T_3^b$, such that the performance index of equation 15 is minimized, while satisfying the slab heating dynamics and the process constraints specified by equations 14 and 16. Many techniques can be used to solve the optimization problem formulated by equations 14–16, such as dynamic programming, genetic algorithms, neural networks, heuristic optimization, etc. [14–16].

A heuristic optimization strategy (HOS) is developed by incorporating domain expertise and operating experience into a rule-base, and using rule-guided heuristic search techniques to find the optimal (or more precisely the sub-optimal) zone temperature setpoints [7]. This HOS algorithm is outlined below:

STEP 1: Obtain the reheat furnace production information, such as the slab size and steel grade, the slab moving rate, the expected slab discharging temperature, etc. to set up the steady-state conditions for optimization.

STEP 2: Set the iterative number $i = 0$, and select the initial furnace zone temperature vector $\mathbf{T}_F^{(0)} = \mathbf{T}_{F0}$ according to the steady-state condition. Here rules for initial setpoints might be applied.

STEP 3: Calculate the furnace temperature distribution based on the zone temperature setpoint vector $\mathbf{T}_F^{(i)}$, and execute the slab-following model, equation 14, to obtain the estimation of the slab temperature distribution $\mathbf{X}^s(k), k = 1, 2, \ldots, k_f$.

STEP 4: Check the estimated slab discharging temperature $\mathbf{X}^s(k_f)$ and the cost function, equation 15, to see if the optimization criteria have been satisfied. If all the constraints expressed in equation 16 are satisfied and further cost reduction is impossible or if the iteration number reaches the maximal iteration number, $I_{\max}$, go to step 6, otherwise go to step 5.

STEP 5: Based on the estimated slab discharging temperature and the desired one, update the zone temperature setpoint vector $(\mathbf{T}_F^{(i+1)} = \mathbf{T}_F^{(i)} + \Delta\mathbf{T}_F^{(i)})$ for the next iteration by heuristic search guided by the heuristic rule base. Let $i = i + 1$ and return to step 3.

STEP 6: If $i < I_{\max}$ then get the optimized zone temperature setpoint vector for the given steady-state conditions by setting $\mathbf{T}_F^* = \mathbf{T}_F^i$, otherwise inform the operator that the optimized temperature setpoint vector has not been reached before the maximum number of iterations. Terminate the heuristic optimization procedure.

The critical part in the HOS algorithm is to obtain the heuristic rules (HRs) and to design the search strategies for updating the zone temperature setpoint vector based on the HRs. Information on the estimated discharging slab temperature and the desired one, along with the related operation cost history, is used to guide the search of the optimal heating pattern. HRs are expressed in production rule format, with two simplified rules listed below for illustration:

$$\text{if } \overline{X}^s(k_f) > T^* + \varepsilon_A \cap \Delta X^s(k_f) > \Delta T^* + \varepsilon_D$$
$$\text{Then } T_3^t(k+1) = T_3^t(k) - \Delta T_3^t(\mathbf{X}^s(k_f), T^*, \Delta T^*)$$
$$T_3^b(k+1) = T_3^b(k) - \Delta T_3^b(\mathbf{X}^s(k_f), T^*, \Delta T^*)$$

$$\text{if } \overline{X}^s(k_f) < T^* - \varepsilon_A \cap \Delta X^s(k_f) < \Delta T^* - \varepsilon_D$$
$$\text{Then } T_3^t(k+1) = T_3^t(k) + \Delta T_3^t(\mathbf{X}^s(k_f), T^*, \Delta T^*)$$
$$T_3^b(k+1) = T_3^b(k) + \Delta T_3^b(\mathbf{X}^s(k_f), T^*, \Delta T^*)$$

where ε_A and ε_D are the tolerance thresholds for the average slab temperature and slab temperature difference respectively, and ΔT_3^t and ΔT_3^b are the increments for the top and bottom soaking zone temperature setpoints

Table 2. A typical optimization result and the corresponding empirical setpoints.

Heating Specification	$d(m)$	$v(m/s)$	T_2^{t*}	T_2^{b*}	T_3^{t*}	T_3^{b*}
$T^* = 1200°C$		0.0010	1020	1020	1210	1230
$\Delta T = 20°C$	0.22	0.0015	1080	1100	1230	1250
$\varepsilon_A = 25°C$						
$\varepsilon_D = 10°C$		0.0020	1180	1200	1230	1250
Empirical Setpoints			1220	1250	1280	1300

respectively. The value of these increments can be determined by the difference between the estimated slab discharging temperatures and the desired ones in order to speed up the optimization procedure, or it can be left as a constant adjustment step determined by the process dynamics and operating experience.

When the estimated slab discharging temperature profile (given by the slab-following model) is within the tolerance of the desired one, and the operation cost (performance index) cannot be further reduced by the HOS, it is believed that the (near) optimal heating pattern has been reached. Obviously, the optimized zone temperature setpoint vector, $\mathbf{T}_F^*$, is a function of the total heating time, the slab size, the steel grade and the desired slab discharging temperature. Using the HOS algorithm, the optimal furnace zone temperature setpoints and the furnace temperature distribution can be obtained by off-line simulation. The maximum iteration number, I_{max}, is used to limit the maximum computation for a single optimization process. Table 2 gives a typical steady-state optimization result and its corresponding conventional empirical setpoints.

From Table 2 it is obvious that the steady-state optimization setpoints will introduce significant improvement in reheat furnace performance compared with the traditional empirical setpoint control due to the fact that lower furnace temperature setpoints are adopted which will result in savings on energy consumption as well as reduction in oxidization and burn up loss during the entire slab heating process.

5.4.3. Dynamic Setpoint Control

In the previous section we have discussed the steady-state optimization of reheat furnace zone temperature setpoints. However, such setpoints are hardly realistic in a real production environment since the assumption of steady-state cannot be justified in most production tasks. It is unavoidable that quite often slabs with different sizes and steel grades are heated in the same furnace zone due to the shift of production order and different slab resources, and there exists no unique steady-state as the basis for setpoint optimization. Process disturbances, such as the fluctuation of the mill rolling rate, the deviation between the actual furnace zone temperature and its setpoint, etc., will also deteriorate the control performance if the steady-state optimization strategy is used. Mismatch of the estimated slab temperature and its real slab temperature, caused by model error, and the furnace heating characteristics changing over production time due to wear out and other performance deterioration, are other important factors to penalize the steady-state optimization. Steady-state only represents the equilibrium condition under idealised situation, hence the setpoint should only be applied in a similar way. Dynamic setpoint control must be introduced to pursue better control performance under changing production environments. In the following, a feed-forward/feedback style dynamic setpoint control scheme is introduced.

When different slabs are heated in a furnace zone, the zone temperature setpoint is calculated by the weighted summation of the steady-state optimized setpoints corresponding to each slab in the furnace zone. Figure 5 shows the situation when three different kinds of slabs are heated in the heating zone. For this situation, the temperature setpoint for the heating zone is calculated by the following equation:

$$T_{20}^t = \sum_{i=1}^{N_0} a_i T_{2i}^{t*}, \quad T_{20}^b = \sum_{i=1}^{N_0} a_i T_{2i}^{b*}$$

$$\sum_{i=1}^{N_0} a_i = 1, \quad a_i \geq 0, \quad i = 1, 2, \ldots, N_0 \tag{17}$$

where T_{20}^t and T_{20}^b are the weighted steady-state temperature setpoints for the top and bottom heat zones respectively, T_{2i}^{t*} and T_{2i}^{b*} are the optimized top and bottom heat zone temperature setpoints corresponding to the ith slab, a_i is the weighting factor for the ith slab, and N_0 is the total number of slabs heated in the heating zone. Similar equations can be used for different kinds of slabs which are heated in the soaking zone to obtain the weighted soaking zone temperature setpoint.

For the fluctuation in the slab moving rate, a feed-forward compensation strategy is introduced:

$$\delta \mathbf{T}_{Fv} = \mathbf{k}_v (v(t) - v_0)$$

$$\delta \mathbf{T}_{Fv} \overset{\Delta}{=} [\delta T_{2v}^t, \delta T_{2v}^b, \delta T_{3v}^t, \delta T_{3v}^b]^t \tag{18}$$

$$\mathbf{k}_v \overset{\Delta}{=} [k_{v1}, k_{v2}, k_{v3}, k_{v4}]^t$$

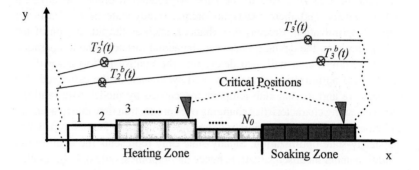

Figure 5. Different kinds of slabs in the heating zone.

where $\delta \mathbf{T}_{Fv}$ is the setpoint compensation vector for the slab moving rate, v_0 and $v(t)$ are the steady-state slab moving rate and the real slab moving rate, respectively, and $\mathbf{k}_v$ is the feed-forward gain vector. The feed-forward gain vector can be determined by means of parameter sensitivity analysis and the principle of feed-forward, which is given by:

$$
\begin{aligned}
k_{v1} &= a_{v1} \frac{\partial T^s(N_y, k_f)}{\partial v} \Big/ \frac{\partial T^s(N_y, k_f)}{\partial T_2^t} \\
k_{v2} &= a_{v2} \frac{\partial T^s(1, k_f)}{\partial v} \Big/ \frac{\partial T^s(1, k_f)}{\partial T_2^b} \\
k_{v3} &= a_{v3} \frac{\partial T^s(N_y, k_f)}{\partial v} \Big/ \frac{\partial T^s(N_y, k_f)}{\partial T_3^t} \\
k_{v4} &= a_{v4} \frac{\partial T^s(1, k_f)}{\partial v} \Big/ \frac{\partial T^s(1, k_f)}{\partial T_3^b}
\end{aligned}
\tag{19}
$$

where α_{vi}, $i = 1, 2, 3, 4$, are adjusting coefficients.

The partial differentials contained in equation 19 can be obtained via the slab-following model using numerical approximation.

Other process disturbances in the operation condition, such as the difference between the setpoint and its corresponding real zone temperature, and the changing of thermal physical properties of the reheat furnace, etc., will be reflected eventually in the slab temperature distribution. The aim of the dynamic setpoint control is to keep the slab temperature distribution close to its optimal distribution (obtained via the corresponding steady-state optimization) while minimizing the operation cost (energy consumption) under varying working conditions. For all those process disturbances, the following feed-forward compensation is introduced based on the difference between the optimized slab temperature distribution and the current estimated slab temperature distribution at the critical positions along the furnace (see Figure 5 for the concept of critical positions):

$$\delta \mathbf{T}_{FT} \triangleq [\delta T_{2T}^t, \delta T_{2T}^b, \delta T_{3T}^t, \delta T_{3T}^b]^t$$

$$\delta T_{2T}^t = k_{T1}(T(i_2, N_y, k) - T_{k2t^*})$$

$$\delta T_{2T}^b = k_{T2}(T(i_2, 1, k) - T_{k2b^*}) \tag{20}$$

$$\delta T_{3T}^t = k_{T3}(T(i_3, N_y, k) - T_{k3t^*})$$

$$\delta T_{3T}^b = k_{T4}(T(i_3, 1, k) - T_{k3b^*})$$

where $\delta \mathbf{T}_{FT}$ is the setpoint compensation vector for slab temperature distribution at the key positions, $T(i_m, j, k)$ is the slab temperature distribution at the mth key position ($m = 2$ for heating zone, $m = 3$ for soaking zone), T_{kmt^*}, and T_{kmb^*} are the optimized slab top and bottom temperatures at the mth key position respectively, k_{Ti}, $i = 1, 2, 3, 4$, are the feed-forward gains which can be calculated by the following equation, based on the same concept used in equation 19.

$$k_{Tv1} = a_{T1}\left[\frac{\partial T(i_2, N_y, k)}{\partial T_2^T}\right]^{-1}, \quad k_{T2} = a_{T2}\left[\frac{\partial T(i_2, 1, k)}{\partial T_2^b}\right]^{-1}$$

$$\tag{21}$$

$$k_{T3} = a_{T3}\left[\frac{\partial T(i_3, N_y, k)}{\partial T_3^t}\right]^{-1}, \quad k_{T4} = a_{T4}\left[\frac{\partial T(i_3, 1, k)}{\partial T_3^b}\right]^{-1}$$

where α_{Ti}, $i = 1, 2, 3, 4$, are adjusting coefficients.

To account for the model error and the time varying characteristics of the reheat furnace, the slab surface temperature measured before it enters the roughing mill for rolling is used for feedback control of the furnace zone temperature setpoints. The following algorithm is used for such a feedback compensation strategy:

$$\delta \mathbf{T}_{FP} = \mathbf{k}_P(T_P - T_{P^*})$$

$$\delta \mathbf{T}_{FP} \triangleq [\delta T_{2P}^t, \delta T_{2P}^b, \delta T_{3P}^t, \delta T_{3P}^b]^t \tag{22}$$

$$\mathbf{k}_P \triangleq [k_{P1}, k_{P2}, k_{P3}, k_{P4}]^t$$

where $\delta \mathbf{T}_{FP}$ is the setpoint compensation vector for slab surface temperature before rolling at the roughing mill. T_p is the slab top surface temperature measured before rolling, and T_{p^*} is the ideal slab top surface temperature before rolling, which can be calculated from $T^s(i, k_f)$ and the elapsed time between discharging and primary rolling, and $\mathbf{k}_P$ is the feedback gain vector.

The final zone temperature setpoints, after the above compensation, are given by:

$$\mathbf{T}_F(k) = \mathbf{T}_{F0} + \delta\mathbf{T}_{Fv}(k) + \delta\mathbf{T}_{FT}(k) + \delta\mathbf{T}_{FP}(k)$$
$$\mathbf{T}_{F0} = [T_{20}^t, T_{20}^b, T_{30}^t, T_{30}^b]^t \tag{23}$$

5.4.4. Slab Tracking System

It is clear from the previous section that if we want to implement the setpoint control strategy we must have real-time information on the reheat furnace operation, including the real-time slab moving rate, the Dynamic Slab In-Furnace Map (DSIF Map) which shows what kind of slabs are heated in the furnace, which slab is currently charged into the furnace and which slab is being discharged, as well as information related to the mill production and the production plan. Such information on the slab movement and arrangement in the furnace, as well as the whole production line, is provided by the slab tracking system. In the following we will confine our attention to the slab tracking around the reheat furnace area. The allocations of slab movement detecting sensors and other process instrumentation are shown in Figure 6.

The task of the slab tracking system (for the furnace area) is to obtain information about the slab allocation in the furnace and other relevant information, based on the slab flow detection signal and the current production plan. The slab allocation and other important information are stored in the DSIF Map, and Table 3 gives an simplified example of the DSIF Map for 1# furnace at a specific time.

Table 3. Dynamic Slab In-Furnace (DSIF) Map for 1# Furnace 1.

No	Steel Grade[2]	Slab Size (mm)	Position (m)	Enter Time	Heating Time (hr)	Slab Temp	Setpoints
1	A	200	26.4	3:05	7.87	Computed	Computed
2	A	200	25.8	3:16	7.69	by Reheat	by Steady
3	A	200	25.2	3:24	7.56	Furnace	State
4	B	240	24.48	3:34	7.39	Model	optimization
...	...	...	...	...	...	Equations	Algorithm:
i	B	240	17.53	5:48	5.15	(9–10),	HOS,
...	...	...	...	...	...	updated	updated
N_{F-1}	C	180	0.81	10.42	0.25	every	every
N_F	C	180	0.27	10.52	0.09	60 Seconds	5 Minutes

[1] Many attributes of the DSIF Map are not listed in this table, such as the slab Id, the expected heating time, the heating requirement, etc.

[2] A = Carbon steel, B = High Carbon Steel, C = Alloy Steel.

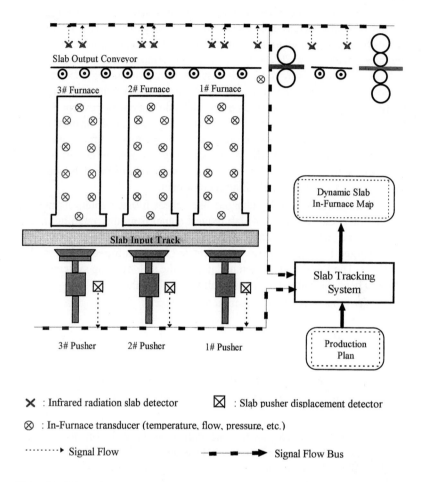

X : Infrared radiation slab detector   ☒ : Slab pusher displacement detector

⊗ : In-Furnace transducer (temperature, flow, pressure, etc.)

········▶ Signal Flow   ▬ ▬ ▬▶ Signal Flow Bus

Figure 6. Allocation of slab detection Sensors and other instrumentation around the furnaces area.

The initial DSIF Map is constructed by the slab pusher movement (displacement) signal and the corresponding production plan. Whenever a slab is charged into the reheat furnace by a slab pusher, attributes such as the slab Id, the slab size, steel grade, the expected heating time of the charged slab, etc., are extracted from the production plan file to form the corresponding attributes of the DSIF Map for the charged slab. The enter furnace time is set as the current time when the slab pusher movement is detected by the pusher displacement detector. The position of the charged slab is set to half the width of the slab. For all the other slabs already in the furnace, the position of those slabs are increased by the width of the charged slab. The N_0 attribute of the charged slab is set to the total number of slabs in the furnace, including the latest charged one. The setpoints (of zone temperature) for the charged slab are computed using the steady-state optimization strategy described in section 5.4.2, according to the slab size, steel grade, and the total expected heating time. All other attributes of the slabs do not change except the slab temperature distribution, which is calculated by the reheat furnace model on a real-time basis. The above process continues until the furnace is fully charged, when the initial DSIF Map is set up and the reheat furnace is ready for discharging once the slab temperature distribution of the slab located at the furnace exit satisfies the rolling requirement.

After the furnace is fully charged, the reheat furnace enters its normal production cycle. When a hot slab request signal from the roughing mill is received, the slab pusher starts to charge slabs into the furnace until a slab is discharged from the furnace exit (detected by the infrared radiation slab detector). The DSIF Map is then updated by the slab tracking system after a slab has been discharged, using the following algorithms:

For slabs that have already been in the furnace:

$$
\begin{aligned}
N_0(k^+) &= N_0(k^-) - 1 \\
x_1(k^+) &= L_c/2 \\
x_1(k^+) &= x_{i+1}(k^-) + L_{\text{out}}, \quad i = 2, 3, \ldots, N_F - 1 \\
\mathbf{P}_i(k^+) &= \mathbf{P}_{i+1}(k^-), \quad i = 1, 2, \ldots, N_F - 1
\end{aligned}
\tag{24}
$$

where k is the time instant when the slab is discharged from the furnace, k^- and k^+ are the time immediately before and after k respectively, $\mathbf{P}_i$ is the attribute vector belonging to the ith slab in the furnace which does not change with time, such as the steel grade, the slab size, etc. x_i is the slab position in the furnace, N_0 is the slab sequence number counted from the furnace exit with the slab to be discharged as $N_0 = 1$, L_c is the width of the charged slab, and N_F is the total number of slabs in the furnace before discharging.

For slabs which are being charged into the furnace during the discharging process, the following procedure is used to determine how many slabs are charged into the furnace for the discharging operation:

(1) Take the attributes of the slab to be charged from the production plan $\mathbf{P}_c(k^+) = \mathbf{P}_{\text{plan}}$, and set $N_{F_0} = N_f - 1$.

(2) if $x_{N_{F_0}}(k^+) - L_c \leq 0$ then go to Step 4 else go to Step 3. Here L_c is the width of slab being charged.

(3) Add the charged slab into the DSIF Map, and set the relevant attributes by the following equation:

$$x_{N_{F0}+1}(k^+) = x_{N_{F0}}(k^+) - L_c$$
$$\mathbf{P}_{N_{F0}+1}(k^+) = \mathbf{P}_c(k^+) \tag{25}$$
$$N_{F0} = N_{F0} + 1$$

and take the attributes of the next slab to be charged from the production plan, $\mathbf{P}_c(k^+) = \mathbf{P}_{\text{plan}}$, then repeat Step 2.

(4) Set the total number of slabs in the furnace to $N_F = N_{F0}$, and end this procedure.

The average slab moving rate is calculated by the following fixed length moving average algorithm based on the DSIF Map:

$$v(k) = \frac{1}{N_v} \sum_{i=0}^{N_v-1} \frac{L_{N_f-1}}{t_0(N_f - i) - t_0(N_F - i - 1)} \tag{26}$$

where N_v is the fixed data length, $t_0(i)$ is the enter furnace time of the ith slab, and L_i is the width of the ith slab. $t_0(i)$ and L_i are taken from the DSIF Map at time k. With the slab tracking system and other on-line instrumentation, the reheat furnace model and the furnace zone temperature control described in the previous sections are integrated and can be executed on a real production environment.

5.5. DDC CONTROL STRATEGIES FOR REHEAT FURNACES

The DDC level control of reheat furnaces consists of zone temperature control, gas and air flow control, furnace pressure control, burner control (on/off), air/gas ratio control, etc. The objective of zone temperature control is to adjust the air flow rate and the gas flow rate such that the zone temperature is kept at its setpoint level to achieve an optimal (ideal) combustion. The air/gas ratio is determined by the ideal combustion condition plus the feedback of oxygen level of the flue gas. For the zone temperature control, the following double cross limit cascade plus ratio control as shown in Figure 7 is used.

The advantage of this double cross limit cascade control scheme is that the air/gas ratio can be kept close to it setpoint value K_u, in a dynamic sense, thus leading to an ideal combustion process. The air/gas ratio is determined based on the theoretical value plus feedback compensation of the oxygen level in the flue gas. The burner on/off control is introduced to keep the satisfactory combustion pattern under different gas flow conditions. There are six burners in each furnace zone located parallel along the direction of furnace width.

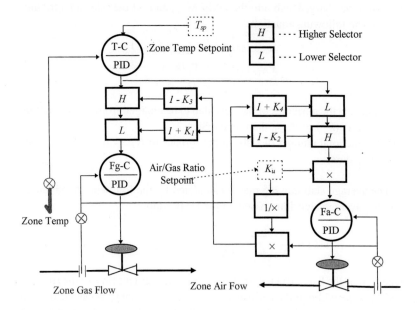

Figure 7. Double cross limit cascade plus ratio temperature control.

The basic idea is that when the zone gas flow is small, some of the burners of that zone, starting from the edge of the furnace, should be turned off in order to keep the remaining burners working at their normal (gas flow) range. Before this new scheme is implemented, the burners are turned off manually when the zone gas flow is too small to keep all the burners working at their normal operation condition. This is a very hard task for the operators due to the high temperature environment around the furnace burner area. Figure 8 shows the block diagram of the burner on/off control as well as the air/gas ratio control scheme.

Furnace pressure is another important control variable in the reheat furnace operation. Ideally, the furnace pressure should be controlled at zero pressure (relative pressure) in order to minimize the heat loss through the furnace surrounding and to increase the heat efficiency of the combustion process within the furnace. Negative furnace pressure will cause cool air penetration into the furnace, which consumes heat energy generated by combustion and decreases furnace temperature, while positive furnace pressure will increase the heat loss of the furnace due to the non-seamless nature of the furnace wall. Conventional PID control here is not sufficient for the furnace pressure control since this process has a very large time delay due to the fact that the

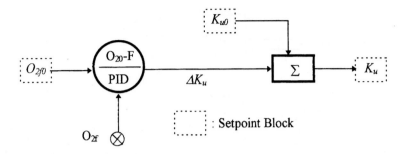

(a) Air/gas ratio setpoint control

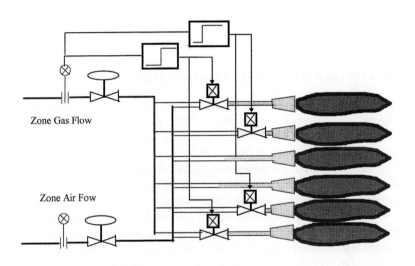

(b) Burner on/off control

Figure 8. Air/gas ratio and burner on/off control.

furnace pressure is adjusted by the position of the damper in the flue stack which is far away from the furnace pressure measurement and the furnace has a very large volume. Another difficulty is that the furnace pressure has a very narrow control tolerance (5 ± 2 mmH$_2$O) beyond which the heat loss will be significant. In our work, a feed-forward (through the overall gas flow) and feedback (through the furnace pressure) control algorithm is used, with its block diagram shown in Figure 9.

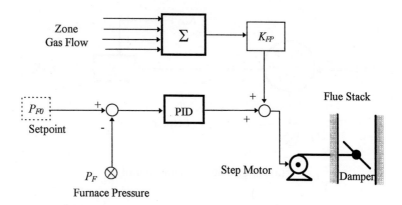

Figure 9. Furnace pressure control.

5.6. CONTROL SYSTEM IMPLEMENTATION

The integrated reheat furnace control, instrumentation and optimization scheme is implemented on a Micro VAX II computer and a WDPF (Westinghouse Distributed Processing Family) computer system, with the Micro VAX II computer in charge of slab tracking, setpoint optimization, and fault diagnosis, while the WDPF distributed computer system is mainly for the DDC level control, plus process monitoring and alarming. The hardware structure of the integrated reheat furnace control system is shown in Figure 10.

WDPF is a sophisticated group of process control technologies that provides powerful control, data acquisition, process management and information technology capabilities. Since its introduction in 1982, WDPF has continued to evolve — incorporating new technologies, greater capabilities, and innovative enhancements. With WDPF, one can fully integrate plant process control, local and wide-area networks, business and engineering systems, etc., in a single unified architecture.

At the heart of the WDPF distributed control system is the powerful, high-speed Westnet data highway. This broadcast-mode highway provides deterministic communication of process data among as many as 254 drops on a WDPF system. A drop is a functional unit consisting of a Data Highway Controller subsystem and one or more functional processors which perform an assigned task(s). Tasks such as modulating and sequential control, operator/engineer functions, historical data collection and logging are physically distributed among drops on the highway, all operating in parallel. This distributed architecture eliminates the need for a host computer

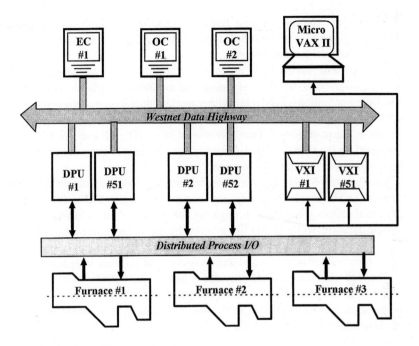

Figure 10. Hardware Structure.

or traffic director, as well as the potential for data bottlenecks that are often associated with such request/response highway schemes. With WDPF, process values are continuously broadcast onto the highway, at least once every second — all the time, regardless of plant conditions. Any drop on the system can instantly access process data from any other drop, regardless of where the data originated or how many drops are on the system. With the continuous broadcast of process values, the data highway essentially serves as a distributed global data base. WDPF supports redundant configuration, which offers full "bumpless" transfer, with no data loss, delay or re-booting required in the event of cable damage or component failure. In Figure 10, DPU #51, DPU #52 and VXI #51 are redundant units (drops).

The WDPF Distributed Processing Unit (DPU) executes simple or complex sequential and modulating control strategies. The DPU features a custom array co-processor to resolve ladder logic, sequence of events, and five control areas with configurable loop speeds. Additional features include Local or Remote I/0 capability, on-line graphic monitoring of control and ladder execution. DPU #1 (and its redundant unit DPU #51) is used for the control

264 Y.Y. YANG and D.A. LINKENS

of reheat furnace #1 and #2, while DPU #2 (and its redundant unit DPU #52) is used for the control of reheat furnace #3 and other overall control and monitoring variables of the reheat furnace process, such as the overall gas pressure, the total gas consumption, etc.

The Operator Station (OS) provides a real-time graphical interface for control, diagnostics, trending, alarming, and monitoring of plant conditions. The Operator Station is available in either single or dual screen configurations, with each screen capable of displaying up to four simultaneous process views, trends or alarm windows. Two operation stations (OS #1 and OS #2) are equipped for the reheat furnace control system, in order to carry out efficient modulating and monitoring functions.

The Engineer Station (EC) provides the interface to build and edit WDPF control logic, process graphics and system databases. Fill-in-the-blank graphical editors guide users through the definition of control logic and the system database, while a CAD-like graphics builder reduces the time required to develop process graphics. The Engineer Station provides multiple levels of security to control access and ensure system and software integrity.

The VAX interface (VXI) provides a link to the WDPF Westnet data highway from external Micro VAX computers via industry standard communication protocols, here TCP/IP compliant devices. The reheat furnace process data required for slab tracking and setpoint optimization within the Micro VAX II computer are obtained through the VXI interface. Also, the optimized dynamic setpoint for individual reheat furnace are transferred to the Westnet Date Highway via the VXI interface. The corresponding DPU will pick the right setpoints for its DDC level control.

The functions of the Micro VAX II computer are slab tracking along the whole production line, the temperature profile estimation of all slabs heated in the reheat furnaces, and the dynamic setpoints calculation for zone temperature of all the reheat furnaces. The VMS operating system is selected in order to facilitate event management and scheduling, to obtain a good real-time performance, and to attain a high system security. Figure 11 shows a functional diagram of the Micro VAX II computer.

The main programme acts as the coordinator of the overall program execution. It supports the data communication among all program modules, runs the appropriate module at a proper time and according to its priority, deals with interruptions from both the reheat furnaces and the human operators. The real-time slab temperature estimation module is based on the reheat furnace model developed in Section 5.4.1. Its main function is to estimate the current temperature distribution of all the slabs in a reheat furnace, by using the current Dynamic Slab In-Furnace (DSIF) Map and the measurable process variables, such as the reheat furnace zone temperatures, the average slab moving rate, and the accumulated heating time of individual slabs.

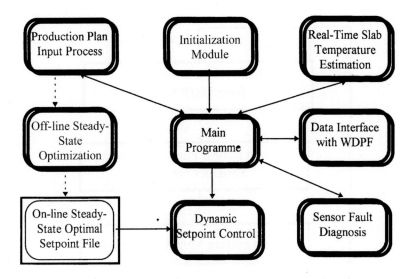

Figure 11. Functional Diagram of the Micro VAX II Computer.

The thermal and physical parameters of the slabs, such as the heat conductivity and specific heat capacity, the density of the slab, the slab thickness and width, the expected slab discharging temperature, etc., are obtained through the corresponding data files related to the production plan. Model parameters can be adjusted by the control engineer through the man-machine interface. Temperature estimates are updated at a fixed cycle of 60 seconds. Figure 12 shows the flowchart of the slab-temperature estimation module.

The dynamic setpoint control module is based on the setpoint control algorithm developed in Sections 5.4.2 and 5.4.3. Its main function is to give the most suitable zone temperature setpoints for individual reheat furnaces, based on the current process condition and the states of the slabs heated in the furnaces. Zone temperature setpoints obtained via the steady-state optimization algorithm, which have been stored in the steady-state optimal setpoint file for on-line use, is the basis for the dynamic setpoint control. The actual setpoint is calculated (or adjusted) around this optimal setpoint according to the deviation of the steady-state and the current state of the reheat furnaces. Here again, the corresponding steady-state and the deviation from this steady-state are determined by the original production plan and the current production state, which is easily available via the DSIF Map. Control process dynamics are considered for outputting the final zone temperature setpoints such that the setpoint trajectory are smoothed and will not cause

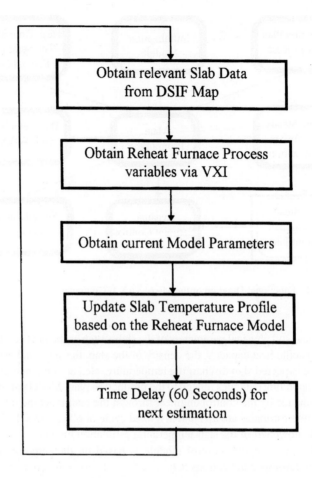

Figure 12. Flowchart of the slab temperature estimation module.

severe fluctuations in the zone temperature control loop. The final setpoints are transferred across the VXI interface and will be picked up by the associated combustion control loops residing in one of the DPUs. The dynamic setpoint control module is executed every 5 minutes. This cycle time is larger than that of the slab temperature estimation in order to obtain satisfactory DDC control behaviour and to maintain the stability of the setpoint control strategy. Figure 13 shows the flowchart of the dynamic setpoint control for the reheat furnaces.

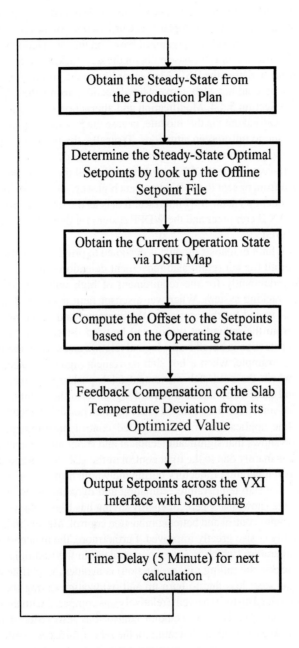

Figure 13. Flowchart of the dynamic setpoint control.

The production plan input process module is designed for inputting the production plan (short-term shift plan). The productions of the reheat furnace and the rolling mill are strictly organized based on this production plan. The production plan also acts as the base point for the slab tracking system which is responsible for updating the DSIF Map according to the production progress via a series of cold/hot slab movement detection, using the algorithms developed in Section 5.4.4. Modification and alternation of production plan can be easily carried out via this module, in case the production plan needs to be changed due to unforeseen situations. Typically, the production plan are typed in by the process operators before the shift begins.

Data interface with the WDPF module identifies all these data which need to be fetched from or sent to the Westnet data highway, along with its desired updating frequency via the VXI interface. It acts as the image copy between the Micro VAX II computer and the WDPF system for those data which need to be transferred.

The sensor fault diagnosis module is designed to provide some basic fault detection for all the hot (cold) slab movement detection sensors, which are configured redundantly for the requirement of high reliability demanded by the slab tracking system. Whenever a sensor fault is detected, an alarm message will issued for the caution of the operators and in the mean time, the signal from the redundant sensor will be used for slab tracking. Such sensor fault detection is based on the logical relation and correlation between sensors. For example, when a hot slab movement signal is detected from a furnace which the cold slab movement (slab pusher movement) signal is absent and there is no hot slab request signal form the roughing mill, it indicates that some fault has occurred to the relevant hot slab detecting sensor.

The on-line application of the integrated control, instrumentation and optimization shows that significant technical and economical benefit can be achieved. It is mainly due to the tight control of the slab discharging temperature and its temperature trajectory during the whole heating period, thanks to the real-time slab temperature estimation and dynamic setpoint control. Statistical data show that 9 percent energy saving has been achieved via the dynamic setpoint control and better combustion control. Meanwhile, the slab heating quality is also greatly improved. Furthermore, the unnecessary high furnace temperature (which is a common strategy in the traditional control scheme to secure the rolling mill production) is avoided. Experimental data show that the steel loss due to burn up and oxidization during the heating period has reduced up to 30 percent (relatively), as compared with the original control scheme. The dynamic setpoint control has also reduced the mill waiting time due to insufficient heating in the reheat furnaces, while on-line fault diagnosis has helped to reduce the production shut down time. As a result, the productivity of the overall production line has also been improved.

5.7. CONCLUSIONS

In this chapter an integrated control, instrumentation, and optimization scheme for reheat furnace operation has been developed by using mechatronic system techniques. A reheat furnace model has been developed which provides a real-time estimation of the slab temperature distribution based on the measurable process variables and the physical data related to the heated slabs, such as the slab size and the steel grade, the heat conduction parameters of the steel, and radiation parameters of the furnace and the slabs etc. Real-time slab temperature estimation is vital for better reheat furnace operation control, and makes it possible to change from empirical control to slab temperature guided optimal control in the reheat furnace operation. Optimization strategies for furnace zone temperature setpoints are also developed under steady-state operation conditions, using heuristic search techniques and the simplified slab-following model. The ultimate furnace zone temperature setpoints are calculated dynamically based on the steady-state optimal setpoints and the deviation of the current working condition from its corresponding steady-state. Additional instruments and measurements are added and a slab tracking system is designed for the purpose of collecting the necessary dynamic information required for the above slab temperature estimation and dynamic furnace setpoint control. The slab tracking system has provided an efficient link between instrumentation, control and optimization, and has made the application of a mechatronic approach possible by integrating different parts into a synergetic unity. The DDC level control has also been upgraded to achieve better dynamic performance of the reheat furnace operation, such as the double cross limit cascade control of the furnace zone temperature, the feed-forward/feedback furnace pressure control, and the automatic on/off control of the burners according to the different zone gas flows. To improve the reliability of the integrated control scheme, redundant instrumentation and computer-based fault diagnosis has also been carefully considered. The proposed integrated scheme has been applied successfully to a real production environment, and the resulting performance is very satisfactory, with significant energy saving and heating quality improvement. The key benefit comes from the integration of control and optimization via mechatronic system principles and techniques. Although the study carried out here is directly concerned with the reheat furnace operation, we believe that the basic principles and techniques developed in this chapter could be applied easily to other similar processes, such as the soaking pit/rolling mill systems, paper mills, etc.

As implied by King [17] mechatronics does not suggest an entirely new subject. Multi-disciplinary engineering involving mechanical, electronic and control techniques has been going on for a long time in areas such as the aviation industry. This kind of integration has become more attractive technically and economically due to the rapid advent of microprocessor and information techniques. Bradley *et al.* [11] also mention that the foundations of a mechatronic approach to engineering design are considered to lie in information and control. They further argued that it may be objected in some quarters that mechatronics is 'only control engineering' in another guise. From the above arguments it seems that the name 'mechatronics' itself is not so important as its key concept of *multi-disciplinary integration*. It is to be hoped that the term 'mechatronics' will assist in highlighting the existence of this type of engineering and attracting more people to experience its advantages first-hand.

The integrated control, instrumentation and optimization scheme proposed in this chapter has efficiently improved the overall system performance of reheat furnace operation on one hand, while at the other hand it also requires a higher level of instrumentation, communication and operation skills. As levels of automation increase, so does the need for communication and instrumentation, which are vital parts in any mechatronic system. For example, considering the case in this chapter, the instrumentation has been significantly intensified by adding all the slab movement detection sensors (via infrared radiation-based sensors for hot slabs and displacement detection for the slab pusher movement), the slab surface temperature detection before a slab enters the roughing mill, and the on/off control valves for burners. Without this kind of instrumentation, it is impossible to carry out the integrated control and optimization strategy. Also, the integrated scheme demands a much higher reliability on the instrumentation, as compared with the traditional control scheme. For example, failure of the thermocouple will result in wrong furnace temperature signals, hence causing the zone temperature setpoint control to be meaningless and the proper zone temperature control to be impossible. In the original control system, the effect of a thermocouple failure is not so dramatic since in that case the zone temperature acts only as a reference variable while the combustion is controlled via direct gas and fuel flowrate. For this reason, a properly designed system should take precautionary measures against instrument failures. One common approach, as adopted in this chapter, is to build in fault diagnosis and prepare several control strategies which require different levels of instrumentation support, along with hardware redundancy for critical measurements. When an instrument (or actuator) failure occurs, the system can automatically detect it and transfer to the proper control strategy if necessary (causing a slight performance degrade). Another important factor of the successful application is the quality and skill of the operators. Basic knowledge of how the system works and what could be wrong under different situations will be helpful in a real operating environment. Attention should be paid to raising the operators awareness and skill to an adequate level demanded by the integrated system.

ACKNOWLEDGEMENTS

The authors wish to thank the invaluable discussions and advice provided by Mr. X.X. Zhou, Mr. B.Y. Li, Mr. X.B. Zheng, Mr. L. Chen, Mr. S. Chen of the Chongqing Iron and Steel Corporation, Shichuan, China during the development and implementation of the advanced reheat furnace control system. Thanks are also given to Professor Y. Z. Lu for his encouragement, advice and support during the development of this project.

REFERENCES

1. Hollander, F. and Huisman, R.L., 1972, *Iron and Steel Engineer*, **48**, 43–56.
2. Hollander, F. and Zuurbier, S.P.A., 1982, *Iron and Steel Engineer*, **59**, 44–52.
3. Iwahashi, Y. and Takanashi, K., 1981, *Proceedings of the 8th World Congress of the International Federation of Automatic Control*, **5**, 2625–2631.
4. Lu, Y.Z. and Williams, T.J., 1983, *Computers in Industry*, **4**, 1–18.
5. Pike, H.E. and S.J., 1972, *Automatica*, **6**, 41–50.
6. Yang, Y.Y. and Lu, Y.Z., 1986, *Computers in Industry*, **7**, 145–154.
7. Yang, Y.Y. and Lu, Y.Z., 1988, *Computers in Industry*, **10**, 11–20.
8. Yabuuchi, K., Shiraishi, K., Kusumoto, K., Kamata, M., Takekoshi, A. and Murakami, H., 1980, *Proceedings of International Conference on Steel Rolling*, Tokyo, Japan, Sept. 29–Oct. 4, **1**, pp.105–116.
9. Miu, D.K., 1993, *Mechatronics: Electromechanics and Contromechanics*, Springer-Verlag, New York.
10. MacConail, P.A., Drews, P. and Robrock, K.H., 1991, *Mechatronics and Robotics*, **I**, IOS Press, Oxford.
11. Bradley, D.A., Dawson, D., Curd, N.C. and Loadre, A.J., 1991, *Mechatronics – Electronics in Products and Processes*, Chapman and Hall, UK.
12. Schweitzer, G., 1990, *Mechanics: Designing Intelligent Machines*, The Institute of Mechanical Engineers, pp.239–246.
13. Chapman, K.S., Ramadhyani, S. and Viskanta, R., 1994, *Combustion Science and Technology*, **97**, 99–120.
14. Goldberg, D.E., 1989, *Genetic Algorithms in Search, optimization and Machine Learning*, Addison-Wesley.
15. Warwick, K., Irwin, G.W. and Hunt, K.J. (Eds.), 1992, *Neural Networks for Control and Systems*, Peter Peregrinus, London.
16. Jiang, J., 1989, *Computers in Industry*, **13**, 253–259.
17. King, T.G., 1995, *Mechatronics*, **4**(2/3), 95–115.

6 NONLINEAR CONTROL OF A GAS FIRED FURNACE

CHRISTOPHER EDWARDS and SARAH K. SPURGEON

Control Systems Research, Department of Engineering,
University of Leicester, Leicester LE1 7RH, UK

6.1. INTRODUCTION

This chapter explores the application of recent nonlinear control and observer theory to the problem of temperature control in a gas fired furnace. The furnace under consideration can be thought of as a gas filled enclosure containing a heat sink (such as a load), bounded by insulating walls. Heat input is achieved via a burner located in one of the end walls and the combustion products are evacuated through a flue positioned in the roof. Although this represents the simplest design possible — a single burner arrangement — such a plant could legitimately represent an industrial furnace for the firing of ceramics. This design has been chosen as a starting point for the study of controllers for high temperature heating plant principally because a prototype single burner furnace is available for controller evaluation purposes.

The control problem considered here involves manipulation of the fuel flow rate so that the measured temperature at some point in the furnace follows some specified temperature/time profile. It is assumed that a controller already exists which regulates the air flow so that for a given fuel flow rate an appropriate air flow rate is maintained to ensure combustion efficiency and an appropriate concentration of unburnt oxygen in the flue products. Currently, in an industrial context, temperature control of such plant would be achieved through the use of traditional Proportional plus Integral plus Derivative (PID) controllers. The gains of such controllers can be tuned on-line and thus, unlike most 'modern control' approaches, no mathematical model of the system is required for controller synthesis. The established technique for modeling of industrial furnaces generates a collection of nonlinear finite difference and algebraic equations which do not easily lend themselves to the problem of controller design. Consequently, for control purposes, one approach is to treat the system as an *uncertain* linear system. In this way the nonlinearities are combined with the true uncertainties such as hysteresis in the valve and disturbances caused by changes in the fuel characteristics.

One effective approach for controlling uncertain systems is the so-called Variable Structure Control Systems methodology. The technique seeks to drive the states of the differential equations representing the system onto a region of the state-space known as the *sliding surface* and then ensure that the states remain there for all subsequent time. The resulting reduced order behavior is termed a sliding motion and the resulting dynamics are specified by the designer's particular choice of sliding surface.

Whilst sliding, the system exhibits certain robustness properties – notably, complete insensitivity to any uncertainty which is implicit in the input channels of the process to be controlled; this class of uncertainty is called *matched uncertainty*. Inherent in this traditional approach to design is a discontinuous control structure. In a mechanical system such as the one comprising the valves that regulate the fuel and air flow to the furnace, this type of control action is not possible and complete insensitivity to the uncertainty is not possible. However close approximation to an ideal sliding motion may still be obtained by the use of a suitable nonlinear controller.

The early sections of this chapter describe the necessary theoretical background. A detailed case study is then presented using a nominal linear model identified from furnace input output data. Quantitative measures of performance which are typically used for closed-loop analysis within the industry are described. The performance of the proposed nonlinear control scheme during both the nonlinear simulation and on-site plant trials can then be explored. For comparative purposes, the typical performance of a sophisticated commercial PID controller is also presented.

The following notation will be used throughout: $\mathbb{R}$ and $\mathbb{C}$ will denote the field of real and complex numbers respectively; $\mathbb{R}_+$ the set of strictly positive real numbers; $\mathbb{C}_-$ the open left half of the complex plane and $\mathbb{R}^{n \times m}$ the set of real matrices with n rows and m columns. For a given matrix A, the transpose will be written A^{T} and the range space and null space will be denoted by $\mathcal{R}(A)$ and $\mathcal{N}(A)$ respectively. For a square matrix $\det(A)$ will represent the determinant; A^{-1} the inverse; $\lambda(A)$ the spectrum i.e. the set of eigenvalues; $\lambda_{max}(A)$ the largest eigenvalue and $\kappa(A)$ the spectral condition number (i.e the ratio of the maximum and minimum singular values). Finally I_n will represent the $n \times n$ identity matrix. The notation $sgn(\cdot)$ will be used for the signum function and $\| \cdot \|$ will denote the Euclidean norm for vectors and the spectral norm for matrices.

6.2. AN INTRODUCTION TO VARIABLE STRUCTURE CONTROL

The focus of much of the research in the area of control systems theory during the seventies and eighties has addressed the issue of *robustness* — i.e. designing controllers with the ability to maintain performance/stability in the presence of discrepancies between the plant and model. One nonlinear approach to robust controller design which emerged during this period is the Variable Structure Control Systems methodology. Variable Structure Control Systems evolved from the pioneering work in Russia of Emel'yanov and Barbashin in the early 1960's. The ideas did not appear outside the Soviet Union until the mid 1970's when a book by Itkis [12] and a survey paper by Utkin [18] were published in English. Variable Structure Systems concepts have subsequently been utilized in the design of robust regulators, model-reference systems, adaptive schemes, tracking systems and state observers. The ideas have successfully been applied to problems as diverse as automatic flight control, control of electrical motors, chemical processes, helicopter stability augmentation, space systems and robotics; for references pertaining to the various applications see the survey paper by Hung *et al.* [11].

Variable Structure Control Systems comprise a collection of different — usually quite simple — feedback control laws and a decision rule. Depending on the 'state' of the system, the decision rule, often termed the *switching function*, determines which of the control laws is 'on-line' at any one time. Unlike, for example, a gain-scheduling methodology, the decision rule is designed to force the system states to reach, and subsequently remain on, a pre-defined surface within the state-space. The dynamical behavior of the system when confined to the surface is described as the *ideal sliding motion*. The advantages of obtaining such a motion are twofold: firstly there is a reduction in order (and for nonlinear systems, by an appropriate choice of surface, the reduced order motion may be chosen to be linear); secondly the sliding motion is insensitive to parameter variations implicit in the input channels. The latter property of invariance towards so-called matched uncertainty makes the methodology an attractive one for designing robust controllers for uncertain systems.

The pertinent properties of Variable Structure Systems can be demonstrated by considering the double integrator given by

$$\ddot{y}(t) = bu(t) \tag{1}$$

where b is a positive scalar whose value is not precisely known. Consider the control law

$$u(t) = \begin{cases} -1 & \text{if } s(y, \dot{y}) > 0 \\ 1 & \text{if } s(y, \dot{y}) < 0 \end{cases} \tag{2}$$

where the *switching function* is defined by

$$s(y, \dot{y}) = my + \dot{y} \tag{3}$$

for some positive scalar m. This clearly fits the description of a Variable Structure Control system given earlier. The reason for the use of the term switching function is also apparent, since it represents the rule governing which control structure is in use at any point $(y, \dot{y})$ in the phase plane. For large values of $\dot{y}$ the phase portrait, obtained from joining the parabolic components of the constituent laws, is shown in Figure 1.

However for values of $\dot{y}$ satisfying the inequality $m|\dot{y}| < b$ then

$$s\dot{s} = s(m\dot{y} + \ddot{y}) = s(m\dot{y} - b\,\text{sgn}(s)) < -|s|(b - m|\dot{y}|) < 0$$

or equivalently

$$\lim_{s \to 0^+} \dot{s} < 0 \quad \text{and} \quad \lim_{s \to 0^-} \dot{s} > 0$$

Consequently when $m|\dot{y}| < b$ the system trajectories on either side of the line

$$\mathcal{L}_s = \{(y, \dot{y}) : s(y, \dot{y}) = 0\} \tag{4}$$

point towards the line. Intuitively high frequency switching between the two control structures will take place as the system trajectories repeatedly cross the line $\mathcal{L}_s$. This high frequency motion is described as *chattering*. If infinite frequency switching were possible, intuitively at least, the motion would be trapped or constrained to remain on the line $\mathcal{L}_s$. The motion when confined to the line $\mathcal{L}_s$ satisfies the first order differential equation obtained from rearranging $s(y, \dot{y}) = 0$, namely

$$\dot{y}(t) = -my(t) \tag{5}$$

This represents a first order decay and the trajectories will 'slide' along the line $\mathcal{L}_s$ to the origin (Figure 2). Such dynamical behavior is described as an *ideal sliding mode* and the line $\mathcal{L}_s$ is termed the *sliding surface*. This simple example demonstrates the key advantages offered by Variable Structure Control Systems, namely:

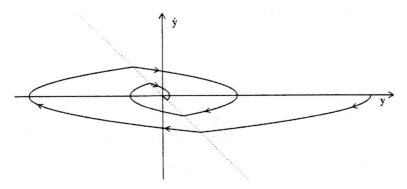

Figure 1. Phase Portrait of the System for large $\dot{y}$.

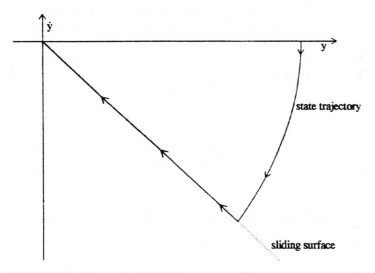

Figure 2. Phase Portrait of a Sliding Motion.

- the sliding motion is independent of the value of b and consequently if this parameter is time varying or uncertain, provided sufficient control energy is available to maintain sliding, the control system completely rejects this variation;
- whilst sliding the system behaves as a first order free motion with all the control effort expended in ensuring that $s(y, \dot{y}) = 0$. The reduced order dynamics can be seen to depend only on the choice of the gradient of the line $\mathcal{L}_s$;

- the control action required to bring about such a motion is discontinuous about the sliding surface.

If a new variable

$$x \overset{s}{=} \begin{bmatrix} y \\ \dot{y} \end{bmatrix}$$

is introduced then equation 1 can be written in *state-space* form as

$$\dot{x}(t) = \begin{bmatrix} 0 & 1 \\ 0 & 0 \end{bmatrix} x(t) + \begin{bmatrix} 0 \\ b \end{bmatrix} u(t) \tag{6}$$

The switching function can also be conveniently expressed in matrix terms as

$$s(y, \dot{y}) = [\, m \quad 1 \,] x(t) \tag{7}$$

This suggests that differential equations written in state-space form constitute a natural framework in which to explore the properties of Variable Structure Control Systems for multi-input linear systems. Without any apparent increase in complexity consider the nth order linear time invariant system with m inputs given by

$$\dot{x}(t) = Ax(t) + Bu(t) \tag{8}$$

where $A \in \mathbb{R}^{n \times n}$ and $B \in \mathbb{R}^{n \times m}$ with $1 \leq m \leq n$. Without loss of generality it can be assumed that the input distribution matrix B has full rank. Define a switching function $s : \mathbb{R}^n \to \mathbb{R}^m$ to be the linear map

$$s(x) = Sx \tag{9}$$

where $S \in \mathbb{R}^{m \times n}$ is of full rank and let $\mathcal{S}$ be the hyperplane defined by

$$\mathcal{S} = \{x \in \mathbb{R}^n \; : \; s(x) = 0\} \tag{10}$$

Suppose $u(s(x), x)$ represents a Variable Structure Control law where the changes in control strategy depend on the value of the switching function. The control law considered as a map $u : x \mapsto u(x)$ is therefore discontinuous. It is natural to explore the possibility of choosing the control action and selecting the switching strategy so that an *ideal sliding motion* takes place on the hyperplane, i.e. there exists a time t_s such that

$$Sx(t) = 0 \qquad \text{for all } t > t_s \tag{11}$$

For the purpose of this discussion, the definition of a sliding motion given above will suffice. (A more general notion and a more rigorous definition of sliding is given in DeCarlo [5].)

Suppose at time $t = t_s$ the systems states lie on the surface S and an ideal sliding motion takes place. This can be expressed mathematically as $Sx(t) = 0$ and $\dot{s}(x) = S\dot{x}(t) = 0$ for all $t \geq t_s$. Substituting for $\dot{x}(t)$ from equation 8 gives

$$S\dot{x}(t) = SAx(t) + SBu(t) = 0 \quad \text{for all } t \geq t_s \tag{12}$$

Suppose the matrix S is designed so that the square matrix SB is nonsingular (in practise this is easily accomplished since B is full rank and S is a free parameter). The *equivalent control*, written as u_{eq}, is defined to be the unique solution to the algebraic equation 12, namely

$$u_{eq}(t) = -(SB)^{-1}SAx(t) \tag{13}$$

This represents the control action which is required to maintain the states on the switching surface. The ideal sliding motion is then given by substituting the expression for the equivalent control into equation 8 which results in a free motion

$$\dot{x}(t) = \left(I_n - B(SB)^{-1}S\right) Ax(t) \quad \text{for all } t \geq t_s \text{ and } Sx(t_s) = 0 \tag{14}$$

It can be seen from equation 14 that the sliding motion is a control independent free motion which depends on the choice of sliding surface, although the precise effect is not readily apparent. A convenient way to shed light on the problem is to first transform the system into a suitable canonical form. In this form the system is decomposed into two connected subsystems, one acting in $\mathcal{R}(B)$ and the other in $\mathcal{N}(S)$. Since by assumption the rank$(B) = m$ there exists an orthogonal matrix $T \in \mathbb{R}^{n \times n}$ such that

$$TB = \begin{bmatrix} 0 \\ B_2 \end{bmatrix} \tag{15}$$

where $B_2 \in \mathbb{R}^{m \times m}$ and is nonsingular. In practice such a matrix can be found by so-called 'QR' decomposition which is an inherent feature within most matrix manipulation environments; for further details see [6]. Let $z = Tx$ and partition the new coordinates so that

$$z = \begin{bmatrix} z_1 \\ z_2 \end{bmatrix} \tag{16}$$

where $z_1 \in \mathbb{R}^{n-m}$ and $z_2 \in \mathbb{R}^m$. The nominal linear system 8 can then be written as

$$\dot{z}_1(t) = A_{11}z_1(t) + A_{12}z_2(t) \tag{17}$$

$$\dot{z}_2(t) = A_{21}z_1(t) + A_{22}z_2(t) + B_2u(t) \tag{18}$$

which is referred to as *regular form*. Equation 17 is referred to as describing the *null-space dynamics* and equation 18 as describing the *range-space dynamics*. Supposing the matrix defining the switching function (in the new coordinate system) is compatibly partitioned as

$$s(z) = S_1 z_1 + S_2 z_2 \qquad (19)$$

where $S_1 \in \mathbb{R}^{m \times (n-m)}$ and $S_2 \in \mathbb{R}^{m \times m}$. Since $SB = S_2 B_2$ it follows that a necessary and sufficient condition for the matrix SB to be nonsingular is that $\det(S_2) \neq 0$. By design assume this to be the case. During an ideal sliding motion

$$S_1 z_1(t) + S_2 z_2(t) = 0 \qquad \text{for all } t > t_s \qquad (20)$$

and therefore formally expressing $z_2(t)$ in terms of $z_1(t)$ and substituting for $z_2(t)$ in equation 17 gives

$$\dot{z}_1(t) = (A_{11} - A_{12} M) z_1(t) \qquad (21)$$

where $M \stackrel{s}{=} S_2^{-1} S_1$. In the context of designing a regulator the matrix governing the sliding motion $\bar{A}_{11} \stackrel{s}{=} A_{11} - A_{12} M$ must have stable eigenvalues. The switching surface design problem can therefore be considered to be one of choosing a state feedback matrix M to stabilize the reduced order system (A_{11}, A_{12}). Because of the special structure of the regular form it follows that the pair (A_{11}, A_{12}) is controllable if and only if (A, B) is controllable [21]. Several approaches have been proposed for the design of the feedback matrix including quadratic minimization, eigenvalue placement in a region and eigenstructure assignment methods [6,23]. It can be seen from equation 21 that S_2 has no direct effect on the dynamics of the sliding motion and acts only as a scaling factor for the switching function. The choice of S_2 is therefore somewhat arbitrary. A common choice however, which stems from the so-called 'hierarchical' design procedure[1],is to let $S_2 \stackrel{s}{=} \Lambda B_2^{-1}$ for some diagonal design matrix $\Lambda \in \mathbb{R}^{m \times m}$ which implies $SB = \Lambda$. By selecting M and S_2 the switching function in equation 19 it completely determined. The remainder of this section is devoted to the synthesis of a control law which will theoretically bring about ideal sliding. In order to investigate the stability of the resulting nonlinear system a *quadratic stability* approach will be adopted (a brief review of the salient features and some definitions are given in the Appendix).

---

[1] The hierarchical approach was the first Variable Structure Controller for multivariable systems and is described in the summary paper by DeCarlo *et al.* [5].

Of the many different multivariable Variable Control Structures which exist [5,6,11] the one that will be considered here is essentially that of Ryan and Corless [14] and may be described as a *unit vector* approach. Consider an uncertain system of the form

$$\dot{x}(t) = Ax(t) + Bu(t) + f(t, x, u) \tag{22}$$

where the function $f : \mathbb{R} \times \mathbb{R}^n \times \mathbb{R}^n \to \mathbb{R}^m$ which represents the uncertainties/nonlinearities satisfies the so-called *matching condition*, i.e.

$$f(t, x, u) = B\xi(t, x, u) \tag{23}$$

where $\xi : \mathbb{R} \times \mathbb{R}^n \times \mathbb{R}^m \to \mathbb{R}^m$ and is <u>unknown</u> but satisfies

$$\|\xi(t, x, u)\| \le k_1 \|u\| + \alpha(t, x) \tag{24}$$

where $1 > k_1 \ge 0$ is a known constants and $\alpha(\cdot)$ is a known function. After changing coordinates with respect to T defined in equation 15 it follows that

$$\dot{z}_1(t) = A_{11}z_1(t) + A_{12}z_2(t) \tag{25}$$

$$\dot{z}_2(t) = A_{21}z_1(t) + A_{22}z_2(t) + B_2(u(t) + \xi(t, z, u)) \tag{26}$$

and because T is orthogonal

$$\|\xi(t, z, u)\| \le k_1 \|u\| + \tilde{\alpha}(t, z) \tag{27}$$

for some new function $\tilde{\alpha}(\cdot)$. Suppose the matrix M defining the switching function in equation 19 has been chosen by some appropriate design procedure. Define a nonsingular linear change of coordinates by

$$T_s \stackrel{s}{=} \begin{bmatrix} I & 0 \\ S_1 & S_2 \end{bmatrix} \tag{28}$$

If new coordinates are defined as

$$\begin{bmatrix} z_1 \\ s \end{bmatrix} = T_s \begin{bmatrix} z_1 \\ z_2 \end{bmatrix} \tag{29}$$

where z_1 and z_2 are defined as in equation 16 then

$$\dot{z}_1(t) = \bar{A}_{11}z_1(t) + A_{12}S_2^{-1}s(t) \tag{30}$$

$$\dot{s}(t) = S_2\bar{A}_{21}z_1(t) + S_2\bar{A}_{22}S_2^{-1}s(t) + \Lambda(u(t) + \xi(t, z, u)) \tag{31}$$

where $\bar{A}_{11} = A_{11} - A_{12}M$, $\bar{A}_{21} = M\bar{A}_{11} + A_{21} - A_{22}M$ and $\bar{A}_{22} = MA_{12} + A_{22}$. The proposed control law comprises two components; a linear component to stabilize the nominal linear system; and a discontinuous component. Specifically

$$u(t) = u_l(t) + u_n(t) \tag{32}$$

where the linear component is given by

$$u_l(t) \stackrel{s}{=} \Lambda^{-1}\left(-S_2\bar{A}_{21}z_1(t) - (S_2\bar{A}_{22}S_2^{-1} - \Phi)s(t)\right) \tag{33}$$

where $\Phi \in \mathbb{R}^{m \times m}$ is any stable design matrix. The nonlinear component is defined to be

$$u_n(t) \stackrel{s}{=} \begin{cases} -\rho(t, z)\Lambda^{-1}\dfrac{P_2s(t)}{\|P_2s(t)\|} & \text{for } s(t) \neq 0 \\ 0 & \text{otherwise} \end{cases} \tag{34}$$

where $P_2 \in \mathbb{R}^{m \times m}$ is a symmetric positive definite matrix satisfying the Lyapunov equation

$$P_2\Phi + \Phi^T P_2 = -I \tag{35}$$

and the scalar function $\rho(t, z)$, which depends only on the magnitude of the uncertainty, is any function satisfying

$$\rho(t, z) \geq (k_1\|\Lambda\|\|u_l\| + \|\Lambda\|\tilde{\alpha}(t, z) + \gamma)/(1 - k_1\kappa(\Lambda)) \tag{36}$$

where $\gamma > 0$ is a design parameter. In this equation it is assumed that the scaling parameter has been chosen so that

$$k_1\kappa(\Lambda) < 1$$

It can easily be established that any function satisfying equation 36 also satisfies

$$\rho(t, z) \geq \|\Lambda\|\|\xi(t, z, u)\| + \gamma \tag{37}$$

and therefore $\rho(t, z)$ is greater in magnitude than the matched uncertainty occurring in equation 31. It can be verified that applying the nonlinear control law 32 to the uncertain system (43) implies

$$\dot{s}(t) = \Phi s(t) - \rho(t, x)\frac{P_2s(t)}{\|P_2s(t)\|} + \Lambda\xi(t, z, u), \quad s(t) \neq 0 \tag{38}$$

The stable design matrix Φ governs the linear free motion decay of the states of the switching function s to the origin. It is straightforward to verify that $V(s) = s^T P_2 s$ guarantees quadratic stability for the switching states s and in particular

$$\dot{V} \leq -s^T s - 2\gamma\|P_2s\| \tag{39}$$

REMARKS
- The closed-loop dynamical behavior obtained from using a Variable Structure Control law comprises two distinct types of motion. The initial phase, occurring whilst the states are being driven towards the surface, is in general affected by any matched disturbances present. Only when the states reach the surface and the sliding motion takes place does the system become insensitive to all matched uncertainty. Ideally the control system should be designed so that the initial (disturbance affected) phase is as short as possible.
- The two stage nature of the design process has clearly been demonstrated: firstly the switching surface in the state space is designed so that the reduced order sliding motion satisfies the specifications imposed by the designer; secondly the control law, discontinuous about the sliding surface, is synthesized so that the trajectories of the closed loop motion are directed towards the surface. An alternative interpretation of the latter property is that the discontinuous control action renders the sliding surface invariant and attractive;
- the control law guarantees that the switching surface is reached in finite time despite the disturbance or uncertainty and once the sliding motion is attained it is completely independent of the uncertainty.
- the linear component of the control law given in equation (33) can be expressed in the form

$$u_l(t) = u_{eq}(t) + \Lambda^{-1}\Phi s(t) \qquad (40)$$

where $u_{eq}(t)$ represent the equivalent control defined in equation 13. In this way, from the viewpoint of controller synthesis, it not necessary to compute the coordinate transformation T_s from equation 28 since the equivalent control can conveniently be expressed (in the coordinates of the regular form) as

$$u_{eq}(t) = -B_2^{-1}[\, MA_{11} + A_{21} \quad | \quad MA_{12} + A_{22} \,]\, z(t) \qquad (41)$$

- In the single-input case, the unit vector component $\frac{P_2 s}{\|P_2 s\|} = \text{sgn}(s)$ when $s \neq 0$ and the control structure 34 becomes a scaled relay structure.

As a final point, comparing the general analysis given above with the double integrator example descried at the start it follows that the parameter b in equation 1 represents matched uncertainty, hence the invariance obtained in the sliding mode dynamics given in equation 5.

So far the concept of ideal sliding motion has been introduced which required the use of a discontinuous control action. In the case of certain electric

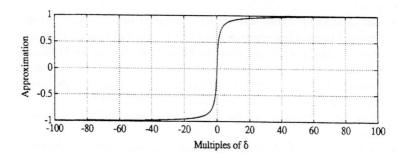

Figure 3. A differentiable approximation of the *signum* function.

motors and power converters the control action is naturally discontinuous
[15]. In other systems however, especially mechanical ones, the introduction
of a discontinuous control action will <u>not</u> induce an ideal sliding motion.
Imperfections in the process such as delays and hysteresis will conspire to
induce a high frequency motion known as *chattering*. This is characterized
by the states repeatedly crossing rather than remaining on the surface. Such
a motion is highly undesirable in practice and will result in unnecessary
wear and tear on the actuator components. It is usual in such situations to
modify the discontinuous control action so that rather than forcing the states
to lie on the sliding surface they are forced to remain within an (arbitrarily)
small boundary layer about the surface. In the literature this is often referred
to as a *pseudo sliding motion*. The total invariance properties associated
with ideal sliding will be lost. However, an arbitrarily close approximation
to ideal sliding can usually be obtained. One approach [3,17] is to replace
the discontinuous control component in equation 32 with the differentiable
approximation

$$u_n^\delta(t) = -\rho(t, x)\Lambda^{-1}\frac{P_2s(t)}{\|P_2s(t)\| + \delta} \tag{42}$$

where δ is a small positive constant. In the single-input case Figure 3
demonstrates the approximation to the relay which results from using
equation 42.

All the ideas presented so far have relied on all the internal system states
being available for use by the control law. In reality, this idealized situation is
quite rare. Usually certain states will be either impossible or prohibitively
expensive to measure. Alternatively the linear model may represent an
approximation of a distributed parameter system or else have been obtained
via system identification; as a result the internal states will have no physical
meaning. In either situation, much of the sliding mode literature is no longer

directly applicable. To circumvent this difficulty the use of Luenberger (linear) observers has been suggested to generate estimates of the unavailable internal states for use in the control law [2,19,22]. More recently Edwards and Spurgeon [7] propose the use of a sliding mode observer and prove that under certain conditions the insensitivity to matched uncertainty provided by sliding mode controllers when full state information is available is retained by the controller/observer pair. This discussion considers a robust output tracking control scheme which utilizes the full state information unit vector controller in conjunction with a sliding mode observer.

6.3. A ROBUST OUTPUT TRACKING CONTROL SCHEME

Consider an uncertain dynamical square system of the form

$$\dot{x}(t) = Ax(t) + Bu(t) + f(t, x, u)$$
$$y(t) = Cx(t) \tag{43}$$

where $x \in \mathbb{R}^n$, $u \in \mathbb{R}^m$ and $y \in \mathbb{R}^p$ with $m = p < n$. Assume that the nominal linear system (A, B, C) is known and that the input and output matrices B and C are both of full rank. The unknown function $f : \mathbb{R}_+ \times \mathbb{R}^n \times \mathbb{R}^m \to \mathbb{R}^n$ which represents any nonlinearities plus any model uncertainties in the system is assumed to have the uncertainty structure described earlier in equations 23 and 24. In addition, it is assumed that the nominal linear system (A, B, C) satisfies

(A1) the pair (A, B) is controllable
(A2) $\det(CB) \neq 0$
(A3) the invariant zeros of (A, B, C) are in $\mathbb{C}_-$ that is

$$\det \begin{bmatrix} sI - A & -B \\ C & 0 \end{bmatrix} \neq 0 \quad \text{for } s \notin \mathbb{C}_-$$

The assumption that the system is square is required since the sliding mode observer formulation requires there to be at least as many outputs as inputs. Conversely, the proposed full information control scheme requires at least as many inputs as outputs — a square system is therefore required. Assumptions A2 and A3 are synonymous with the system being minimum phase and relative degree one.

It can be assumed without loss of generality that the system is already in the so-called 'regular form' usually used in sliding mode design, so that

$$A = \begin{bmatrix} A_{11} & A_{12} \\ A_{21} & A_{22} \end{bmatrix} \quad B = \begin{bmatrix} 0 \\ B_2 \end{bmatrix} \quad C = [C_1 \quad C_2] \tag{44}$$

where $A_{11} \in \mathbb{R}^{(n-m) \times (n-m)}$, $B_2 \in \mathbb{R}^{m \times m}$ and $C_2 \in \mathbb{R}^{p \times p}$. The square matrix B_2 is nonsingular because the input distribution matrix is assumed to be of full rank. Therefore since $CB = C_2 B_2$ is nonsingular by assumption it follows that C_2 is nonsingular.

6.3.1. Controller Formulation

Consider initially the development of a tracking control law for the nominal linear system

$$\dot{x}(t) = Ax(t) + Bu(t) \tag{45}$$

assuming that full state information is available. It is convenient to assume *a priori* that the matrix pair (A, B) is in regular form as in equation 44. The control law described here is based on that described by Davies and Spurgeon [4] and utilizes an integral action methodology. Consider the introduction of additional states $x_r \in \mathbb{R}^p$ satisfying

$$\dot{x}_r(t) = r(t) - y(t) \tag{46}$$

where the differentiable signal $r(t)$ satisfies

$$\dot{r}(t) = \Gamma \left(r(t) - R \right) \tag{47}$$

with $\Gamma \in \mathbb{R}^{p \times p}$ a stable design matrix and R a constant demand vector. The design matrix Γ can be thought of as defining an ideal transient response to the input R; however it also serves to provide a differentiable 'reference signal' r for use in the controller. Augment the states with the integral action states and define

$$\tilde{x} \stackrel{s}{=} \begin{bmatrix} cx_r \\ x \end{bmatrix} \tag{48}$$

The (augmented) nominal system can then be conveniently written in the form

$$\dot{\tilde{x}}_1(t) = \tilde{A}_{11} \tilde{x}_1(t) + \tilde{A}_{12} \tilde{x}_2(t) + B_r r(t) \tag{49}$$

$$\dot{\tilde{x}}_2(t) = \tilde{A}_{21} \tilde{x}_1(t) + \tilde{A}_{22} \tilde{x}_2(t) + B_2 u(t) \tag{50}$$

where $\tilde{x}$ is partitioned as $\tilde{x}_1 \in \mathbb{R}^n$ and $\tilde{x}_2 \in \mathbb{R}^m$ and

$$\begin{bmatrix} \tilde{A}_{11} & \tilde{A}_{12} \\ \tilde{A}_{21} & \tilde{A}_{22} \end{bmatrix} \stackrel{s}{=} \left[\begin{array}{cc|c} 0 & -C_1 & -C_2 \\ 0 & A_{11} & A_{12} \\ \hline 0 & A_{21} & A_{22} \end{array} \right] \tag{51}$$

The input distribution matrix for the 'demand' signal $r(t)$ in the null space dynamics is given by

$$B_r = \begin{bmatrix} I_p \\ 0 \end{bmatrix} \tag{52}$$

The proposed controller seeks to induce a sliding motion on the surface

$$S = \{(\tilde{x}_1, \tilde{x}_2) \in \mathbb{R}^{n+p} \ : \ s(\tilde{x}, r) = 0\} \tag{53}$$

where the switching function is defined by

$$s(\tilde{x}, r) \overset{s}{=} S_1 \tilde{x}_1 + S_2 \tilde{x}_2 - S_r r \tag{54}$$

and the matrices $S_1 \in \mathbb{R}^{m \times n}$, $S_2 \in \mathbb{R}^{m \times m}$ and $S_r \in \mathbb{R}^{p \times p}$ are design parameters which govern the reduced order motion. Let $S_2 = \Lambda B_2^{-1}$ where Λ is a nonsingular diagonal design matrix which satisfies

$$k_1 \kappa(\Lambda) < 1 \tag{55}$$

Equation 54 is a slightly more complicated switching function compared to the one given for the regulator described previously, however choosing $S_r = 0$ recovers the previous case. The choice of S_r and its effects will be discussed in more detail later in the chapter. If a controller exists which induces an ideal sliding motion on S, then using an equivalent control argument as before it follows that the ideal sliding motion is given by

$$\dot{\tilde{x}}_1(t) = (\tilde{A}_{11} - \tilde{A}_{12}M)\tilde{x}_1(t) + \left(\tilde{A}_{12}S_2^{-1}S_r + B_r\right)r(t) \tag{56}$$

where $M \overset{s}{=} S_2^{-1} S_1$. It can be shown [7] that if the nominal linear system (A, B, C) is assumed to have no invariant zeros at the origin, the pair $(\tilde{A}_{11}, \tilde{A}_{12})$ is completely controllable. This is guaranteed by A3 and therefore any of the methods referred to in [6,23] can be used to design the matrix M. For the remainder of this controller description it will be assumed that this switching function design problem has been addressed successfully so that the dynamics of the sliding motion

$$\bar{A}_{11} = \tilde{A}_{11} - \tilde{A}_{12}M \tag{57}$$

satisfy the closed loop performance requirements imposed on the designer; stability is clearly a minimum requirement. Essentially the same unit vector control law described previously will now be used to induce a sliding motion on the surface S defined in equation 53. As before let the $\Phi \in \mathbb{R}^{m \times m}$ be any stable design matrix and let $\bar{P}_2$ be the unique symmetric positive definite solution to the Riccati equation

$$\bar{P}_2 \Phi + \Phi^{\mathrm{T}} \bar{P}_2 + \bar{P}_2 \bar{Q}_2 \bar{P}_2 = 0 \tag{58}$$

for some positive definite design matrix $\bar{Q}_2$. This represents a slight departure from the full information case described earlier and is required for the closed loop analysis of the combined controller/observer scheme when only output information is available. As in the regulator case, the proposed control law will be of the form

$$u = u_l(\tilde{x}, r) + v_c \qquad (59)$$

where the nonlinear component

$$v_c = \begin{cases} -\rho_c(u_l, y)\Lambda^{-1} \dfrac{\bar{P}_2 s(\tilde{x}, r)}{\|\bar{P}_2 s(\tilde{x}, r)\|} & \text{if } s(\tilde{x}, r) \neq 0 \\ 0 & \text{otherwise} \end{cases} \qquad (60)$$

and u_l represents the linear component. In this case

$$u_l(\tilde{x}, r) = L_{eq}\tilde{x} + \Lambda^{-1}\Phi s(\tilde{x}, r) + L_r r + L_{\dot{r}}\dot{r} \qquad (61)$$

where the gains are defined as

$$L_{eq} = -B_2^{-1}[c|cM\tilde{A}_{11} + \tilde{A}_{21} \mid (M\tilde{A}_{12} + \tilde{A}_{22})] \qquad (62)$$

$$L_r = -B_2^{-1}MB_r \qquad (63)$$

$$L_{\dot{r}} = B_2^{-1}S_2^{-1}S_r \qquad (64)$$

and

$$\rho_c(u_l, y) = \|\Lambda\|\frac{\left(k_{b_1}\|u_l\| + \alpha(t, y) + k_{k_1}\gamma_c\|\Lambda^{-1}\| + \gamma_o\right)}{\left(1 - k_{k_1}\kappa(\Lambda)\right)} + \gamma_c \qquad (65)$$

where γ_o and γ_c are two design scalars. The state feedback matrix L_{eq} in equation 62 can again be interpreted as the state dependent part of the equivalent control. The nonlinear gain function in equation 65 however is more complicated than the one given earlier. This is because the reliance on full-state information will be dropped.

Remarks

The seemingly unnecessarily complicated notation $\bar{P}_2$ is used in equation 58 because a Lyapunov function of the form $\tilde{x}_1^T \bar{P}_1 \tilde{x}_2 + s^T \bar{P}_2 s$ will be constructed for the closed loop system resulting from applying the control law 59–65 to the uncertain system 43.

This control law relies on all the internal states being available. An observer will thus be incorporated to provide estimates of the internal states which will be used in place of the true states in equations 62 and 60.

6.3.2. Nonlinear Observer Formulation in Regular Form

Consider the observer structure given by

$$\dot{z}(t) = Az(t) + Bu(t) - GCe(t) + Bv_o \qquad (66)$$

where z represents an estimate of the true states x, and $e \stackrel{s}{=} z - x$ is the state estimation error. The output error feedback gain matrix G is chosen so that the closed loop matrix $A_0 \stackrel{s}{=} A - GC$ is stable and has a Lyapunov matrix P satisfying both

$$PA_0 + A_0^T P = -Q \qquad (67)$$

for some positive definite design matrix Q and the structural constraint

$$PB = C^T F^T \qquad (68)$$

for some nonsingular matrix $F \in \mathbb{R}^{m \times m}$. The discontinuous vector v_o is given by

$$v_o = \begin{cases} -\rho_o(u_l, y) \dfrac{FCe}{\|FCe\|} & \text{if } Ce \neq 0 \\ 0 & \text{otherwise} \end{cases} \qquad (69)$$

where $\rho_o(u_l, y)$ is the scalar function

$$\rho_o(u_l, y) = (\rho_c(u_l, y) - \gamma_c) / \|\Lambda\| \qquad (70)$$

The observer given in equation 66 may be viewed as a Luenberger Observer with an additional nonlinear term and is similar to that proposed by Walcott and Żak [20]. It can be shown [6] that assumptions A2 and A3 are necessary and sufficient conditions for the existence of such an observer which is insensitive to matched uncertainty and induces a sliding motion on

$$\mathcal{S}_o = \{\, e \in \mathbb{R}^n \ : \ FCe = 0 \,\} \qquad (71)$$

Since the system is square and the matrix F is nonsingular, sliding on $\mathcal{S}_o$ implies the output of the observer is identical to that of the plant. Under the assumption of matched uncertainty it follows that the error system satisfies

$$\dot{e}(t) = A_0 e(t) + B\left(v_o - \xi(t, z, u)\right) \qquad (72)$$

For equations 65 and 70 and the uncertainty bounds inequality 24 it can be shown that

$$\rho_o(u_l, y) \geq \|\xi(t, z, t)\| + \gamma_o \qquad (73)$$

Choosing $V_e = e^{\mathrm{T}} P e$ and taking derivatives along the trajectories it follows that

$$\dot{V}_e = P A_0 + A_0^{\mathrm{T}} P + 2 e^{\mathrm{T}} P B \, (v_o - \xi)$$
$$\leq -e^{\mathrm{T}} Q e - 2 \gamma_o \| F C e \|$$

where both the structural constraint 68 and the inequality 73 have been used. The original formulation of Walcott and Zak required the use of symbolic manipulation to synthesize the matrices G and P which completely define the observer. More recently Edwards and Spurgeon [7] proposed an analytic solution which is described below. Let A_{22}^s be a stable design matrix and let

$$G = \begin{bmatrix} A_{12} C_2^{-1} \\ A_{22} C_2^{-1} - C_2^{-1} A_{22}^s \end{bmatrix} \tag{74}$$

It follows that

$$A_0 = \begin{bmatrix} A_{11} - A_{12} C_2^{-1} C_1 & 0 \\ A_{21} - A_{22} C_2^{-1} C_1 - C_2^{-1} A_{22}^s C_1 & C_2^{-1} A_{22}^s C_2 \end{bmatrix}$$

It can be verified [8] that the invariant zeros of the system (A, B, C) are given by $\lambda(A_{11} - A_{12} C_2^{-1} C_1)$ and therefore since $C_2^{-1} A_{22}^s C_2$ is stable, by construction A_0 is stable. If P_2 is a symmetric positive definite matrix satisfying the Lyapunov equation

$$P_2 A_{22}^s + (A_{22}^s)^{\mathrm{T}} P_2 = -Q_2 \tag{75}$$

for some symmetric positive definite design matrix Q_2 then define

$$F \stackrel{s}{=} (P_2 C_2 B_2)^{\mathrm{T}} \tag{76}$$

If P_1 is a symmetric positive definite matrix satisfying the Lyapunov equation

$$P_1(A_{11} - A_{12} C_2^{-1} C_1) + (A_{11} - A_{12} C_2^{-1} C_1)^{\mathrm{T}} P_1 + \mathcal{A}_{21}^{\mathrm{T}} P_2 Q_2^{-1} P_2 \mathcal{A}_{21} + Q_1 = 0 \tag{77}$$

where $\mathcal{A}_{21} \stackrel{s}{=} C_1(A_{11} - A_{12} C_2^{-1} C_1) + C_2(A_{22} C_2^{-1} C_1 + A_{21})$ and Q_1 is a symmetric positive definite matrix, then the symmetric positive definite matrix

$$P = \begin{bmatrix} P_1 + C_1^{\mathrm{T}} P_1 C_1 & C_1^{\mathrm{T}} P_2 C_2 \\ C_2^{\mathrm{T}} P_2 C_1 & C_2^{\mathrm{T}} P_2 C_2 \end{bmatrix} \tag{78}$$

is a Lyapunov matrix for $A_0 = A - GC$ which satisfies $PB = C^{\mathrm{T}} F^{\mathrm{T}}$. For further details see [7].

REMARKS

From the numerical computation viewpoint the Lyapunov equation given in equation 77 need not be solved and the matrix P in equation 78 need not be evaluated – they are only required for analysis purposes.

6.3.3. Closed-Loop Considerations

Although both the controller and observer strategies provide robustness against matched uncertainty it is by no means clear that the overall system, i.e. the uncertain plant with the observer/controller forming a closed loop system, will retain these properties. However, Edwards and Spurgeon [7] prove that quadratic stability of the closed loop is guaranteed. An outline of the construction of the quadratic Lyapunov function is given below:

Let $A_c = \tilde{A} - \tilde{B}L$ represent the closed loop system matrix where the pair $(\tilde{A}, \tilde{B})$ represents the original system (A, B) augmented with the integral action states and L is defined in equation 62. Also define

$$\bar{G} = \begin{bmatrix} I_n & 0 \\ S_1 & S_2 \end{bmatrix} \begin{bmatrix} -I_p \\ G \end{bmatrix} \tag{79}$$

where G is the observer gain matrix in equation 66 and S_1 and S_2 are components of the switching function 54. It should also be noted that the partitions in equation 79 are not compatible although the overall expression is correct. Edwards and Spurgeon [7] demonstrate that for an appropriate class of symmetric positive definite matrices $\bar{Q}$, the Riccati equation

$$\bar{P}A_c + A_c^T\bar{P} + \bar{P}\bar{G}CQ^{-1}C^T\bar{G}^T\bar{P} + \bar{P}\bar{Q}\bar{P} = 0 \tag{80}$$

where Q is defined in equation 67, has a symmetric positive definite solution

$$\bar{P} = \begin{bmatrix} \bar{P}_1 & 0 \\ 0 & \bar{P}_2 \end{bmatrix}$$

where $\bar{P}_1 \in \mathbb{R}^{n \times n}$ and $\bar{P}_2$ satisfies the Riccati equation 58. Finally let P_r be the unique symmetric positive definite solution to the Lyapunov equation

$$P_r\Gamma + \Gamma^T P_r = -\bar{G}_r\bar{Q}^{-1}\bar{G}_r^T - Q_r \tag{81}$$

where Γ is the stable design matrix from equation 47, the matrix

$$\bar{G}_r \overset{s}{=} [\,(B_r + \tilde{A}_1 S_2^{-1} S_r)^T \quad 0_{p \times p}\,] \tag{82}$$

and Q_r is any symmetric positive definite design matrix. Then it can be shown [7] that the function

$$
\begin{aligned}
V(\tilde{x}_1, s, e, r) = {} & \left(\tilde{x}_1 + \bar{A}_{11}^{-1}(B_r + \tilde{A}_1 S_2^{-1} S_r) R \right)^{\mathrm{T}} \\
& \times \bar{P}_1 \left(\tilde{x}_1 + \bar{A}_{11}^{-1}(B_r + \tilde{A}_1 S_2^{-1} S_r) R \right) \\
& + s^{\mathrm{T}} \bar{P}_2 s + e^{\mathrm{T}} P e + (r - R)^{\mathrm{T}} P_r (r - R) \qquad (83)
\end{aligned}
$$

is a Lyapunov function for the uncertain system and controller/observer pair combined and that quadratic stability is guaranteed. Furthermore it can be shown that sliding motions take place on the controller and observer sliding surfaces S and S_o respectively.

REMARKS
In establishing quadratic stability it is sufficient to ensure that equation 70 which relates the scalar functions $\rho_c(\cdot)$ and $\rho_o(\cdot)$ holds. There is no other dependence between the observer and the controller and a form of separation principle is seen to hold.

6.4. DESIGN AND IMPLEMENTATION ISSUES

The remainder of this chapter considers the application of the theoretical results presented earlier to a high temperature furnace control problem. The Gas Research Center at Loughborough, UK has an experimental furnace which is representative of a kiln for the firing of pottery. The furnace can be thought of as a gas filled enclosure bounded by insulating surfaces. For convenience, the effect of a load is simulated by a network of pipes covering the interior surfaces of the furnace through which water is circulated. Heat input is achieved by a burner located in one of the end walls, and the combustion products are evacuated via a flue in the roof. The burner is fed by a fuel and air supply and the flow rate of these gases can be modulated independently by means of motorized 'butterfly' valves present in the respective flow lines.

From a control systems perspective, the outputs of the 'system' are the furnace temperature (as measured by the thermocouple) and the percentage of oxygen present in the combustion products (as measured by the oxygen analyzer in the flue). In an industrial situation, the internal furnace temperature would be required to exhibit a specific time/temperature profile comprising, say, a period of low fire, a ramp to a higher temperature, a period of soak and finally a return to ambient temperature. During normal furnace operation, efficient fuel combustion is desirable. For a given mass of fuel, a theoretical mass of oxygen is required to completely oxidize the hydrocarbons (so called *stoichiometric combustion*). An inadequate air supply will result in incomplete combustion with a corresponding loss in thermal energy release. Conversely excess air, whilst guaranteeing complete combustion, will give rise to unnecessary enthalpy losses through the flue due to the increased flue flow rate. Therefore efficient combustion is ensured by minimizing the amount of excess oxygen present in the combustion products in the flue. In addition to the potential energy savings, it can be argued that accurate control of both temperature and excess oxygen is of paramount importance from the point of view of reducing pollutant emissions.

For safety reasons, such furnaces usually operate at a fixed fuel/air ratio as a result of an Electronic Ratio Controller (ERC) [9]. The fuel and air flows to the burner are measured and fed-back to the ERC which makes appropriate adjustments to the air valve position. The device also provides an additional safety feature in the form of a 'shut down' alarm which isolates the fuel supply when the furnace persistently operates 'off ratio', i.e. away from the required fuel/air ratio set-point. This framework is usually described as 'gas led' since the air flow is modulated as a result of changes in the fuel flow to maintain the appropriate fuel/air ratio necessary for efficient combustion. An additional input to the ERC exists, referred to as the 'trim signal', which allows the fuel/air ratio set-point to be adjusted. The control problem considered here however involves the manipulation of the fuel flow rate so that the measured temperature at some point in the furnace follows some specified temperature/time profile. This can be represented schematically as shown in Figure 4. Currently, temperature control in single burner furnaces is usually achieved through the use of traditional PID controllers. Such controllers are commercially available with appropriate protective logic to prevent undesirable effects such as integral wind-up. The rest of this chapter uses the theoretical results described earlier to develop a temperature controller.

The dominant mechanism of heat transfer in high temperature furnaces is radiation. The established method of mathematically modeling the radiation exchange within such an enclosure is the 'Zone Method' [10]. In this approach the surfaces and interior volumes are divided into sub-surfaces and sub-volumes small enough to be considered isothermal. The integro-differential equations governing radiation exchange are reduced to algebraic and finite-difference equations, which can be solved numerically. This technique, in conjunction with models of the dynamics of the valves, can be used to build up a detailed nonlinear simulation of the process. One such simulation [13] which has been developed and validated on extensive input output data has been used as a test-bed for evaluating the performance of the controllers prior to implementation. The nonlinear finite difference and algebraic equations that comprise the model do not lend themselves easily to

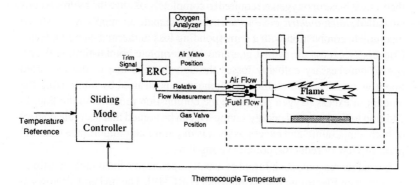

Figure 4. Schematic of proposed control scheme.

the problem of control law design. Instead a system identification approach has been used to obtain a nominal linear model. A low order linear model relating valve position to temperature, which provides good agreement with input-output data, may be represented as

$$A = \begin{bmatrix} 0 & 1 & 0 \\ 0 & 0 & 1 \\ -0.0001 & -0.0082 & -0.1029 \end{bmatrix}$$

$$B = \begin{bmatrix} 0 \\ 0 \\ 1 \end{bmatrix} \quad C = [\,0.0001 \quad 0.0022 \quad 0.0053\,] \tag{84}$$

The inherent nonlinearities present will be treated as bounded uncertainty and will be incorporated into the (true) uncertainties associated with the valve dynamics and external disturbances. As a result of the identification approach adopted, measurement of internal states of the linear model will not be possible — indeed the states will not necessarily have any physical meaning.

6.4.1 A Quantitative Measure of Controller Performance

To compare the performance of the nonlinear controller/observer pair with other control schemes, a quantitative measure of performance will be used which takes account of how accurately the output tracks the demand signal, the amount of control action used, and the degree to which high frequency components appear in the control action. If $\{y_i\},\{u_i\}$ and $\{r_i\}$ are finite sequences which represent the output, input and demand signals respectively which are sampled at a fixed rate, then formally the index comprises:

(1) *Mean Absolute Error* — representing the accuracy of the tracking performance, is defined by:

$$\bar{e} = \sum_{i=1}^{N_s} |y_i - r_i|/N_s \quad \text{where } N_s \text{ is the number of samples}$$

(2) *Mean Control Action* — reflecting the amount of control action used, is defined in the obvious way as

$$\bar{u} = \sum_{i=1}^{N_s} u_i/N_s$$

(3) *Measure of Excitation* — which assesses the degree to which high frequency components appear in the control signal, is defined as

$$u_{ex} = \sum_{i=1}^{N_s} |u_i - (u_f)_i|/N_s \quad \text{where } \{u_f\} = \mathcal{F}(\{u\})$$

and $\mathcal{F}$ is a linear low pass filter designed to remove the high frequency components from the control action (which are unrepresentative in terms of the movement of the valve).

From a sliding mode perspective the control signal u_f may be viewed as the equivalent control action necessary to maintain a sliding motion. The choice of filter is therefore somewhat arbitrary. Here a discrete linear filter with no phase distortion has been used which is similar to the one suggested by Backx [1]. Formally the filter comprises:

(1) A linear first order filter defined by

$$z_i = \alpha z_{i-1} + (1 - \alpha)x_i$$

where $\alpha = 0.2$ which generates an intermediate sequence $\{z_i\}_{i=0}^N$.
(2) Since the filtering takes place off-line it is possible to define a new sequence $\{z_i'\}_{i=0}^N$ according to $z_i' = z_{N-i}$.
(3) Applying the same first order filter to $\{z_i'\}_{i=0}^N$ results in another intermediate sequence $\{y_i'\}_{i=0}^N$ satisfying

$$y_i' = \alpha y_{i-1}' + (1 - \alpha)z_i'$$

(4) Finally define $\{y_i\}_{i=0}^N$ according to $y_i = y'_{N-i}$

The overall effect $\{x_i\} \rightarrow \{y_i\}$ is a linear (low pass) filter with no phase distortion.

6.4.2. Design of the Controller Observer Pair

This section discusses the design of a nonlinear observer/controller pair for the furnace system based on the theory described earlier. From the results described there the controller and observer can be designed independently. The observer design is discussed first. Conveniently the realization given in equation 84 is already in regular form. The invariant zeros of this linear system are given by $\{-0.3749, -0.0358\}$ and therefore all the preceding theory is valid.

6.4.3. Controller Design

Because the system is already in regular form, the augmented system from equation 51 can easily be identified to be

$$
\begin{bmatrix} \tilde{A}_{11} & \tilde{A}_{12} \\ \tilde{A}_{21} & A_{22} \end{bmatrix} =
\left[\begin{array}{ccc|c}
0 & -0.0001 & -0.0022 & -0.0053 \\
0 & 0 & 1 & 0 \\
0 & 0 & 0 & 1 \\
\hline
0 & -0.0001 & -0.0082 & -0.1029
\end{array}\right]
$$

Since (A, B, C) has stable invariant zeros, as predicted from the theory, the pair $(\tilde{A}_{11}, \tilde{A}_{12})$ is completely controllable. The poles of the ideal sliding motion dynamics have been chosen to be $\{-0.025, -0.03 \pm 0.025i\}$, which represent dynamics marginally faster than the dominant pole of the open-loop plant. The unique M such that $\lambda(\tilde{A}_{11} - \tilde{A}_{12}M) = \{-0.025, -0.03 \pm 0.025i\}$ is given by

$$ M = [-0.5372 \quad 0.0019 \quad 0.0822] $$

For single-input single-output systems no additional design freedom is provided by the scalar Λ. For simplicity it has been chosen so that $S_2 = 1$. The remaining pole, associated with the range space dynamics, has been assigned the value -0.1 by selecting the design matrix $\Phi = -0.1$. Because of the single-input single-output nature of the system there is no need to calculate the Lyapunov matrix $\bar{P}_2$ associated with the unit vector controller. The nonlinear component of the controller is given by

$$ v_c = -\rho_c(u_l, y) \, \text{sgn}(S\tilde{x} - S_r r) \tag{85} $$

and the switching function is given by

$$[\, S_1 \quad S_2 \,] = [\, -0.53717 \quad 0.00186 \quad 0.08216 \quad | \quad 1.00000 \,]$$

The hyperplane defined in section 6.2 is given by

$$\mathcal{S} = \{\tilde{x} \in \mathbb{R}^{n+p} \, : \, s(\tilde{x}, r) = 0\} \tag{86}$$

where $s(\tilde{x}, r) = S_1 \tilde{x}_1 + S_2 \tilde{x}_2 - S_r r$ which is now completely defined except for the design parameter $S_r \in \mathbb{R}^{m \times m}$. One possible choice is of course to let $S_r = 0$; alternatively S_r may be viewed as affecting the values of the integral action components since from equation 56 at steady state

$$(\tilde{A}_{11} - \tilde{A}_{12} M)\tilde{x}_1 + \left(\tilde{A}_{12} S_2^{-1} S_r + B_r\right) r = 0 \tag{87}$$

where B_r is defined in equation [52]. One possibility is therefore to choose a value of S_r so that, for the nominal system at steady state, the integral action states are zero. If $K_s \stackrel{s}{=} -B_r^{\mathrm{T}} \tilde{A}_{11}^{-1} \tilde{A}_{12}$ then provided it is nonsingular, choosing

$$S_r = S_2 K_s^{-1} B_r^{\mathrm{T}} \tilde{A}_{11}^{-1} B_r \tag{88}$$

implies $x_r = 0$. It can be shown [7] that because (A, B, C) does not have any zeros at the origin K_s is nonsingular. Substituting the appropriate values into equation 88 gives $S_r = 26.1642$. From equations 62–64 the gains are given by

$$L_{eq} = [\, 0.00000 \quad 0.00002 \quad -0.00521 \quad -0.01792 \,]$$
$$L_r = 0.53717$$
$$L_{\dot{r}} = 26.16416$$

The control law is given by

$$u(t) = L_{eq}\tilde{x}(t) + L_r r(t) + \Lambda^{-1}\Phi s(\tilde{x}, r) + L_{\dot{r}}\dot{r}(t) + v_c \tag{89}$$

where v_c is defined as in equation 85. The stable matrix Γ from equation 47 affects the closed loop performance. Here it has been chosen to tailor the step response of the nominal closed loop.

Figure 5 shows the step response for different values of Γ equally spaced in the interval $[-0.04, -0.015]$. In practice, overshoot in a thermal process is very undesirable. As a consequence, the value of -0.025 has been chosen as a compromise between the conflicting objectives of overshoot and rise-time.

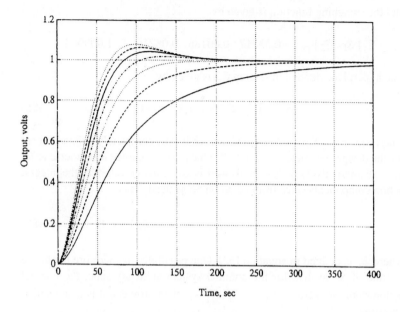

Figure 5. Selection of the stable design matrix Γ.

6.4.4. Design of an Observer

Before discussing the design procedure adopted it should be noted that because of the structure of the realization, any variation in the elements in the last row of the system matrix (which are the coefficients of the characteristic polynomial) occurs in $\mathcal{R}(B)$, and so can be considered as matched uncertainty. The observer is therefore insensitive to changes in the poles of the system. Utilizing the notation used earlier it follows that

$$\begin{bmatrix} A_{11} & A_{12} \\ A_{21} & A_{22} \end{bmatrix} = \left[\begin{array}{cc|c} 0 & 1 & 0 \\ 0 & 0 & 1 \\ \hline -0.0001 & -0.0082 & -0.1029 \end{array} \right]$$

and

$$[\, C_1 \quad C_2 \,] = [\, 0.0001 \quad 0.0022 \mid 0.0053 \,]$$

If the stable design matrix $A_{22}^s = -0.2$ it follows immediately from equation 74 that

$$G = \begin{bmatrix} A_{12} C_2^{-1} \\ A_{22} C_2^{-1} - C_2^{-1} A_{22}^s \end{bmatrix} = \begin{bmatrix} 0 \\ 188.8498 \\ 18.3328 \end{bmatrix}$$

Again because the system is single-input single-output there is no need to compute the sliding surface matrix F because the discontinuous component simplifies to

$$v_o = -\rho_o(t, y)\, \text{sgn}(Ce)$$

The observer is thus complete except for the nonlinear scalar gain function given in equation 70. This is discussed in detail in the following subsection.

6.4.5. Design of the Nonlinear Gain Function

Formally, the nonlinear gains are related to the magnitudes of the uncertainty bounds, which in this situation are not available. In this case, the approach that has been adopted is to let

$$\rho_o(u_l, y) = r_1|y| + r_2|u_l(\cdot)| + \gamma_o \tag{90}$$

where $u_l(\cdot)$ represents the linear component of the control action and the positive scalars r_1, r_2 and γ_o are to be chosen empirically. An estimate of the design constants in the nonlinear gain function can be obtained by considering the range of allowable inputs. For the system under consideration, the input is restricted to the interval $[0, 4]$ Volts. Consequently it is reasonable to require that the inequality $r_1|y| < 4$ is satisfied. In practice, as a result of the chosen temperature profile to be tracked, it was found that y is of the order 1. In this way, a sensible upper bound on r_1 can be obtained. Figure 6 represents simulation tests with the nonlinear gain

$$\rho_o(u_l, y) = r_1|y| \quad \text{for } r_1 \in \{0, \tfrac{1}{2}, 1, 1\tfrac{1}{2}, 2\}$$

This demonstrates the increase in performance obtained as a result of increasing r_1 and hence the nonlinear component of the control action. The final value of r_1 was chosen as a trade-off between the tracking error and the control effort required. The remaining parameters were chosen in a similar way.

6.4.6. Furnace Simulations

Under typical operating conditions, the controller is required to take the furnace from one operating temperature to another, along a specified trajectory. A typical temperature demand signal, which will subsequently be used for the simulations and plant trials, is given in Figure 7. It can be seen that this function is piecewise linear and there exist a finite number of

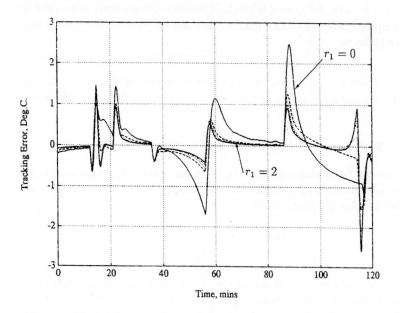

Figure 6. Simulations with different nonlinear components.

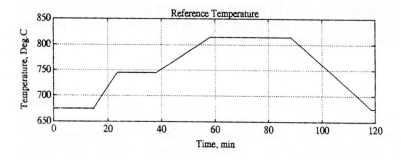

Figure 7. Typical temperature reference signal.

points at which it is not differentiable. To ensure that the differentiable signal $r(t)$ tracks the profile at points that comprise the ramps between the different steady state operating points, it is necessary to modify equation 47. Consider the (noncausal) differential equation

$$\dot{r}(t) = \Gamma \left(r(t) - R(t - \tfrac{1}{\Gamma}) \right) \tag{91}$$

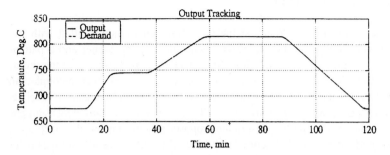

Figure 8. Nonlinear simulation under sliding mode control.

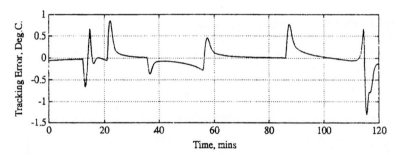

Figure 9. Tracking error using the sliding mode controller.

It is assumed that the intervals which define the piece-wise linear components of the reference are large enough compared to the time constant of Γ so that 'steady state' occurs. On each interval $\ddot{R}(t) = 0$ and therefore $\dot{R}(t) = \alpha$ for some fixed scalar. If $e_r \overset{s}{=} r(t) - R(t)$ then from equation 91 it follows immediately that

$$\dot{e}_r(t) = \Gamma r(t) - \Gamma R(t - \tfrac{1}{\Gamma}) - \alpha$$
$$= \Gamma e_r(t) + \Gamma R(t) - \Gamma R(t - \tfrac{1}{\Gamma}) - \alpha$$
$$= \Gamma e_r(t)$$

Therefore $e_r(t) \rightarrow 0$ and the solution $r(t)$ to equation 91 follows the profile in Figure 7 asymptotically.

Figure 8 represents the response of the furnace simulation under nonlinear control with the tracking error given in Figure 9. This performance is very good even on the parts of the demand profile that comprise transients between steady state operating points, for which asymptotic tracking is not guaranteed theoretically.

6.5. PLANT TRIALS

The experimental furnace at the Gas Research Center is fitted with five thermocouples — one on the end wall opposite the burner and a pair on each side wall. This section initially considers the performance of the sliding mode observer/controller pair with regard to the end-wall thermocouple; this represents the 'nominal' situation as the linear model in equation 84 relates valve position as input to the end-wall thermocouple measurement as output. To examine the effectiveness of the proposed scheme it is sensible to compare its performance with that of a 'well-tuned' PID, the controller settings being found via an auto tuning algorithm.

6.5.1. End Wall Thermocouple Trials

Figure 10 represents a typical furnace response under PID control. The corresponding control action is given in Figure 11.

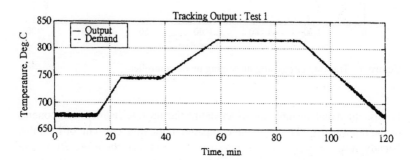

Figure 10. Experimental furnace under PID control.

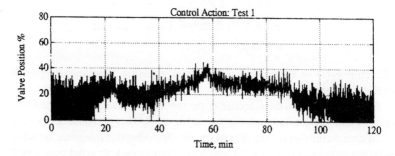

Figure 11. PID control action.

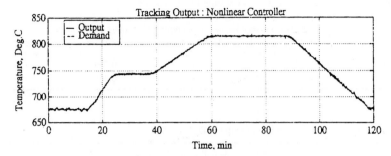

Figure 12. Response of end-wall thermocouple under nonlinear control.

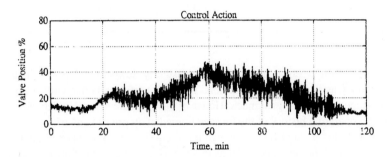

Figure 13. Nonlinear control action.

The controller/observer pair was implemented for trial purposes using a portable PC. An input/output card provided the interface between the PC and the analogue thermocouple and valve positioner signals. The controller/observer algorithm was coded in Turbo Pascal, compiled and run as an executable file. The temperature signal was sampled, the output voltage updated and transmitted 10 times a second. This left sufficient CPU time to 'sample' the observer at a rate of 100 times as second. The controller sampling rate is comparable with that of the PID which has a sampling rate of 8 Hz. Figure 12 gives the response of the furnace, under nonlinear control, to the reference signal used earlier. Here the average tracking error is less than 1° which represents a level of accuracy probably greater than the measurement precision. The control excitation, as shown in Figure 13, is higher than that of the simulation but is considerably less than that of the PID.

Table 1 given below presents the performance indices for trials of the nonlinear controller whilst Table 2 shows the performance indices of PID controller trials conducted on the same furnace during early 1993.

Table 1. Performance indices for sliding mode scheme.

| Test | Performance Indices for Nonlinear Controller | | | |
	Absolute Error	Mean Control	Excitation	Overall
1	0.893	1.085	0.169	2.148
2	0.901	1.048	0.205	2.155

Table 2. Performance indices for the PID controllers.

| Test | Performance Indices for PID Trials | | | |
	Absolute Error	Mean Control	Excitation	Overall
1	1.333	1.022	0.287	2.643
2	1.433	1.001	0.313	2.757
3	1.440	0.971	0.301	2.712

The absolute error measure is consistently lower for the nonlinear controller as is the valve excitation. The PID performs better with regard to the mean control signal index. However this measure is heavily dependent on the operating history of the furnace immediately prior to any trial. This is demonstrated by the PID tests which represent three back-to-back trials; the mean control measure decreases as less fuel is used, because of the retention of heat by the furnace. The overall measure indicates that the nonlinear controller/observer pair is performing at least as well as a commercial PID controller for the nominal test. Robustness issues will now be addressed.

6.5.2. Robustness — Side Wall Thermocouple Trials

In order to explore the robustness of the nonlinear controller, trials were undertaken where the temperature signal was supplied from one of the side-wall thermocouples. This represents a seriously perturbed system from the point of view of control since it is known that under normal operating conditions while the end wall thermocouple operates in the region of 650–820°, the side-wall temperatures are in the range 450–650°. It is therefore not possible to use the original demand signal since the side walls cannot attain such elevated temperatures. Figure 14 demonstrates the results of a typical side-wall test using the same nonlinear controller/observer pair but a modified demand. Results of a comparable PID test are presented in Figure 15; as in the case of the nonlinear controller the original parameters have been left unaltered. These test results, summarised in Table 3, indicate

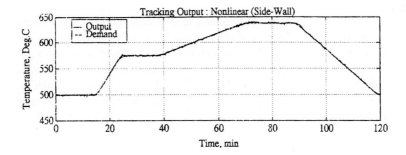

Figure 14. Response of side-wall thermocouple under nonlinear control.

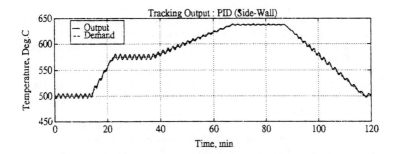

Figure 15. Response of side-wall thermocouple under PID control.

that the nonlinear controller performs better from the point of view of robustness. In an industrial setting, returning of the PID controller would be necessary. Here, this is circumvented by the use of a single robust nonlinear controller.

Table 3. Performance indices for robustness trials.

| Controller | Performance Indices for Robustness Trials | | | |
	Absolute Error	Mean Control	Excitation	Overall
VSC	0.703	0.985	0.105	1.793
PID	2.138	1.060	0.218	3.414

6.6. CONCLUDING REMARKS

This chapter has attempted to present a tutorial description of some recent advances in the area of nonlinear sliding mode control. The theoretical results have been applied to the practical problem of temperature control in a gas fired furnace. The practical choice of parameters for the proposed controller/observer pair has been described. It should be noted that although the case study presented is single-input-single-output, the theoretical approach is multivariable.

References

1. Backx, T., 1987, *PhD thesis*, Dept. of Electrical Engineering, Eindhoven University of Technology.
2. Breinl, W. and Leitmann, G., 1987, *App. Math. Comput.* **22**, 65–87.
3. Burton, J.A. and Zinober, A.S.I., 1986, *Int. J. Sys. Sci.*, **17**, 876–885.
4. Davies, R. and Spurgeon, S.K., 1993, In *Proc. IFAC World Congress, Sydney*, **vol. 10**, 43–46.
5. DeCarlo, R.A. Żak, S.H. and Matthews, G.P., 1988, *Proc. IEEE.*, **76**, 212–232.
6. Edwards, C. and Spurgeon, S.K., 1996, *Sliding Mode Control: Theory and Applications* (Taylor and Francis, 1998).
7. Edwards, C. and Spurgeon, S.K., 1996, *Int. J. Contr.*, **64**, 967–983.
8. El-Ghezawi, O.M.E., Billings, S.A. and Zinober, A.S.I., 1983, *Proc. IEE, Part D.*, **130**, 1–5.
9. Hammond, P.S., 1992, In *Proc. International Gas Research Conference*, Florida.
10. Hottel, H.C. and Cohen, E.S., 1958, *J. A. I. Ch. E.*, **4**, 3–33.
11. Hung, J.Y., Gao, W. and Hung, J.C., 1993, *IEEE Trans. Ind. Electron.*, **40**, 2–22.
12. Itkis, U., 1976, *Control Systems of Variable Structure*, Wiley, New York.
13. Palmer, M.R., 1989, *Proc. of the International Gas Research Conference*, Tokyo.
14. Ryan, E.P. and Corless, M., 1984, *IMA J. Math. Contr. Information*, **1**, 223–242.
15. H. Sira-Ramírez, H. and Lischinsky-Arenas, P., 1991, *Int. J. Contr.*, **54**, 111–134.
16. Slotine, J.J.E. and Li, W., 1991, *Applied Nonlinear Control*, Prentice-Hall.
17. Spurgeon, S.K. and Davies, R., 1993, *Int. J. Contr.*, **57**, 1107–1023.
18. Utkin, V.I., 1977, *IEEE Trans. Automat. Contr.*, **22**, 212–222.
19. Utkin, V.I., 1992, *Sliding Modes in Control Optimization*, Springer-Verlag.
20. Walcott, B.L. and Żak, S.H., 1986, In *Proc. CDC, Athens*, pp.961–966.
21. Young, K.-K.D., Kokotović, P.V. and Utkin, V.I., 1977, *IEEE Trans. Automat. Contr.*, **22**, 931–937.
22. Young, K.-K.D. and Kwatny, H.G., 1982, *Automatica*, **18**, 385–400.
23. Zinober, A.S.I., 1994, In A.S.I. Zinober, editor, *Variable Structure and Lyapunov Control*, Springer-Verlag, chap. 1, pp.1–22.

APPENDIX: QUADRATIC STABILITY

Consider the nonlinear system given by

$$\dot{x}(t) = F(x, t) \qquad (92)$$

where $x \in \mathbb{R}^n$ and $F : \mathbb{R}^n \times \mathbb{R}_+ \to \mathbb{R}^n$ with $F(0, t) = 0$ i.e. the system has an equilibrium point at the origin. If a generalized energy function — known as a Lyapunov function — can be found, which is nonzero except at an equilibrium point and whose total time derivative decreases along the system trajectories, then the equilibrium point is stable (for rigorous details see for example [16]). The key point is that this approach obviates the need to obtain an analytical solution to the nonlinear differential equation when assessing its stability properties. Unfortunately no systematic way exists to synthesis Lyapunov functions for nonlinear systems. Define a scalar function $V : \mathbb{R}^n \to \mathbb{R}$ to be the *quadratic form* given by

$$V(x) = x^{\mathrm{T}} P x \qquad (93)$$

where $P \in \mathbb{R}^{n \times n}$ is some symmetric positive definite matrix. By construction the function is nonzero except at the origin.

The (origin) of the system (92) is said to be *quadratically stable* if there exists a symmetric positive definite matrix $Q \in \mathbb{R}^{n \times n}$ such that the total time derivative satisfies

$$\dot{V}(x) = 2x^{\mathrm{T}} P f(x, t) \leq -x^{\mathrm{T}} Q x \qquad (94)$$

This implies $\|x(t)\| < e^{-\alpha t}$ where $\alpha = \lambda_{min}(P^{-1} Q)$ and hence the origin is asymptotically stable. If $F(x, t) = Ax(t)$ then it is well known that A has stable eigenvalues if and only if given any symmetric positive definite matrix Q there exists an unique symmetric positive definite matrix P satisfying the *Lyapunov Equation*

$$P A + A^{\mathrm{T}} P = -Q \qquad (95)$$

Consequently any stable linear system is quadratically stable. A symmetric positive definite matrix P satisfying (95) will be referred to as a *Lyapunov Matrix* for the matrix A.

7 MECHATRONIC SYSTEMS TECHNIQUES IN TELEOPERATED CONTROL OF HEAVY-DUTY HYDRAULIC MACHINES

N. SEPEHRI[1] and P.D. LAWRENCE[2]

[1]*Experimental Robotics and Teleoperation Laboratory, Department of Mechanical Engineering, University of Manitoba, Winnipeg, Manitoba, Canada R3T-5V6*
[2]*Robotics and Control Laboratory, Department of Electrical Engineering, University of British Columbia, Vancouver, British Columbia, Canada V6T-1Z5*

The human interface for controlling heavy-duty hydraulic equipment, for example excavators, has not changed significantly over many years. Traditionally two joysticks, each with two degrees of freedom, have been used by operators to control the joint rates of the machine arms independently. Motivated by potential benefits such as reduction in learning/adaptation time, productivity gains, reduction in fatigue and enhanced safety, a project has been initiated to develop, implement and examine new control interfaces in teleoperated control of such heavy-duty hydraulic machines [1]. This chapter presents a demonstration of benefits and performance evaluation of the recently developed mechatronic systems and techniques through direct application to real-world machines or through computer simulations. The necessary implementation to retrofit these machines into teleoperated systems is also outlined. The goal is to reduce the required level of operator skill, while maintaining machine performance and safety, by controlling the machine in task coordinates. In a task coordinate system, the machine implement direction and speed are specified by the deflection of a 3D joystick handle, and the appropriate control signals to the actuators are determined by a computer control system.

7.1. INTRODUCTION

The development of telerobotic applications for the operation of heavy-duty hydraulic machines has recently received particular attention. Agriculture, forest, mining and construction industries are currently utilizing many heavy-duty machines, such as excavators or log loaders. The environments in which these machines operate are unstructured and potentially hazardous. Consequently, the operation and control of these machines are very much operator dependent and require significant visual feedback, judgment and skill. The operator must remain constantly alert in order to accomplish the work efficiently and at the same time protect his/her safety and that of others. Our recent research effort has been to develop the means for converting these machines into teleoperated control systems.

Teleoperation is the control of a machine by a human through some form of barrier in a way that naturally extends the human capabilities of sensing and actuation beyond that barrier. The terminology in the area has been reviewed by Sheridan [2]. In the context of a powerful manipulator controlled by a human, the barrier is a force barrier — the human is not capable of lifting large loads. There could also be a distance barrier if the manipulator is remotely operated. Since the human is involved in the control loop during at least some phases of tasks, human factors studies have also been important elements in the research in this area [3–6].

The application of teleoperating systems becomes more challenging when applied to heavy-duty, hydraulic machines. The actuation system is complex, coupled and difficult to control. They are designed to do heavy-duty tasks such as pick and placing loads, or excavation. A Caterpillar 215B log loader (Figure 1) and a Unimate Hydraulic manipulator (Figure 2) have been used in this investigation. These machines incorporate many aspects of typical robotic systems and are the basis for most heavy duty hydraulic machines. Thus, the analyses and development reported in this chapter can be employed in other similar systems.

7.2. DESCRIPTION OF THE CANDIDATE MACHINES

7.2.1. Caterpillar 215B Machine

Machine description

The Caterpillar 215B excavator-based machine (Figure 1) is a mobile three-degree-of-freedom manipulator with an additional moveable implement. The implement is a grapple for holding and handling objects such as trees. It can be a bucket to dig and carry loads.

The upper structure of the machine rotates on the carriage by a "swing" hydraulic motor through a gear train. 'Boom' and 'stick' are the two other links which, together with the 'swing' serve to position the implement. Boom and stick are operated through hydraulic cylinders. The cylinders and the swing motor are activated by means of pressure and flow through the main valves.

Figure 1. Caterpillar excavator-based log loader.

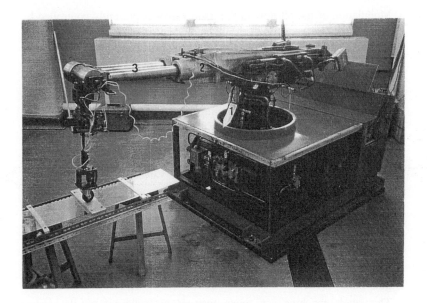

Figure 2. Unimate MKII industrial hydraulic manipulator.

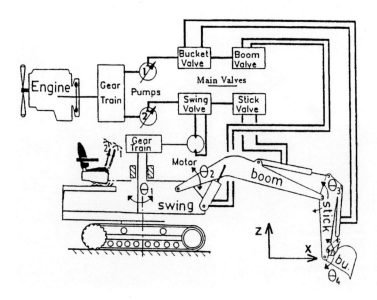

Figure 3. Schematic of hydraulic drive unit.

Modulation of the oil flow in the main valves is presently controlled by the pilot oil pressure through the manually operated pilot control valves. Joysticks 1 and 2 (Figure 3) are used to control these valves; forward or backward movement or side to side movement of these two joysticks provide individual control to the link motion. Figure 3 also shows the main hydraulic circuit. The output of the engine is used to turn three hydraulic pumps; two axial piston variable displacement pumps and one gear pump (the latter not shown in the diagram). The output flow from the axial piston pumps is used to operate the hydraulic cylinders as well as the swing motor. Oil from pump 2, for example, goes through swing and stick main valves to the tank. Movement of the stick link, controlled by the stick main valve, cannot be achieved if the swing main valve is fully open. If the latter is partly open, the stick can operate but not at maximum rate. In some models, the motion of the boom and the stick are also coupled by the cross-over valves (not shown in the figure). This will cause a faster movement of one when the other is at low speed. When the total (summing) of the pressures in the implement circuits become high enough, the axial piston pumps reduce their outputs to prevent engine stall.

The whole machine can move forward or backward on its tracked undercarriage. An additional pair of valves (which should be superimposed onto the diagram shown in Figure 3 perform the mobility as well as the steering. They are controlled by two foot pedals and an additional joystick. Due to the mobility, theoretically the machine has infinite work-space.

Retrofitting the machine

In order to retrofit the excavator-based machine into a teleoperated system, the original manually operated pilot valves were bypassed by a new set of servovalves. The servovalves used (Figure 4) were typical commercially available two-stage electrohydraulic servovalves with direct feedback [7]. They were originally designed to control the flow-rate in proportion to the input current. An attachment was designed and built (marked A in Figure 4a) which converts the above flow control servovalves into pressure control ones. This part connects the two outlet ports 1 and 2 to the tank through a fixed orifice, a_0. Figure 4b shows the new servovalve orifice arrangement around one of the main valves. A certain given voltage to the servovalve opens orifices a_1 and a_3. Port 2 is connected to the return tank through two orifices a_3 and a_0 which releases the pressure on one side of the main valve spool. The orifice arrangement on the side of port 1 is sketched as in Figure 4c which is similar to the one in the original pilot valves. The pressure P1 can be shown to be as below:

$$P_1 = \frac{\left(\frac{a_1}{a_0}\right)^2 P_s + P_e}{\left(1 + \left(\frac{a_1}{a_0}\right)^2\right)} \tag{1}$$

a_0 is the fixed orifice area; a_1 is the variable orifice area controlled by the servovalve through an input voltage. Depending on the value of a_1, P_1 can change from the tank pressure, P_e, to almost the pilot main line pressure, P_s. The pilot servovalves provide the interface between the computer and the hydraulics part. They take voltages coming from a digital-to-analog (D/A) card and produce pilot pressures proportional to the voltages. These pressures in turn are applied across the spools of the main valves.

The retrofitted excavator-based machine is located at the University of British Columbia. The machine is instrumented with a flexible VME-bus based computer system for machine control system development and human factors study. The computer interface system has been designed to provide a real-time control using a 20 MHz Inmos-T800 transputer-based processor board and a UNIX-based SPARC-1E VME-bus board. The machine hydraulic system has been modified to include electrohydraulic pilot valves. The maximum control input signal to each pilot valve is 6 volts. In order to keep track of the position of the machine, resolvers (with the resolution of

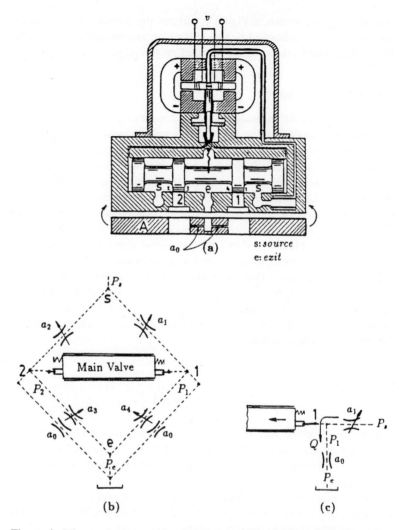

Figure 4. Pilot servovalve used to retrofit the excavator-based machine.

0.1 deg) are used in conjunction with the resolver-to-digital (R/D) card. The resolvers are located at the pivot points of the links and their output gives the angular position of each link. An on-line low-pass 5-point differentiator software package is used to calculate the joint velocities from the joint angle measurements. Pressure transducers (0 to 5000 psi with accuracy of 0.5%) are installed to measure the pressures at the cylinders. Also, pressure transducers (0 to 500 psi with accuracy of 0.5%) are used to measure the pilot pressures produced by the new servovalves. Their output is proportional to the pressure, and is between 0 to 5 volts. The analog-to-digital (A/D) card converts these signals into digital values.

7.2.2. Unimate MKII-2000 Manipulator

The Unimate MKII industrial hydraulic manipulator, shown in Figure 2, is based on a spherical geometry, with links 1, 2, and 3 enabling left-right, up-down and in-out motions, respectively. The actuators are supplied by a 7000 ± 500 kPa constant pressure source. Spool valves are closed-center types and have parabolic orifice area-to-displacement functions, and a 6% deadband. Position feedback is obtained using digital encoders with resolutions 0.028 deg, 0.028 deg, and 0.105 mm on links 1, 2, and 3, respectively. The control signals are the current inputs to the servovalves. The force sensor has a stiffness of 350 kN/m, and could measure forces along the vertical Cartesian direction with ± 10N of noise. No attempt was made to reduce the level of noise in the readings, because of the intended industrial application of the controller, where such idealizations may not be feasible. The environment with which the manipulator interacts, is a simply-supported aluminum plate. By varying the support spacing, the environment stiffness could be varied (7 kN/m to 53 kN/m).

The Unimate MK-II robot which was built in 1972 is located at the University of Manitoba. When received in 1991, all the original control hardware was removed with the intention of establishing a research platform for testing different control algorithms. The manner in which the computer has been integrated into this system is best described with reference to Figure 5. Two three-degree-of-freedom joysticks are used to control the five axes of the manipulator. Joystick 1 with left-right, up-down and forward-backward motions is used to control rotate-left-right (link one), up-down (link two) and in-out (link three) movements of the manipulator, respectively. Up-down motion of joystick 2 controls the pitch, while rotation about the shaft controls the roll of the gripper. The voltage outputs from both joysticks are read by the computer through an A/D card. Depending on the strategy chosen, analog voltages are sent to the servovalve current amplifiers. The D/A card allows the computer to output a valve current proportional voltage (± 5 volts to produce ± 600 mA current). The valves regulate the flows to and from the actuators in proportion to the input currents.

The encoders generate digital-like gray code signals. The gray code translation to binary requires a cascade of exclusive OR (XOR) operation at the bit level. A circuit board has been devised for this operation to reduce the computational time overhead. The digital data from the encoders to the

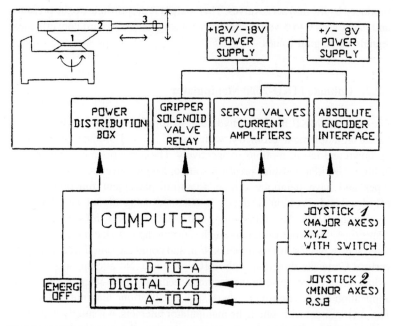

Figure 5. Unimate hardware configuration of sensory and control system.

computer are multiplexed. The interface board first selects the encoder signals (one at a time), translates them digitally into binary and presents them to a digital I/O card. Referring to Figure 5, the arrow pointing to the absolute encoder interface indicates digital input for encoder selection while the arrow pointing to the digital I/O card indicates digital output from this card to the computer. Velocity measurements are obtained using a 20-point linear regression on digital encoder readings, collected at ≈ 200 Hz.

7.3. HYDRAULIC FUNCTIONS

A typical hydraulically actuated system consists of a hydraulic actuator, connected to a valve through flexible hoses. The valve monitors the flow to and from the motor. A pump provides sufficient flow at sufficient pressure to provide the motion. Depending on the type of valve used, the pump could be a constant pressure (pressure-compensated), or constant flow (torque-limited) type. Valves, connecting hoses and actuators are the main components. Other components such as check valves and relief valves are for the machine safety. In the following, the basics of these main components are described.

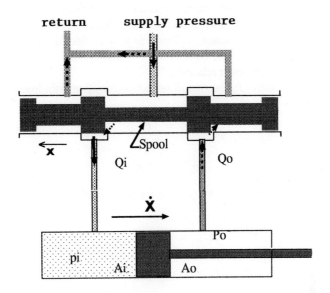

Figure 6. Hydraulic closed-center valve.

7.3.1. Valves

Valves are devices that control the fluid flow. The most widely used valve is the sliding valve with spool type construction. Two types of valves, commonly used in manipulators, are discussed here:

Closed-center spool valves

Figure 6 shows a typical critical center (zero-lapped) symmetric valve. This type of valve is usually used along with a constant supply pressure system. In a multi-link manipulator, this arrangement allows each valve to function independently which simplifies the control scheme. Most indoor hydraulic manipulators such as the Unimate robots use the above valves.

The inputs required to model a valve are the spool displacement (x), the supply pressure (P_s), the return pressure (P_e) and the line pressures $(P_i$ and $P_o)$. The governing nonlinear equations are written using the continuous equations for orifices;

$$Q_i = kwx\sqrt{P_s - P_i} \qquad (2)$$

$$Q_0 = kwx\sqrt{P_o - P_e} \qquad (3)$$

where k is the orifice coefficient and w is the area gradient [7]. In the above equations, the orifice areas were assumed to be linearly proportional to the normalized spool displacement, x. This was done to simplify the derivation of the transfer function. The model used in the simulation study, however, applies the actual nonlinear relations.

The pump pressure (P_s) is normally selected according to the load pressure (P_l) as follows [7]:

$$P_s = \frac{3}{2}P_l \tag{4}$$

where

$$P_l = P_i - P_0 \tag{5}$$

Further, for a symmetric, critical center valve, the following equations can be shown to exist

$$P_s = P_i + P_0 \tag{6}$$

$$Q_l = kwx\sqrt{\frac{P_s - P_l}{2}} \tag{7}$$

where

$$Q_l = \frac{(Q_i + Q_0)}{2} \tag{8}$$

The nonlinear equation 7 can be linearized using a Taylor series expression around an operating point and neglecting the higher order terms.

$$\Delta Q_l = \frac{\partial Q_l}{\partial x}\Delta x + \frac{\partial Q_l}{\partial P_l}\Delta P_l \tag{9}$$

where

$$\frac{\partial Q_l}{\partial x} = kw\sqrt{\frac{P_s - P_l}{2}} = K_x = \text{ flow gain} \tag{10}$$

and

$$-\frac{\partial Q_l}{\partial P_l} = kwx\frac{1}{4\sqrt{\frac{P_s - P_l}{2}}} = k_p = \text{ flow-pressure gain} \tag{11}$$

Q_l is usually measured around the neutral (critical) point where $x = 0$ and $P_l = 0$. Thus,

$$Q_l = K_x x - K_p P_l \tag{12}$$

where

$$K_x = kw\frac{P_s}{2}$$
$$K_p = 0 \tag{13}$$

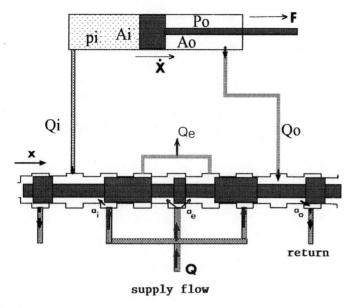

Figure 7. Hydraulic open-center valve.

K_p contributes to the damping ratio. Therefore for critical center valves, it makes sense to carry out the analysis around the neutral point where damping is minimum.

Open-center spool valves

Figure 7 shows a schematic of a typical open-center flow control valve. When the spool is in neutral position, the flow passes through the valve and returns to the tank, or the next valve. As the spool moves to the left or the right, the flow to the tank is restricted and the supply pressure increases until it exceeds the load pressure, and, at the same time the passage to the load opens. The flow is distributed to the load and the tank, depending on the orifice arrangement, the spool displacement and the load. In general the equations governing the flow pattern are:

$$Q_i = kwx\sqrt{P_s - P_l} \tag{14}$$

$$Q_0 = kwx\sqrt{P_0 - P_e} \tag{15}$$

$$Q_e = Q - Q_i = kw(1 - x)\sqrt{P_s - P_e} \tag{16}$$

where Q_e is the remaining pump flow, back to the tank having an exit pressure P_e.

Open-centre valves are normally used for outdoor, mobile heavy-duty manipulators (see Figure 1), along with a variable-flow torque-limited pump system. In such a system, the highest load is seen by the system changes the pump output flow to meet the engine output requirement. For symmetric open-centre valves, the same linearized equation as in closed-center valves can be used for qualitative and basic analysis. However, compared to closed-center valves, the pump pressure, in open-center valves, is variable depending on the load and the displacement of the valve spool.

7.3.2. Actuators and Connecting Hoses

The line pressures (P_i and P_o) and the cylinder linear acceleration ($\ddot{X}$) are related as below:

$$F = (P_i A_i - P_o A_o) = J\ddot{X} + f_d \ddot{X} \tag{17}$$

where f_d is the equivalent viscous damping coefficient and J is the inertia driven by the actuator. A_i and A_o are the piston areas. Note that for the case of a rotary actuator, such as the excavator cab swing motor, $\ddot{X}$ is taken as the angular acceleration and $A_i = A_o$. Using lumped parameter theory [7], the first order transmission line model is

$$\dot{P}_i \frac{V_i}{\beta} = Q_i - A_i \dot{X} \tag{18}$$

$$\dot{P}_o \frac{V_o}{\beta} = A_o \dot{X} - Q_o \tag{19}$$

V_i and V_o are the volumes of fluid trapped at the sides of the actuator and is the effective bulk modulus of the hydraulic circuit.

7.3.3. Frequency-Domain Analysis

Knowing the following relations,

$$Q_i = K_x x - K_p P_i \tag{20}$$

$$Q_0 = K_x x + K_p P_o \tag{21}$$

and using a Laplace transformation, the input-output relationship between the actuator displacement, X, and the main valve spool displacement, x, in an explicit form is written as follows:

$$\frac{X(s)}{x(s)} = \frac{K_x(A_i + A_o)}{(JC)s^3 + (JK_p + f_d C)s^2 + (f_d K_p + A_i^2 + A_o^2)s} \tag{22}$$

where $C = \frac{V_i}{\beta} \approx \frac{V_o}{\beta}$ is the hydraulic compliance [7]. Equation 22 is a third order system and has the following form:

$$\frac{X(s)}{x(s)} = \frac{\kappa}{s\left[\frac{s^2}{\omega_n^2} + \left(\frac{2\zeta}{\omega_n}\right)s + 1\right]} \tag{23}$$

where

$$\kappa = \frac{K_x(A_i + A_o)}{K_p f_d + (A_i^2 + A_o^2)} \tag{24}$$

$$\omega_n = \sqrt{\frac{K_p f_d + (A_i^2 + A_o^2)}{JC}} \tag{25}$$

$$\zeta = \frac{K_p J + f_d C}{2\sqrt{JC(K_p f_d + (A_i^2 + A_o^2))}} \tag{26}$$

The inclusion of $\frac{1}{s}$ in equation 23 means that the actuator is pushed by fluid flow. The oscillatory roots are lying in the second order part and they are functions of ζ (damping) and ω_n (hydraulic natural frequency). The increase in compliance, C, also decreases damping which together with low natural frequency may cause an undesirable response. The effect of damping in stability is quite important. The sources of damping excluding J, C, A_i and A_o are load damping which appears in the form of K_p in the equations, structural damping included in f_d, dry friction and leakage. Leakage effect can be included in the linearized model by changing K_p to $(K_p + K_l)$, where K_l is the leakage coefficient.

For the heavy-duty machines, the fluid flow is controlled through a two-stage valve arrangement. The first stage is the pilot servovalve system which, given a control signal, provides enough pressure to move the main valve spool. For such an arrangement, a first order relation between the input (control signal to the pilot valve) and the output (main valve spool displacement) proved to be accurate enough [1]. Therefore, the response of the whole system can be described as a fourth order system;

$$\frac{X(s)}{x(s)} = \frac{\kappa}{s(\tau s + I)\left[\frac{s^2}{\omega_n^2} + \left(\frac{2\zeta}{\omega_n}\right)s + 1\right]} \tag{27}$$

where τ is the response time of the main valve dynamics.

7.4. MACHINE SIMULATORS

Simulations are often used as a tool to help answer questions experimentation cannot answer. There are a few benefits for utilizing simulations as a tool: (i) to avoid equipment damage when trying new designs, (ii) to accelerate the engineering evaluation of many parameter settings and, (iii) to carry out human factors evaluation of a new system. This section outlines the development of the machine models, with computational modifications towards real-time simulations.

Most heavy-duty manipulators use hydraulic actuation with a complex arrangement of coupled hoses and pressure sources. The actuation dynamics are highly nonlinear and interconnected, with a mixture of fast and slow changing state variables, causing the system to be mathematically stiff. Using integration routines specifically developed for stiff systems could be useful in terms of computing time, however, they are prone to failure or becoming inefficient when they reach a point of discontinuity [8] as occurs quite frequently in the machine under investigation. The model, explained here, handles the complex and coupled actuators used in Caterpillar 215B machine. We showed that efficient simulation was possible by combining the best properties of the transient and steady-state solutions in the modeling phase. It is evident that since humans are included in the control loop, it is important to have real-time response to users' inputs. Therefore, the aim is to ultimately develop a real-time simulator which is implemented as a 'flight-type' simulator.

In this study, the Caterpillar 215B machine was considered to be a large-scale system [9] consisting of two subsystems — a complex interconnected hydraulic actuation system and the linkages. A partitioned hierarchical modeling technique was developed for fast and accurate simulations of the hydraulics [10]. Analytical, steady-state and numerical techniques were combined, using large-scale systems analysis [11]. Dynamics of the linkages was studied using the Lagrangian approach; it led to an explicit and compact mathematical model which allowed for further computational arrangement and customization towards fast simulations [12].

The primary simulation study of the hydraulic dynamics showed that the state variables of the hydraulic circuit could be divided into two groups; fast-response states and slow-response states. An example of slow-response states is the pressures in the flexible hoses connecting the main valves to the actuators, which have a considerable effect on the dynamics and the vibrational behavior of the structure [13]. An element involving a restriction linking two rigid pipes (e.g., in the main valve system) is a typical example of a component with a high-frequency response [14].

To achieve efficient simulations, the equations were studied and grouped with respect to different criteria. The main stiffness was found to originate from the connection between the valves and localized in one part. The differential equations responsible for the stiffness were then converted to steady-state equations. This allowed a larger integration step size and a reduced number of states to be integrated.

Figure 8 shows the whole system in the form of subsystems and their input/output connections. Referring to Figure 8, the main valve system (level 2) with its corresponding cross-overs and the connections is considered as one subsystem which is activated by the pilot subsystem (level 1), and connected to the links (level 4) by means of flexible hoses (level 3). Each subsystem has its own set of dynamic characteristics. This arrangement is the result of the study based on the initial formulation as well as experimental observations. The subsystems are outlined as follows.

The input to the pilot system is the voltage applied to the servovalves. The relationship between the input voltage and the spool displacement can be modeled as a first-order differential equation. The servovalves operate independently and their inputs are external. Thus, an analytical solution was written for each valve separately. The connection between the valves and the actuators are localized in subsystem 2. Lumped parameter theory for the transmission lines (level 3) and the multibody dynamics method mentioned previously (level 4) were used to model these two subsystems. The resulting differential equations contain only states with similar response times, which dictates the time-interval for numerical integration. The Lagrangian approach [15] was used for the dynamics of the linkages. Although the configuration of hydraulic manipulators, due to the attachment of the cylindrical actuators, implies that they belong to the class of closed-kinematics chain manipulators, however, compared with the main structure, the actuator linkage dynamics is not significant and thus the simpler serial link dynamic analysis was used here.

Figure 9 shows the simulator constructed for this study. It is comprised of a silicon Graphics 3130 workstation with a VME bus analog/digital interface card to which were connected two spring-centered PQ [Model M220] joysticks. The joysticks are located on each side of a stationary chair, facing the display. The display replicated the three dimensional graphical view through the front window of the excavator cab.

7.5. CONTROL

7.5.1. Joint Control

The controls found on excavator-based machines today consist of two, two-degree-of-freedom joysticks with a one-to-one mapping between the

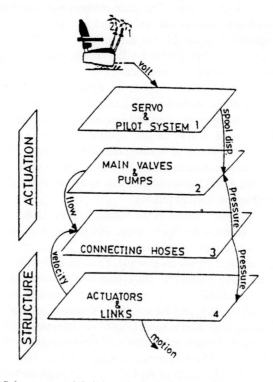

Figure 8. Subsystems and their input/output arrangement.

joystick motions and the links. Each link of the machine is controlled by a specific motion of one of the joysticks; 'in-out' or 'left-right'. A motion of the joystick corresponds to a velocity command. This is called a joint-speed control. This method of control is easy to implement as it does not require any external computation. However, it is accompanied by the drawback that this mode requires much coordination on the part of the operator. For instance, suppose the operator desires to bring the implement towards the cab, at a constant height and a constant angle with respect to the ground such as in scraping the ground. This necessitates the simultaneous coordination of three links, i.e., three simultaneous motions of the joysticks. The efficiency and accuracy is then directly dependent on the human operator's judgment and skill.

This section describes the recently developed methods for closed-loop control of each joint through the computer. Each link in a heavy-duty machine, moves through the application of a torque, based on a pressure differential

Figure 9. Real-time excavator-based machine simulator.

in the hydraulic actuator. The pressure differential is achieved by means of a spool displacement in a main valve that controls the fluid flow into and out of the corresponding actuator. In the retrofitted machine, the main valve spools are pushed by special electrohydraulic pilot valves (see Section 7.2.1). The control signals then activate the pilot valves.

Feedforward control of joint speeds. Based on the model developed for the hydraulics, an algorithm has been developed which is applied in conjunction with the closed-loop components [16]. It is basically a feedforward load-compensating scheme [17] which uses the measured hydraulic line pressures along with an appropriate portion of the hydraulics model to control the joint speeds. The scheme incorporates a logic system which, according to the states relevant to the machine task, decides on the minimum portion of the hydraulics model to be used. Inclusion of this algorithm eliminates most of the effects of loading and coupled actuation, which allows easier implementation of the closed-loop part. A knowledge of some hydraulic parameters is the only requirement of this scheme. No knowledge about the structure or loading is necessary. The algorithm has been tested on the machine and has been proven to successfully compensate for the effects of load variation in joint speed of heavy-duty machines within the range of their operating speed and accuracy. The basic idea of the algorithm is explained in the following.

Referring to Figure 7, the three orifices a_i, a_o and a_e, in the main open-centre valve are controlled by a single spool displacement, x. Line pressures P_i and P_o are shown in Figure 7 and are related to the flow-rates as was described in Section 7.3.1. The algorithm presented here assumes no knowledge about the structure or load. It uses the information contained in the hydraulic line pressures which reflects the dynamics of the structure, load or interaction forces. Knowing that the joysticks operate in the velocity mode, the measured values of P_i and P_o are used and the spool displacement (and consequently the appropriate input voltage) is determined such that the fluid-flow provided by the hydraulic system, aims at bringing the link to the desired velocity, i.e.,

$$Q_i^d = A_i \dot{X}^d \tag{28}$$

$$Q_o^d = A_o \dot{X}^d \tag{29}$$

where the superscript 'd' denotes a desired quantity. This way, only the above two equations along with Equations 14 to 16 are needed to be solved for the required spool displacement to achieve the desired velocity, $\dot{X}^d$. Since the three orifices, a_i, a_o and a_e are related to a single spool displacement, depending on whether the pump to cylinder flow (Equations 14 and 16) or cylinder to tank flow (Equation 15) is intended to be controlled, two different solutions may be obtained for the spool displacement. This is because the number of equations are more than the number of unknowns and thus a unique solution is not guaranteed. A criterion is thus required to choose the appropriate set of equations.

Referring to Figure 7, when the acting force F is in the opposite direction to the desired motion, $\dot{X}^d$, the drain side (cylinder-to-tank orifice area) is controlling the load and therefore the flow from the cylinder to the tank needs to be regulated. Equation 15 is therefore used to determine the spool displacement, x:

$$x = \frac{Q_o^d}{[kw\sqrt{(P_o^m - P_e^m)}]} \tag{30}$$

where Q_o^d is the desired flow from the actuator back to the tank from Equation 29. P_o^m and P_e^m are the measured return line and tank pressures.

When the desired velocity, $\dot{X}^d$, and the acting force, F, are in the same direction. The orifices on the pump side control the motion, Equations 14 and 16 are thus rewritten as below:

$$Q - Q_i^d = kw(1 - x)\sqrt{P_s - P_e^m} \tag{31}$$

$$Q_i^d = kwx\sqrt{P_s - P_i^m} \tag{32}$$

where Q_i^d is the desired flow into the actuator from Equation 28. Unknowns are P_s and x. Pressures P_i^m and P_e^m are known through measurement.

Figure 10 shows the velocity control of the bucket when it moves out-in with a constant velocity towards its joint limits. As is seen, after a certain delay, the bucket reaches the desired out velocity (10 deg/s). The delay is due to the time required for the pressure to be built up in the pilot system to move the main valve spool. Also, the spool has to travel a bit (valve deadband) before any orifice can open. The same delay is also seen when the bucket changes its direction. The valve spool has to move to the other direction and has to pass through its deadband. Figure 11 shows the variation of the net force on the actuator piston, during the experiment; when it is positive, it compensates for gravity. Thus, the 'pump side' should control the motion.

Figure 12 shows how the measured line pressures (shown as P_i^m and P_o^m) change during the motion. Pump 1 is connected to the bucket main valve and is thus active; it senses one of the line pressures. During bucket-out motion, the line with the pressure shown as P_i^m is connected to the pump; it is connected to the reservoir in bucket-in operation and thus is noted as P_o^m.

Fuzzy logic position control

One characteristic of fuzzy logic is its applicability to control systems with model uncertainty or even with unknown models. This section demonstrates, how fuzzy logic can be directly applied to the control of nonideal heavy-duty hydraulic functions. The goal was to examine the behavior of fuzzy controller during step and ramp inputs looking for improved tracking ability.

In spite of many articles on the successful application of fuzzy control to electrically driven actuators [18,19], the application of fuzzy control to hydraulic manipulators is sparse and can only be found in a few research papers. Zhao and Virvalo [20] applied a fuzzy rule evaluation method to adjust the position, velocity and acceleration gains of a linear state controller. Chou and Lu [21] developed a more direct fuzzy control for a class of indoor hydraulic robots.

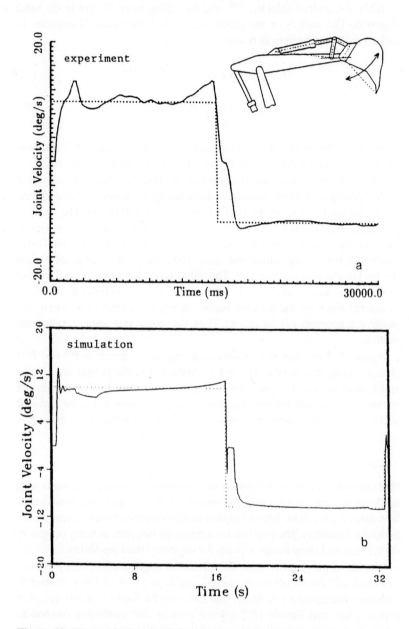

Figure 10. Bucket out-in velocity control; (a) experiment; (b) simulation.

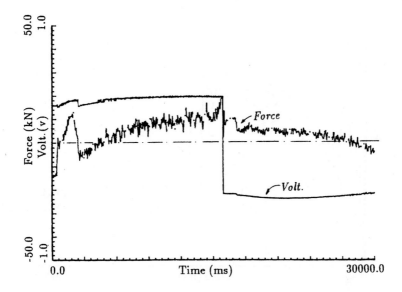

Figure 11. Actuator force in bucket out-in velocity control operation.

The basic fuzzy controller used here is a simple two-input controller. The inputs are the scaled angular position error, G_a (Desired Angle–Actual Angle), and the scaled angular velocity error, G_v (Desired Velocity–Actual Velocity). G_a and G_v are the scaling factors that put the error values within the controller's universe of discourse. The fuzzy controller then fuzzifies these input quantities and applies a series of rules to these variables which are now converted to linguistic notions. It then outputs a control action based on a weighted result of the rule evaluation. Rules are normally written based on experience, common sense, observations and understanding how the system responds and the attributes it must contain.

Five degrees of position errors and three degrees of velocity errors are the inputs to the fuzzy controller having three degrees of control output (Figure 13). The membership functions that represent the values of the input and output degree of truth for each set of linguistic variables are simple symmetric triangular functions. They have sufficient overlap to produce smooth controls. The 5-output regions of control signals are also shown in Figure 13; MIN being the centroid of the negative large control action and MAX the centroid of the positive large control action. The width of the zero output membership function is chosen to be narrow to produce large control values for the position errors close to zero. This is done in view

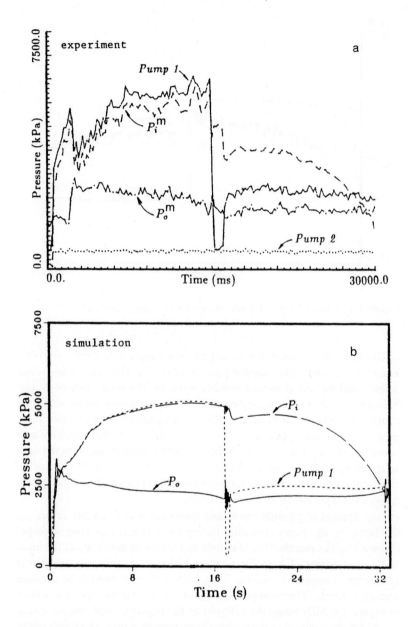

Figure 12. Pressures in bucket out-in motion; (a) experiment; (b) simulation.

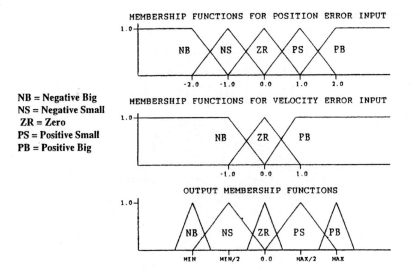

NB = Negative Big
NS = Negative Small
ZR = Zero
PS = Positive Small
PB = Positive Big

Figure 13. Membership functions.

of the existence of nonidealities inherent to heavy-duty manipulators such as friction, leakage, slow valve response, pump pressure changes and valve deadbands. Also from Figure 13, the output membership functions contain narrow large (positive or negative) regions. This allows large control signals (i.e., full spool travel) to be mostly effective in the presence of absolutely large errors; they are less active when the position error has membership in both small and large error zones.

For any set of scaled position and velocity errors, an output is produced based on the rules summarized in a matrix notation shown in Figure 14. The membership functions and rules have been designed and adjusted based on simulations, testing and a knowledge of the characteristics and response for the class of hydraulic robots under investigation.

The reasoning method used in this work is based on Mamdani's 'Minimum Operation Rule (MOR)' and the 'Center Of Area (COA)' defuzzification technique [22,23]. As a demonstration, consider the case depicted in Figure 15. The position error of 0.75 (after being scaled) falls into two regions of its fuzzy set-zero (ZR) and positive small (PS). The degree of truth in each statement is determined by membership functions. For example the degree of truth in the statement of "position error being a small error" is 75% which is graphically displayed in Figure 15. The velocity error of −1.5, on the other hand, falls within one single region, i.e., negative large (NB).

		Position Error				
		NB	NS	ZR	PS	PB
Velocity Error	NB	NB	NB	NS	**PS**	PB
	ZR	NB	NS	ZR	PS	PB
	PB	**NB**	**NS**	PS	PB	PB

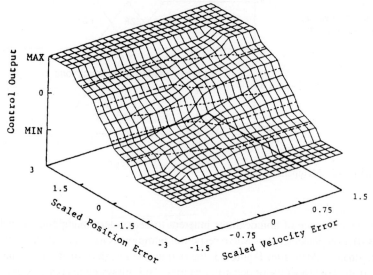

Figure 14. Fuzzy Associative Memories and Fuzzy control surface.

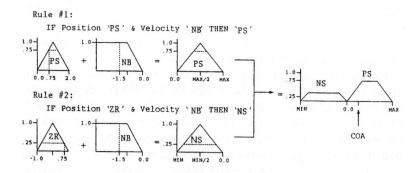

Figure 15. Example of fuzzy reasoning and rule defuzzification.

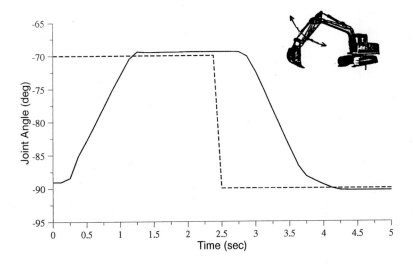

Figure 16. Typical step input response of fuzzy PD controller (experiment).

This combination causes two rules to fire. Referring to the rule table shown in Figure 14, the first rule implies sending a positive small (PS) control signal, for a small positive position error and a large negative velocity error. The second rule demands a negative small (NS) control output for a zero position error and a large negative velocity error. These two rules have, somehow, to be combined to form a single output. A logical AND is first applied to each rule which is equivalent to taking the lowest of the two membership functions ('MOR'). The output control areas for both rules are then combined, and the center of the resultant control area ('COA') provides a crisp control action.

Two sets of experiments were performed. The first set was to test the positioning ability of the controller. The second test was to examine the tracking ability. Figure 16 shows the step input response of the fuzzy controller to link three (stick) of the log loader, shown in Figure 1. The response for well tuned gains was fast with an acceptable steady-state error and with no overshoot or oscillations about the set-point. The steady-state error is due to the main valve deadband, dry friction as well as the open-center characteristic of the main valve system. This is due to the fact that the effect of the velocity error on the fuzzy control action is considered in union with the position error. This nonlinear characteristic of the controller is of great advantage.

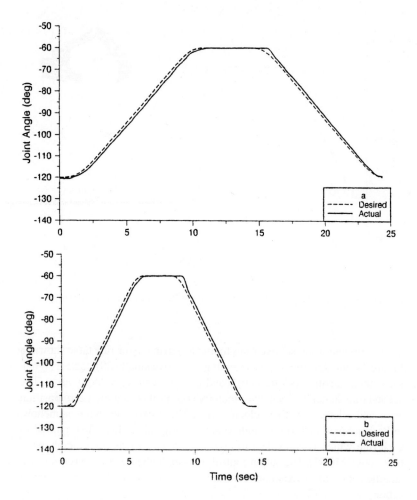

Figure 17. Tracking response of fuzzy PD controller (experiment): (a) low speed; (b) increased speed.

The same control gains as in the step input were then used and kept fixed in the trajectory tracking experiment. The trajectory profile (Figure 17a) included a gradual change in the required velocity at the start and at the end. A constant error is seen when following the ramp. This is expected and is due primarily to the nature of the fuzzy PD controller as applied to hydraulics. Zero position and velocity errors would result in the controller closing the valve and stopping the motion; thus, some error must be present

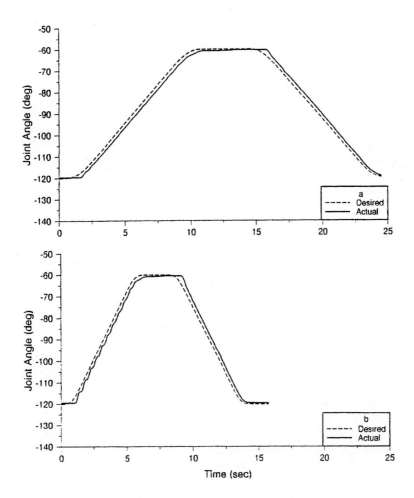

Figure 18. Tracking response of non-fuzzy PD controller (experiment): (a) low speed; (b) increased speed.

in the response. The response of the fuzzy controller for a similar trajectory but at a faster speed is shown in Figure 17b.

Similar tracking tests were performed this time with a well-tuned non-fuzzy PD controller. The results are summarized in Figure 18. The PD controller responded well for slow tracking; with the same gains it failed to produce a satisfactory response at a higher speed. The non-fuzzy PD controller showed substantial oscillations during the stick-out motion. Reducing the gains to improve the response, produced a very large steady-state error. The fuzzy controller displayed characteristics which were difficult to achieve with a conventional linear PD controller.

Nonlinear PI position control

A good position controller is required to meet a number of stringent criteria. It must (i) have excellent tracking and regulating abilities, (ii) respond quickly to changes in set-point, in spite of stiction and hydraulic flow deadband, (iii) reverse directions quickly without overshoot, and (iv) retain the above properties for both large and small changes in set-point. To meet these criteria, the following control law has been developed:

$$u_t = K_P e_t + K_I I_t \tag{33}$$

where K_p and K_I are the proportional and integral gains, respectively. The control signal u, at time t, is obtained using a nonlinear PI control law, acting on the generalized joint angle θ_t, and its associated error e_t. The nonlinearity enters from the calculation of the error integral I_t.

$$I_t = (I_{t-\Delta t} + e_t \Delta t + K_a \ddot{\theta}_t^{\text{braking}} \Delta t) \frac{\alpha}{\alpha + \dot{e}_t^2} \tag{34}$$

Δt is the sampling time interval.

The first modification is to iteratively multiply the accumulated error integral by the velocity error factor $\frac{\alpha}{\alpha + \dot{e}^2}$. This factor is unity at zero velocity error ($\dot{e} = 0$), and approaches to zero as the velocity error goes to infinity. The rate at which this factor goes to zero depends on the value of the scalar parameter α. This method improves the controller's regulating and tracking abilities over a wide range of inputs, and eliminates the problems of actuator saturation and integral windup [24,25].

The second modification to the integral is to include the braking acceleration $\ddot{\theta}^{\text{braking}}$, weighted by gain K_a, in the integration, where

$$\ddot{\theta}^{\text{braking}} = \begin{cases} \ddot{\theta}^d & \text{if } \ddot{\theta}^d \dot{\theta}^d < 0 \\ 0 & \text{if } \ddot{\theta}^d \dot{\theta}^d \geq 0 \end{cases} \tag{35}$$

where the superscript 'd' refers to a desried (set-point) quantitiy. This modification is made in order to reduce the overshoot observed with large changes in set-point velocity. Integrating the joint set-point braking acceleration is equivalent to adding a braking rate feedforward term contained in the derivative of a PID controller, but not including the rate feedback term. The set-point braking acceleration term allows the necessary control signals to be added in an on-off manner without sudden signal discontinuities. Deceleration is therefore enhanced without affecting the response time or stability.

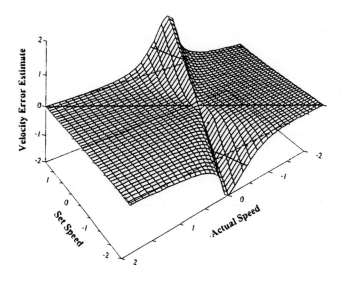

Figure 19. Velocity error estimate plot ($\beta = 50$).

One further improvement was made in order to eliminate some of the problems observed due to stiction and actuator deadband. The improvement was based on the realization that, when the manipulator is not following the set-point because of stiction, the greatest of the three usual state error signals available (position error, error integral and velocity error) is the velocity error. Therefore, in order to detect the occurrence of stiction, the following function was developed:

$$\dot{e}^{\text{striction}} = (\dot{\theta}^d - \dot{\theta}) \frac{(\dot{\theta}^d)^2}{(\dot{\theta}^d)^2 + \beta(\dot{\theta})^2} \tag{36}$$

where β is a constant. The estimate of velocity error due to stiction, $\dot{e}^{\text{striction}}$, that is generated by this function is shown in Figure 19. The estimate is close to zero unless there is a velocity error at low actual velocities, i.e., the estimate is large when the set-point is moving and the manipulator is not. For high values of β, this function sharply discriminates between velocity errors due to sticking and those due to other causes.

When the velocity error estimate detects the occurrence of stiction, the integral control signal can simply be augmented at the onset of sticking in order to overcome it. The control signal augmentation takes the form of

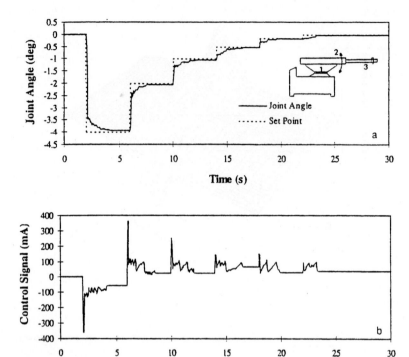

Figure 20. Experimental evaluation of the NPI controller; step inputs.

a one-time addition to the error integral;

$$I_{t-\Delta t} = \begin{cases} \bar{I}\,\mathrm{sgn}(\dot{e}^{\text{striction}}) & |\dot{e}^{\text{striction}}| > \bar{\dot{e}} \\ I_{t-\Delta t} & |\dot{e}^{\text{striction}}| < \bar{\dot{e}} \end{cases} \qquad (37)$$

Referring to Equation 37, if the velocity error due to stiction, $\dot{e}^{\text{striction}}$, exceeds a threshold $\bar{\dot{e}}$, the integral $I_{t-\Delta t}$ is set to $\bar{I}$. In short, when sticking is detected, the integral is immediately brought to the level necessary to overcome it, instead of waiting for the position error to accumulate.

Figures 20 and 21 present an illustration of the set-point regulating and tracking, resulted from the control action on link two of the Unimate manipulator (see Figure 2) with up-down motion. The control provided good tracking in step inputs (Figure 20) as well as a 1 Hz sine wave of diminishing amplitude (Figure 21). Figure 21 also shows that the controller still has difficulty following the sine wave once the peak-to-peak amplitude

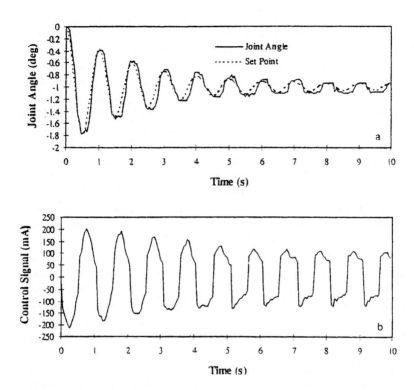

Figure 21. Experimental evaluation of the NPI controller; sine wave test.

drops to ≈ 0.2 deg. This amplitude corresponds to about 7 times the encoder resolution. Tracking such small changes accurately while retaining stability in the face of large changes in set- point is a very difficult task because the effects of stiction and actuator deadband are most apparent in this region of operation.

7.5.2. Endpoint Control

Coordinated-motion control

With reference to Figure 3, the standard method of operating a 4 degree-of-freedom excavator utilizes a pair of spring-centered joysticks with two-degrees-of-freedom (joint axes) each. The deflection of a joystick about one of its joint axes away from the neutral position, results in an increase in velocity of the controlled joint. This is called joint-space velocity-control, and has a very desirable advantage in that if the hand controls are released, all motion stops. Also, the gain of the system defined as "G_v = (slave joint velocity)/(joystick deflection from zero)" is adjustable. The disadvantage of this system is that it requires an operator to learn what joint speeds are necessary at each moment in time in order to obtain the desired direction and speed of the implement. For example for the implement to move straight across the ground at the same level, the operator must first cause the boom to rise until the implement is directly under the stick joint, and then the operator must cause the boom to fall all while maintaining a constant stick motion. The relation between the motions that one desires to make in a Cartesian workspace, and the joint angle velocities required to achieve those motions is given by the Inverse Jacobian [26]. The Inverse Jacobian is often a complex nonlinear function of joint angles. It is this mapping that the operator must learn in order to do the tasks.

The solution proposed here is to use Cartesian-space velocity-control instead. Such a control system has been studied for space manipulators [27]. In the work reported here however, the operator specifies endpoint velocities in a rotating Cartesian space via joystick positions. An on-board computer determines the necessary velocities of all the joints required to attain the specified Cartesian endpoint velocity [28]. It then decides on the control inputs that are required for each link, and at each moment in time (see Section 7.5.1), in order to obtain the desired direction, speed and/or position of the implement. We have termed this control "coordinated-motion control" — a term that appears to be meaningful to operators.

Figure 22 illustrates different steps involved to control an excavator-based machine in the proposed coordinated motion control mode. First, the joysticks are sampled to get the user commands in terms of $V_x = \frac{\Delta(\text{Radius})}{\text{sec}}$, $V_z = \frac{\Delta(\text{Height})}{\text{sec}}$, $V_y = \frac{\Delta(\text{Arc})}{\text{sec}}$. Note that the right-left motion of the implement is actually done along an arc with its center at the swing pivot point. Multiplying these numbers by the sampling period, and adding the result to the previous desired position/orientation of the implement, the current desired position/orientation is obtained. By solving the inverse kinematics of the machine, the desired position/orientation in terms of the individual joint angles and joint velocities is determined. Finally, passing these values to a control algorithm, appropriate voltages are sent to the electrically activated pilot valves which were previously operated manually.

Human factors studies

The coordinated-motion control system completely conforms to the spirit of the ISO 1503-1977 standard which states "*Movements in control elements*

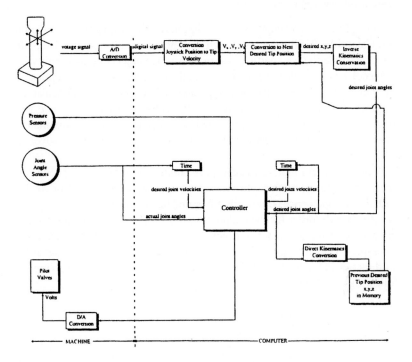

Figure 22. Coordinated-motion control of excavator-based machine.

and the intended changes in considered object should be logically coordinated." In other words if the "considered object" is a bucket, when the hand controller is pushed forward, one would expect the bucket to move in the same direction (i.e. forward). Similarly when an upwards motion is desired of the bucket, one would naturally expect a corresponding upwards motion of the hand controller to cause that motion.

The development of the coordinated-motion control for excavator-based machines has been through many distinct phases. In phase one, the real-time graphical simulator, described in Section 7.4, was utilized and human factors studies were performed to assess potential benefits. Phase two consisted of implementation on a Caterpillar 215B machine, configured as a log loader, and a human factors evaluation using two joysticks for coordinated control. Phase three consisted of improvements in hardware and software, and human factors studies using only a single joystick for coordinated control.

Simulation Experiments

The first phase of experiments was based on the real-time graphical simulation of a Caterpillar 215B excavator machine. There were 20 right-handed male subjects (students) who participated in the study. Subjects sat in a chair with a joystick on each side of the chair, facing the display. The display replicated the view through the front window of the machine cab. Three dimensional graphical information was obtained by the subjects from perspective and other cues. The subjects were presented with nine tasks over approximately a two hour period. The details of the task and complete results are given in [29].

In an industrial robot, Cartesian space commands are usually specified in terms of world coordinates. In an excavator, however, the operator rotates with the cab and hence commands are issued in the operator's rotating frame of reference. In order for the endpoint to move in the same 3D direction as the handle of the joystick is currently deflected, the gain "G_v = (endpoint workspace velocity)/(joystick deflection from zero)" must be the same for all directions of deflection of the joystick. This was achieved by solving the following equations

$$x = \Delta x + x_{\text{old}}$$

$$z = \Delta z + z_{\text{old}}$$

$$d = \sqrt{x^2 + z^2}$$

$$\theta_1 = \frac{\Delta y}{x} + \theta_{1\text{old}} \tag{38}$$

$$\theta_2 = \tan^{-1}\left(\frac{z}{x}\right) + \cos^{-1}\left[\frac{d^2 + b^2 + s^2}{2bd}\right]$$

$$\theta_3 = \cos^{-1}\left[\frac{d^2 - b^2 - s^2}{2bs}\right]$$

Δx, Δy, Δz are the increments in workspace endpoint position implied by the current joystick position; x, y, z are the new coordinates of the endpoint; x_{old} and z_{old} are the previous coordinates of the endpoint; θ_1, θ_2 and θ_3 are the current joint angles of the swing, boom and stick, respectively; $\theta_{1\text{old}}$ is the previous value of θ_1; finally, 'b' and 's' are the length of the boom and stick respectively.

The net effect of these equations is that the operator observes the direction of a line between the current endpoint and a new desired location for the endpoint. The operator pushes the joystick in that direction. As the cab rotates, the operator rotates with it and adjusts the joystick to keep pushing it "toward" the desired target location. Consequently, control of movement

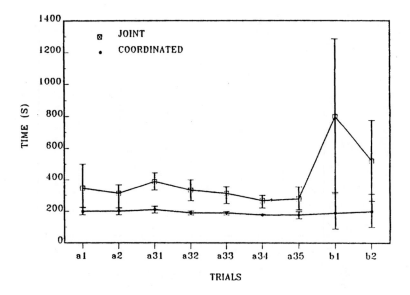

Figure 23. Mean time to complete each trial (simulation).

no longer requires learning a new coordination task. The results of these experiments on novices (see Figure 23) indicated that the novices were more productive using coordinated control than using the standard controls and that the advantage increased as the tasks were made more difficult (tasks 'b_1' and 'b_2').

Machine Experiments

During the actual experiment, two hand controls were located in place of the original Caterpillar pilot controls. The right hand-controller was a PQ (model M220) three function control (x, y and z). The left hand-controller (also PQ), for implement orientation, had pitch, roll and yaw functions. The yaw function was used to rotate the grapple in either direction. The trigger switch on the right hand-controller closed the grapple and the one on the left hand-controller opened it.

The log loader was located in a field surrounded by three sets of three numbered log cradles (see Figure 1). The subjects' task consisted of moving these logs from cradle to cradle in a designated sequence. The two different control methods (joint control and coordinated control) were compared as a function of time. The human factors experiments were set up to study 10

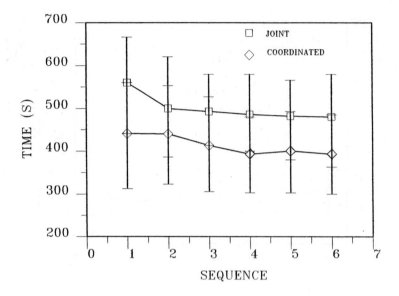

Figure 24. Average sequence time to complete each trial (experiment).

novices and 6 experienced operators in two separate studies. The novices had about 4 hours exposure to both systems and the experienced operators had about 2–3 hours per day total on the machine over a 5-day period for both joint mode (on which baseline data was only taken during day 1) and coordinated-motion mode. The details of the human factors studies are presented in [30]. The results from the human factors tests for novices are shown in Figure 24. It is clear that novices benefited immediately from coordinated control. For experienced operators, however, it took some time to unlearn existing control patterns. The experienced operators' coordinated control performance improved with time. Two operators actually matched their joint control times at the end of day 5 (8–10 hours in coordinated mode).

Simulation versus Experiment

The general perception of coordinated-motion control among the experienced operators tested is that it would reduce learning time for novices and that the experienced operators could adapt to it given sufficient time. The operators in general were quite positive about its potential. One lesson that we learned during the course of these experiments was that achieving results comparable to the real-time simulations studies was not as easily achieved on the actual machine. Although not reported here, some informal studies during phase one were carried out with experienced operators. These studies predicted that coordinated control performance could match or exceed joint control performance. In reality, it took several more years to actually accomplish that goal on the machine itself. One of the reasons for that is the initial simulation only included kinematics. Thus, the dynamics of the grapple the dynamics of pilot and main valves, the interaction between the machine and the operator which disturbed hand control action, and the interaction between logs and the implement were not simulated at that time. However, when the simulator was improved by adding hydraulic and mechanical dynamics, it became valuable in understanding problems, modeling control ideas and determining control system gains.

Inverse kinematics solution of machines with redundancy

One aspect of implementing the coordinated-motion control is the solution to the inverse kinematics for machines with redundant linkages. Figure 25 shows a class of machines containing redundant linkages. Presently, the human operator decides on the individual joint motion in order to provide the desired implement trajectory and at the same time utilizes the redundancy to ensure smooth joint velocities and efficient utilization of power and linkages. Numerous inverse kinematics solutions have been developed to utilize the redundancy in manipulators. In teleoperation, any solution to the inverse kinematics should, to a great extent, resemble human operator performance, i.e., the joint motions are to be determined as close to what an experienced operator produces, in the conventional joint control mode. We suggest that, using artificial neural networks, it is possible to learn the human performance during the joint-motion control and utilize it for the coordinated motion control. The focus of this section is to exercise this concept.

Figure 26 depicts a typical neuron used in artificial neural networks. On the top are the multiple inputs to the neuron, each arriving from another neuron and connected to the neuron shown at the center. Each interconnection has an associated connection strength, given as W_{i0}, W_{i1}, ... The neuron performs a weighted sum on the inputs and uses a nonlinear threshold function (e.g., a sigmoid threshold) to compute its output. The calculated result is transmitted along the output connections to other target cells shown at the bottom of the figure.

One of the most popular 'Artificial Neural Network' (ANN) models, the back-propagation model, is used here to learn the relationship between the inputs (desired end-effector velocity and the current joint angles) and the outputs (desired joint velocities). Back-propagation training is mathematically designed to minimize the mean squared aggregate error

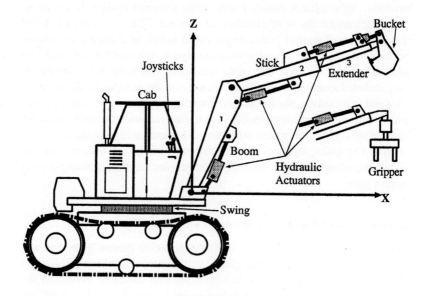

Figure 25. Excavator machine with a redundant link.

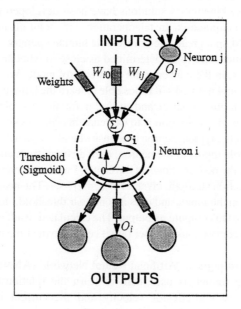

Figure 26. Schematic neuron from a neural network.

across all training patterns. Moreover, it is a supervised training technique, meaning that the network designer can dictate the desired results he/she wants the network to achieve, and the network's performance can always be measured against those results. Back-propagation can be applied to any problem that requires pattern mapping. Given an input pattern, the network produces an associated output pattern. Its learning and update procedure is intuitively appealing because it is based on a relatively simple concept — the network is supplied with both a set of patterns to be learned and the desired system response for each pattern. If the network gives the wrong answer, then the weights are corrected so that the error is lessened and as a result future responses of the network are more likely to be correct [31]. The training process usually takes many cycles before the network converges to a solution.

A gradient projection optimization technique was previously adapted to solve for the joint rates, when the end-effector is directed to move at a certain velocity [32]. The technique, when applied, in simulation, to the class of machines under investigation, effectively prevented the links from reaching their limits, was fast and provided moderately smooth joint velocities, similar to what an experienced operator would achieve. The scheme was then used within the real-time machine simulator, as a test rig, to resemble the human operator performance in a joint control mode.This was done, since a real instrumented machine similar to the one in Figure 26 was not available at the time.

Using the simulator, training data were collected by moving the end-effector in different direction and at different velocities (see Figure 27). The network was supplied with the information to establish the relationship between the inputs and the outputs. The inputs are the current joint angles and the desired end-effector velocity, and the outputs are the desired joint angle velocities (Figure 28). Table 1 shows the input neurons, the hidden layer and the output neurons with their specifications. The training was performed using the Xerion Neural Network Simulator implemented on a SUN workstation [33]. Once the ANN was trained, it was substituted for the gradient projection scheme in the simulator. The network was then tested by having the end-effector to move in straight line segments as shown in Figure 29a. Figures 29b to 29d compare the results (i.e., the desired joint angle velocities) obtained from the network and the simulator. The results are promising indicating that artificial neural networks can be trained to model the human performance in manipulating heavy-duty machines with redundant linkages. Further investigation should include the actual data from a real-world instrumented machine (instead of a simulator) for training the network.

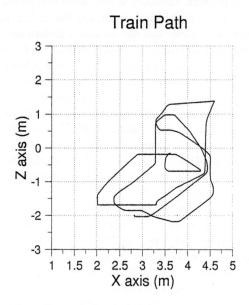

Figure 27. Training path.

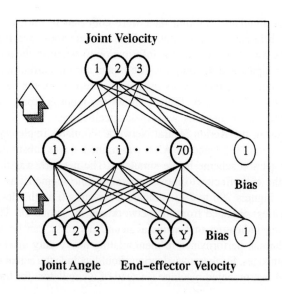

Figure 28. Three layers artificial neural network model.

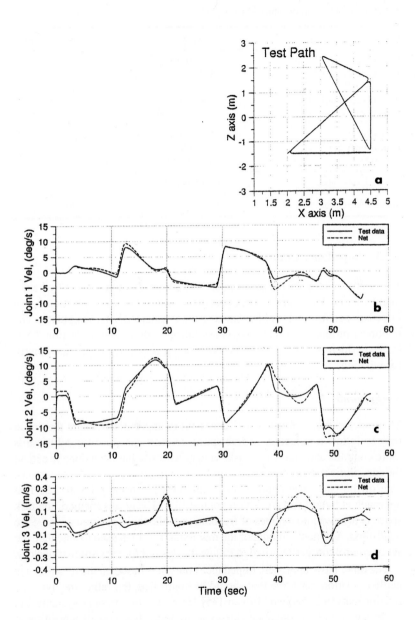

Figure 29. Joint trajectories comparison for an unseen path trajectory.

Table 1. Inputs, outputs and hidden layers.

Inputs	Min–Max	No. of Neurons
Joint angle 1	−20 – +60 deg	1
Joint angle 2	−145 – +35 deg	1
Joint angle 3	−0.1 – +0.1 m	1
End-effector velocity (X)	−0.6 – +0.6 m/s	1
End-effector velocity (Y)	−0.6 – +0.6 m/s	1
Joint velocity 1	−15 – +15 deg/s	1
Joint velocity 2	−20 – +20 deg/s	1
Joint velocity 3	−0.4 – +0.4 m/s	1
Hidden Layer Type	Connection	No of Neurons
Single hidden layer	Fully connected	70

7.5.3. Impedance Control

This section addresses the problem of environmental interaction in hydraulic manipulators. Motivated by the previous works [34–36], a position-based impedance controller is implemented on the Unimate MK-II manipulator. The nonlinear proportional integral (NPI) controller, described in Section 7.5.1, is used which meets the accurate positioning requirements of this control formulation.

Background

Manipulators have proven to be successful in applications where they need only to follow a trajectory in free space. Interest has since turned towards more complex and interactive tasks, and much work has been done to extend the capabilities of robotic manipulators to handle environmental interactions. One of the two basic approaches to this task are usually taken. The first approach is hybrid force/position control [37]. Using this method, the task space is divided into degrees of freedom in which only position or force are controlled. A dynamic model of the manipulator is then used to calculate the set of control torques required to follow the desired force/position trajectory.

The other basic approach, impedance control, is based on the idea that it is neither position nor force that should be controlled, but rather the dynamic relation between the two. Hogan [35] showed how physical systems can be divided into two conjugate classes — systems that produce a position in response to an imposed force, and those that produce a force in response to an imposed position. The former are termed admittances, the latter impedances. Since the manipulator's environment may be constrained, it falls into the category of admittances. Therefore, the controlled manipulator should behave as an impedance [35].

The most common way of controlling the manipulator impedance as seen from the environment is similar to hybrid position/force control. The required accelerations of the end-effector are calculated according to a desired (target) impedance relationship [35]. A model of the manipulator dynamics is then used to calculate the joint torques required to achieve these accelerations. Evidently, the only difference between the implementations of hybrid position/force control and impedance control just presented is in the calculation of the desired accelerations. In fact, it has been shown that, using these implementations, second-order impedance control and proportional force control can be considered equivalent [38].

The technique of computing joint torques from accelerations assumes that these torques are controllable. This assumption is most appropriate for electrically-driven manipulators, and consequently almost all of the work that has been done in the field to date involves this type of manipulator, or simulations thereof. Unlike electric actuators, torque control of hydraulic actuators is a difficult problem [39]. There have recently been a number of reports of an implementation of impedance control based on position control [34,40,41]. Instead of deriving a set of accelerations from the target impedance to be passed to a dynamic model, an end-effector position is calculated and used by a position controller. In the following we investigate the applicability of the position-based impedance control technique to a hydraulically actuated manipulator.

Formulation

The general setup for an impedance controller as implemented on a single hydraulic link is shown in Figure 30. The impedance controller block takes desired trajectories, $\mathbf{X}^d$ and $\mathbf{F}^d$, for both force and position and combines them with feedback information to produce a position set-point, $\mathbf{X}^s$, according to a target impedance. A position controller then uses this modified set-point and feedback to generate a control signal $\mathbf{U}$.

The target impedance used here is that of a linear, second-order system. This class of impedance was chosen because of its familiarity and widespread use [34–36, 41]. With reference to Figure 31, the transfer functions used for the impedance target models are

$$\mathbf{M_X}(s) = \frac{cs + k}{ms^2 + cs + k}$$

$$\mathbf{M_F}(s) = \frac{1}{ms^2 + cs + k}$$

(39)

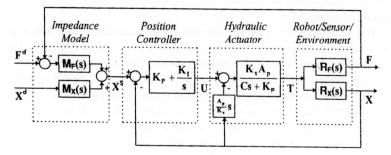

Figure 30. Block diagram of impedance control system.

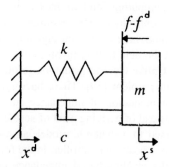

Figure 31. Physical model of target impedance.

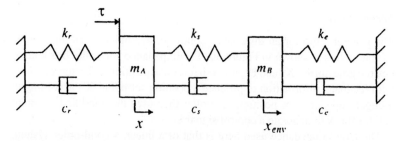

Figure 32. Model of robot-sensor-environment dynamics.

where m, c, and k are the target impedance parameters. Note that in the present work, $\mathbf{M_X}$ is taken from the target impedance relation, and therefore includes the dynamics of changing the reference point $\mathbf{X^d}$.

The set-point, $\mathbf{X^s}$, is passed to a position controller which controls the signal to the actuators. In this work the Nonlinear PI controller, described earlier, is implemented for positioning of the end-effector. The hydraulic actuator then produces a generalized torque/force, **T**, which in turn produces force and position at the end-effector of the manipulator through its interaction with the environment.

The linearized hydraulic functions are also included in Figure 30. Here, A_p is the piston area (assumed equal on both sidess of the piston), C is the hydraulic compliance, and K_x and K_p are the flow gain and flow-pressure coefficients, respectively.

Volpe and Khosla [38] showed how a fourth order model could be used to analyze the dynamics of the coupled system. With reference to Figure 32, this model of the robot, sensor and environment system is shown by the following transfer functions,

$$\mathbf{R_X(s)} = \frac{m_B s^2 + (c_s + c_e)s + (k_s + k_e)}{\Delta_{\mathbf{R}}(s)}$$

$$\mathbf{R_F(s)} = \frac{k_s(m_B s^2 + c_e s + k_e)}{\Delta_{\mathbf{R}}(s)}$$

(40)

where

$$\Delta_{\mathbf{R}}(s) = (m_B s^2 + (c_s + c_e)s + (k_s + k_e)).$$
$$(m_A s^2 + c_r s + k_r) + (m_B s^2 + c_e s + k_e)(c_s s + k_s)$$

Subscripts r, s, and e refer to the robot, sensor and environment, respectively. Subscripts A and B refer to the interfaces between these elements. In the event that contact with the environment is lost, m_B, c_e and k_e become essentially zero.

Assembling the overall transfer function from the individual system components shows that even a linearized model of one link of the controlled system has eighth-order dynamics. Nonlinearities such as actuator saturation and deadband, joint stiction, varying hydraulic supply pressure and nonlinear environmental admittance further complicate the system. Clearly, the efficacy of the position controller is paramount for the viability of this control scheme. This is especially true when stiff environments are encountered, in which case minute position adjustments may be required to make the manipulator response conform to that of the target impedance.

Experiments

A test of static force control ability is shown in Figure 33. The end-effector was placed in contact with the aluminum plate with stiffness $\cong$ 7 kN/m (see also Figure 2). In spite of the large noise in measurement (20N peak-to-peak), the force response followed the behavior required by the target impedance.

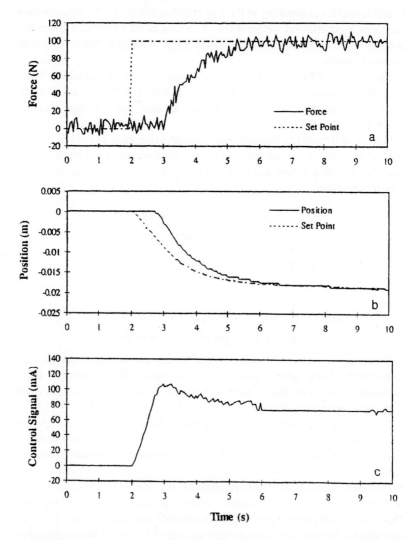

Figure 33. Force regulation of impedance control ($m = 500$ kg, $c = 948$ Ns/m, $k = 50$ kN/m).

During the experiment, it was observed that as the impedance model stiffness k was increased, the steady-state error increased, matching the relation

$$\frac{f}{f^d} = \frac{k_e}{k_e + k} \tag{41}$$

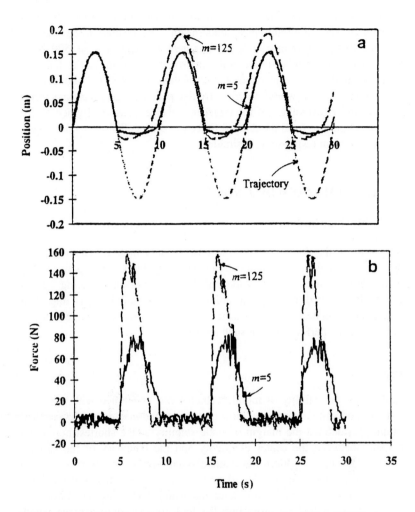

Figure 34. Collision force reduction using impedance control ($k = 500$ N/m).

Evidently, for good force regulation, the target impedance stiffness k must be much less than the environmental stiffness k_e. However, there exist cases in which the target stiffness k is required to be high, for example during a force trajectory tracking. In such cases, one may simply multiply the force set-point by a factor $\frac{\hat{k}_e + k}{k}$, where $\hat{k}_e$ is an estimate of the environmental stiffness.

Figure 34 shows the effect of impedance parameter selection on the manner in which collisions are dealt with. The Unimate robot was commanded to move in a simple sine wave motion, but half of the trajectory was blocked by the aluminum plate underneath. The desired force was set to zero. Target impedance stiffness was held constant at $k = 500$ N/m. Using an experimental estimate of the environmental stiffness, the damping parameter in the target impedance was set so that the composite robot-sensor-environment system would be critically damped. The resulting trajectories are shown for two values of the target impedance mass, m. Evidently, for minimal departure from the desired trajectory after collision, a small mass, i.e. a fast system response, is desirable.

7.6. FAULT-HAZARD DIAGNOSIS

7.6.1. Machine Stability

One issue in the computer control of heavy-duty mobile machines, is to develop a suitable method that can indicate the potential of tipping-over. A heavy-duty mobile machine while carrying a load or experiencing a force needs to maintain its balance. The ability to maintain stability is an important consideration in design, to assure the safety and comfort of the operator and to improve the performance of such machines.

Background

Early work on stability of mobile vehicles was concerned with the static stability and gait generation of slow moving legged machines. McGhee and Frank [42] defined the margin of static stability as the shortest horizontal distance between the center of gravity and the boundary of the support pattern. The support pattern was defined as a minimum area convex polygon containing all contact points. This definition was later modified [43] to include locomotion over rough terrain. Song and Waldron [44] improved McGhee and Frank's method by projecting the foot contacts onto a plane which is positioned to minimize the distance of projection for the foot contacts. Davidson and Schweitzer [45] pointed out cases where both horizontal and sloping projection methods could give misleadingly large values for margin of stability. Additionally, both methods provide a qualitative measure of stability according to Messuri and Klein [46], who demonstrated that McGhee and Frank's definitions predict the same level of stability for both front and rear sides of a leveled body on a gradient. Messuri and Klein then introduced a different measure of stability termed "energy stability margin". Energy stability level for each edge of the support boundary was defined as the work required to rotate the body center of gravity about that edge to the verge of instability. The support boundary was defined as a set of three dimensional line segments connecting the tips of the supporting feet. The method of energy stability appears to be the only method that includes the effect of vertical position of the mass center or top-heaviness.

Methods described so far could deal with cases where the only destabilizing load is due to the gravitational force. However, for moving base manipulators with heavy tools, a large portion of destabilizing forces and moments could be due to the inertial or external loads arising from maneuvering the implement. Since the inclusion of external and inertial loads is important in the stability analysis of the machines subject to this study, we extend the measure of stability by Messuri and Klein [46].

Extended measure of stability

The approach taken was to first develop an algorithm which finds the configuration of the machine on the verge of instability for each potential tipping-over edge. The amount of the work done when the machine is virtually brought to this onset of instability stance, from the current position, is then computed to represent the stability level of the machine. This work is in fact a quantitative measure of the impact energy, through an impact force, that can be sustained by the vehicle without overturning at that instant [47]. The calculation of energy stability level, associated with an edge of potential tipping-over consists of the following three steps: (i) find the equilibrium plane (the equilibrium plane associated with a particular edge of a support boundary is a plane containing the edge and with an angle with respect to the vertical plane, such that if the body rotates around the edge until the center of gravity is placed in this plane, the net moment due to all forces and moments around the edge becomes minimum in the absolute sense); (ii) calculate the work done when the vehicle body (which is subjected to gravitational forces as well as external and inertial loads) is hypothetically rotated about the edge until the center of gravity reaches the equilibrium plane.

Formulation

Figure 35 hows the general coordinate systems used in our formulation. Frame XYZ (also referred to here as $\{\mathcal{M}\}$) is attached to the vehicle body at point P and is called machine coordinate system. For machines subject to this study, the manipulator base is the most convenient choice for P. All reaction forces and moments resulting from manipulating the arms are known in this frame. Coordinates of the contact points, f_i's, are also known in the machine frame.

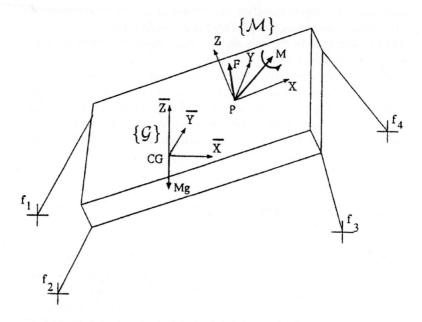

Figure 35. Machine $\{\mathcal{M}\}$ and gravity $\{\mathcal{G}\}$ coordinate systems.

$\bar{X}\bar{Y}\bar{Z}$ (also referred to here as frame $\{\mathcal{G}\}$) is defined with the origin always at the center of gravity of the machine. $\bar{Z}$ is in the direction opposite to the gravitational acceleration. $\bar{X}$ and X are in parallel planes; so are $\bar{Y}$ and Y of frame $\{\mathcal{M}\}$. As the center of gravity moves with respect to frame $\{\mathcal{M}\}$, so does the base of the gravity frame. When the machine is in a horizontal position, the machine frame $\{\mathcal{M}\}$ is parallel to the gravity frame $\{\mathcal{G}\}$. The total force due to gravitational loading is described in this frame.

Vector descriptions of the system are given with reference to Figure 36. Here XYZ is the machine frame and point P is where all the forces, $\mathbf{F}$, and moments, $\mathbf{M}$, other than the weights are transferred to. The gravity frame $\bar{X}\bar{Y}\bar{Z}$, and the location of the center of gravity, CG, of the whole system including the load are also shown. The total weight shown as $\mathbf{mg}$ is described in this frame. f_1 and f_2 are the two contact points forming an edge of the support boundary. $\vec{f}_1$ and $\vec{f}_2$ are the position vectors of these two points, and $\vec{R}$ is a vector perpendicular to line $f_1 f_2$ that connects the contact points f_1 and f_2. In machine frame $\{\mathcal{M}\}$, $\tilde{b}$ is defined as the unit vector parallel to the line $f_1 f_2$. Direction of $\tilde{b}$ is chosen such that a moment vector in the opposite direction of $\tilde{b}$ contributes to the stability of the machine.

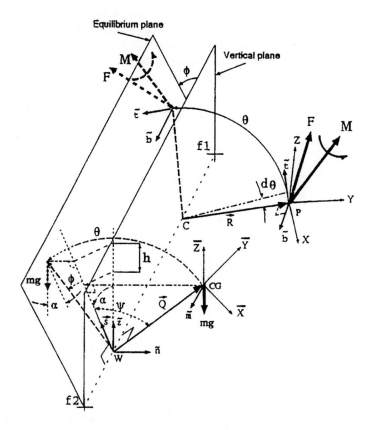

Figure 36. Equilibrium plane and energy stability calculation.

Referring to Figure 36, knowing the position vectors $\vec{f}_1$ and $\vec{f}_2$, $\vec{R}$ could be found from the following relation:

$$\vec{R} = (\vec{f}_1 \cdot \tilde{b})\tilde{b} - \vec{f}_1 \qquad (42)$$

Unit vector $\tilde{t}$ is perpendicular to the plane containing vectors $\vec{R}$ and $\tilde{b}$. Similarly vector is computed in the gravity frame $\{\mathcal{G}\}$ as

$$\vec{Q} = (\vec{f}_1^* \cdot \tilde{m})\tilde{m} - \vec{f}_1^* \qquad (43)$$

where $\vec{f}_1^*$ is the position vector of the contact point f_1 in frame $\{\mathcal{G}\}$. $\tilde{m}$ is the unit vector similar to $\tilde{b}$, defined in frame $\{\mathcal{G}\}$. We further define two unit vectors; vector $\tilde{z}$ in Figure 36 which is in the direction of $\bar{Z}$ and, $\tilde{n}$ to be perpendicular to the vertical plane constructed by $\tilde{z}$ and $\tilde{m}$.

The above vector descriptions are now utilized to determine the equilibrium plane and to calculate the energy stability levels. Referring to Figure 36, the equilibrium plane for a general three dimensional case is described with an angle of ϕ from the vertical plane. Assuming that the machine goes through a rotation which puts the center of gravity, CG, in this plane, the following relation then holds:

$$\sum_{f_1, f_2}^{M} = (\vec{F} \cdot \tilde{t})|\vec{R}| + \vec{M} \cdot \tilde{b} + mg|\vec{Q}| \cos \alpha \sin \phi = 0 \qquad (44)$$

α in the above equation represents the angle that the support boundary edge, $f_1 f_2$, makes with the horizontal plane. Its value is found from the following relation

$$\alpha = \cos^{-1} \left(\tilde{z} \cdot \frac{\vec{s}}{|\vec{s}|} \right) \qquad (45)$$

where $\vec{s}$ is a vector perpendicular to $f_1 f_2$ in the vertical plane (see Figure 36). It is the outer product between $\tilde{m}$ and $\tilde{n}$. From the above equation, the equilibrium angle ϕ can be determined.

Since the weight force is conservative, the work done when the center of gravity rotates around a support boundary edge to the equilibrium plane, depends only on the vertical displacement of the center of gravity h. This work is calculated from the following equation

$$W_1 = -mg|\vec{Q}|(\cos \phi - \cos \psi) \cos \alpha \qquad (46)$$

where the angle ψ between $\vec{Q}$ and the vertical plane is calculated from the following relation

$$\psi = \cos^{-1} \left(\frac{\vec{Q}}{|\vec{Q}|} \cdot \frac{\vec{s}}{|\vec{s}|} \right) \qquad (47)$$

The work done by the net destabilizing force, **F**, and moment, **M**, other than the weight force, is calculated from the following relation

$$W_2 = [(\vec{F} \cdot \tilde{t})|\vec{R}| + (\vec{M} \cdot \tilde{b})]\theta \qquad (48)$$

where $\theta = (\psi + \phi)$ from Figure 36. Note that the vectors in the above equation are all defined in machine frame $\{\mathcal{M}\}$ and therefore directions of these forces and moments do not depend on the orientation of the machine frame with respect to the gravity frame.

The total energy stability level, for edge $f_1 f_2$ of the support boundary, is then

$$\text{Energy Stability Level} = -(W_1 + W_2) \qquad (49)$$

The minimum of all energy stability levels, corresponding to all support boundaries. is called the energy stability margin for that instant.

Simulation studies

The extended measure of stability is now exemplified using the excavator-based log loader simulator. The machine runs on its tracks; thus, the support boundary is a simple planar rectangle. Figure 37 shows a typical swing motion (rotation of the cabin including the manipulator arms). Given a step input control voltage, the cabin rotated from initial position, $\theta = 90°$, and was brought to stop after one full revolution (at $\theta = 450°$). The displacement, velocity and acceleration profiles for the swing joint are shown. Referring to Figure 37a, setting the command voltage to zero (at $t \approx 10$s) resulted in closing the hydraulic valve and stopping the motion due to the oil trapped on both sides of the hydraulic motor. There was a period in which the acceleration was minimum and remained flat. This saturation was the result of the relief valve action. Pressures beyond a certain limit caused the relief valve to open. From the machine stability point of view, the action of the relief valve is favorable but results in overriding of the swing rotation. Figure 37b summarizes the results of the stability analysis based on the machine response. Changes in the dynamic energy stability levels, as a result of changes in both the location of the gravity center and the inertial loads arising from the swing motion, are shown in the figure. Edges 2 and 4 indicate the minimum stability levels.

The previous experiment is now repeated on a sloping ground. Figure 38 shows the energy stability levels for the machine positioned along the slope. Compared with Figure 37b, a slight decrease in the energy stability levels for edges 2 and 4 are seen. The main changes, however, occurred for edges 1 and 3 where the stability level for edge 1 increased because it was on the uphill side and the stability level for edge 3 dramatically decreased. Figure 39 shows the energy stability levels for all edges when the machine was positioned across the slope. It is seen that the stability level for edge 2 was reduced significantly. At certain times such as $t \approx 1$s and $t \approx 10$s, the energy stability level for side 2 was almost zero, meaning that the machine was on the verge of instability and the possibility that the machine to completely overturn. It is therefore desirable to prevent such a situation by keeping the stability level within a comfortable margin at any time during the manipulation.

7.6.2. System Parameter Tracking using Genetic Algorithms

A common characteristic of heavy-duty manipulators is a high hydraulic compliance which indicates the degree of joint flexibility. The identification

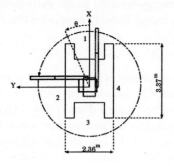

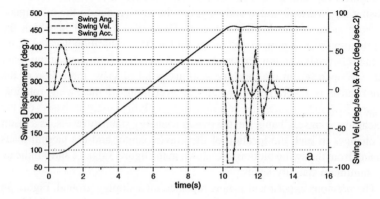

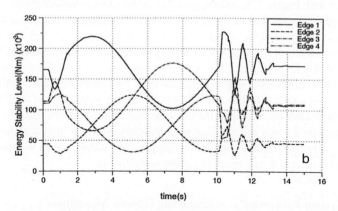

Figure 37. Energy stability analysis pertaining to swivel of manipulator on a level ground: (a) joint angle, velocity and acceleration responses; (b) stability for all edges of the support boundary.

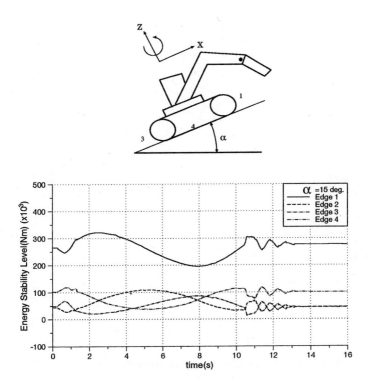

Figure 38. Effect of sloping ground; machine along the slope.

of hydraulic compliance is beneficial towards: (i) developing an exact model, (ii) fault/hazard diagnosis system and, (iii) improving control strategies [48]. In this section we demonstrate the application of a genetic algorithm towards identification of the compliance in the boom hydraulic actuator.

Genetic Algorithms (GAs)

GAs are search procedures based on the mechanics of natural genetics and natural selection [49]. They extract the ideas of survival of the fittest from nature to form a robust search mechanism. Some properties of genetic algorithms are:

1. GAs are parallel, global search techniques that can simultaneously search through many points in the parameter space.

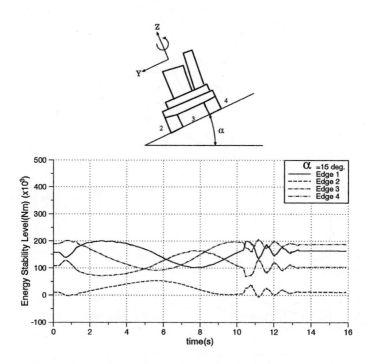

Figure 39. Effect of sloping ground; machine across the slope.

2. GAs use finite length codings of the parameters (usually in binary), instead of the values of the parameters themselves.
3. GAs apply a population-by-population approach rather than point-by-point techniques.
4. GAs need only fitness values (dependent on application). There is no requirement for derivatives or other auxiliary knowledge.
5. GAs use probabilistic transition rules instead of deterministic ones.

Figure 40 shows how a genetic algorithm operates. A simple GA uses three operators: 'Reproduction', 'Crossover' and 'Mutation'. Reproduction is based on "the survival of the fittest" principle; it decides which strings survive and which ones are going to disappear according to their normalized fitness values. Individuals with above average fitness will have more offspring than those with below average fitness. This step directs the search towards the best individuals in a search space population.

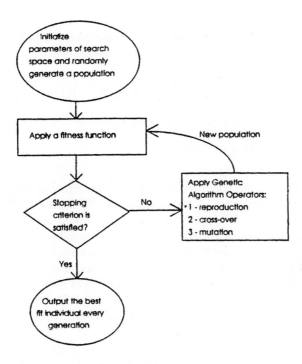

Figure 40. Simple genetic algorithm flow chart.

Crossover is a randomized (with a probabilistic crossover rate), yet structured, operator that takes valuable information from both strings (parents) and combines it in order to find an individual which, hopefully, will be fitter than its immediate predecessor.

The mutation operator is used to introduce new information into the population at the bit (in a binary formatted GA) level. It acts as an insurance against the loss of important genetic material at a particular position. Mutation is a purely randomized operator. It will flip one or more bits in a string with a low probability.

GAs have been shown to work much more subtly than classical random search techniques by preserving the best solution to date. Their ability to handle functions that are nonlinear and discontinuous makes them good candidates in function optimization and in identification of complex systems.

Formulation

Referring to Figure 7, line pressures P_i and P_o are related to the flow out of and into the valve (Q_i and Q_o, respectively) as follows:

$$Q_i = A_i \dot{X} + \dot{p}_i C_i = kwx\sqrt{(P_s - P_i)}$$

$$Q_e = kw(1 - x)\sqrt{(P_s - P_e)} \tag{50}$$

$$Q = Q_i + Q_e$$

These equations are nonlinear and can become highly coupled as the system complexity increases. A unique feature of GAs is that they do not put strict requirements on the model. Their population-based search technique can easily handle discontinuities and nonlinearities in a model. GAs do not necessarily require rearranging the equations that are used for simulations, in terms of their unknowns. Moreover, GAs are nonrecursive algorithms. They store the estimated parameters in a population of solutions, which is independent from changes in the model of the system. To demonstrate the technique, C_i (the compliance at the pump side) is identified as follows [50]:

1. Given the measured values of P_i^m and the measured joint angular velocity $\dot{\theta}^m \propto \dot{X}^m$ at time t, the derivative of the pressure, $\dot{P}_i^m$, is first determined using a piece-wise linear least-squares curve-fitting technique.
2. A population consisting of n values of $C_i(j)$ $j = 1, 2, \ldots, n$ is initially set up. For every $C_i(j)$, equations at the pump side are used recursively as follows:

$$\overline{Q}_i = A_i \dot{X}^m + \dot{P}_i^m C_i(j)$$

$$\overline{P}_s = \left(\frac{\overline{Q}_i}{kwx} \right)^2 + P_i^m$$

$$\overline{Q}_e = kw(1 - x)\sqrt{(\overline{P}_s - P_e)} \tag{51}$$

$$\overline{Q} = (\overline{Q}_i + \overline{Q}_e)$$

$$W(j) = Q - \overline{Q}$$

where $\overline{Q}_i, \overline{Q}_e, \overline{Q}$ and $\overline{P}$ are the intermediate states information. $E(j)$ is the error between the actual pump flow, Q, (considered known) and the calculated one, $\overline{Q}$.

3. Step two is performed over a window of measured values including of the past states. The fitness (payoff) assigned to each $C_i(j)$ is then defined as follows:

$$0 \le \text{fitness } (C_i(j)) = \sum_t^{\text{window}} \{B - E^2(j)\} \le \sum_t^{\text{window}} B \qquad (52)$$

B is a positive bias constant that sets a limit for the GA reproduction operator (only individuals with fitness >0 can reproduce). This bias is determined according to the nature of the application. The use of the window concept is to minimize the variance of the estimated parameters and the effect of noise.

4. Using the fitness values and GA's operators, a new population of $C_i(j)$ is then generated. This produces a population with the best fit which statistically converges to the correct value of C_i.

As is noticed, the mathematical model should be a good approximation of the physical process. Otherwise, it would be difficult for GAs, like other identification methods, to work properly. Figure 41 shows typical results based on experimental data. The measured joint angle, θ^m, as well as the line pressures P_i^m and P_o^m are shown in Figures 41a and 41b. This information was used to identify the value of the hydraulic compliance at the pump side (C_i). Figure 41c shows the calculated value of the best fit. At the termination of identification, the algorithm determined the value of 0.0275 cm³/(kPa) for compliance. This value was shown to be within the expectation [50]. The identification was performed during the pressure rise where the compliance was effective. The population size was 120 with starting values between 0.0 to 0.1 and resolution of 0.0004 (i.e., 8 bits strings). At each time, the moving window incorporated the last ten data points. The crossover rate (i.e., the frequency with which this operator was applied) and the mutation rate were 0.8 and 0.01, respectively.

7.7. SUMMARY

The present control of existing mobile machines that are utilized in primary industry, such as excavators and feller bunchers, is very much operator dependent and requires significant visual feedback, judgment and skill. Application of robotics technology to this class of equipment will bring about enhanced operator safety and improved productivity. The objective of this chapter was to report the results of our recent project in converting such heavy-duty machines into task-oriented human supervisory control systems.

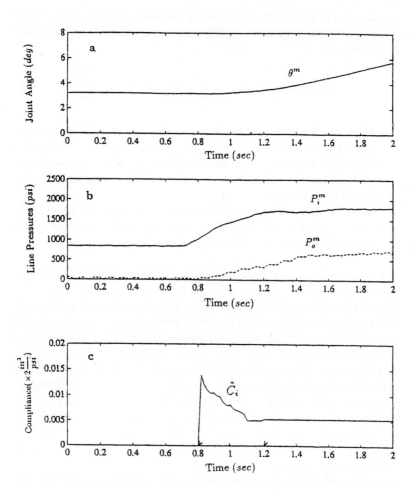

Figure 41. Identification of hydraulic compliance (experiment); (a) measured joint angle; (b) measured line pressures; (c) calculation of the best fit.

A Caterpillar 215B excavator and a Unimate MK-II industrial robot were chosen as target machines. First necessary changes were made to retrofit them into teleoperated control systems. They were then used as testbeds to validate the methods and approaches developed during the course of the research.

One issue in this project was to build accurate and comprehensive real-time simulators. We showed that this was possible by combining the best properties of the transient and steady-state solutions in the modeling phase. The systematic cycle of simulate, experiment, fine-tune and repeat made it possible to isolate the individual effects of various phenomena and consequently converge to models that not only have sufficient degree of predictive accuracy, but also are simple and fast enough for real-time simulations. The results of human factors studies carried out on these simulators proved to be useful in developing hand control interfaces and new approaches to directing the control of the machine implements.

Different methods were then implemented to address the joint control problem in hydraulically actuated links. First a model and sensor-based algorithm was developed to control the joint velocities. The method is basically a feedforward load compensating scheme that uses the measured hydraulic line pressures along with the appropriate portion of the hydraulic model. Next the application of a fuzzy logic control was investigated. A simple, yet effective, set of membership functions and rules were developed which meets the control requirements of hydraulic functions. Experimental studies have proven the promise of both techniques. We also explored the application of an impedance control technique in order to enable hydraulic manipulators to properly interact with the environment. A novel nonlinear PI (NPI) controller was first developed to meet the accurate position tracking and regulating requirements of this application. The NPI was capable of accurate tracking of trajectories upto 0.2 deg, and of regulation upto 0.06 deg. Because of this excellent performance, the impedance controller was capable of masking the actual, highly nonlinear manipulator dynamics and replacing them with those of the target impedance.

Coordinated motion control, in heavy-duty machines with redundant linkages, requires an appropriate inverse kinematics solution. In teleoperation, such a solution should preferably resemble the experienced operator performance, that is, the desired joint motions are to be determined as close to what an experienced operator produces, in a conventional joint control operation. Simulation study indicated that artificial neural networks have the potential to be trained to model the human performance in manipulating heavy-duty machines having redundant linkages.

The issue of stability against tipping-over was also investigated in this project. The ability to maintain stability is an important consideration in design, to assure the safety of the operator and to improve the performance of such machines. A model was introduced which quantifies the stability limits for such machines. The model can be used, in conjunction with the simulator, to provide the designer with an inexpensive, fast and safe method to fine tune the design details and to put necessary limits on loads, velocities and forces in different machine configurations. Presently these limitations only concern the load and radius of the implement position, i.e., static stability of the machine. The velocity adjustments are mainly left to the skill and experience of the operator.

Identification of hydraulic compliance (indication of joint flexibility) is particularly of interest in our research because of its significant effect on the control of the hydraulic functions. A genetic algorithm (GA) was used for robust and fast identification of this important parameter. The GA was able to handle the nonlinear and coupled hydraulic functions with its nonrecursive, population-based search power. The algorithm has been implemented on a SUN workstation, connected to 16 T800 transputers and is presently capable of real-time simultaneous monitoring of hydraulic compliances at two joints.

ACKNOWLEDGMENTS

The authors wish to acknowledge the contributions of Todd Corbet in implementing the fuzzy control software, Ahmad Ghasempoor in simulating the machine stability, Brad Heinrichs and Amir Khayaat in implementing the impedance control, and many other undergraduate and graduate students, at both the University of Manitoba and the University of British Columbia, involved in projects related to different stages of the development described here. The authors are also grateful to Al Lohse of the University of Manitoba, Dan Chan and Simon Bachmann of the University of British Columbia in implementing the hardware and sensor packages. Support for the projects was derived from Natural Science and Engineering Research Council (NSERC) of Canada, the Institute of Robotics and Intelligent Systems (IRIS) and Precarn Associates Inc. The authors are also grateful to Caterpillar Inc., Peoria, and their distributor, Finning Inc., for providing the Cat 215B testbed machine, to CP Rail, Winnipeg, for providing the Unimate robot, and to MacMillan Bloedel Ltd. and RSI Research for their support of this project.

REFERENCES

1. Lawrence, P.D., Sassani, F., Sauder, B., Sepehri, N., Wallersteiner, U. and Wilson, J., 1993, *Computer-Assisted Control of Excavator-Based Machines*, International Off-Highway Powerplant Congress & Exposition, Milwaukee, Wisconsin, SAE Technical Paper No. 932486.
2. Sheridan, T.B., 1989, Telerobotics, *Automatica*, **25**, 487–507.
3. Stark, L. *et al.*, 1987, Telerobotics: Display, Control and Communication Problems, *IEEE Journal of Robotics and Automation*, **3**, 67–75.
4. Sheridan, T.B. *et al.*, 1987, MIT Research in Telerobotics, Proceedings — NASA Workshop on Space Telerobotics (G. Rodriguez, ed.), Massachusetts Institute of Technology, Cambridge, Massachusetts, pp.403–412.

5. Hannaford, B. *et al.*, *Performance Evaluation of a Six-Axis Generalized Force-Reflecting Teleoperator*, Report No. 89–18, Jet Propulsion Laboratory, California Institute of Technology, June 1989.

6. Wicker, S.F., Hershkowitz, E. and Zik, J., 1986, *Teleoperator Comfort and Psychometric Stability: Criteria for Limiting Master-Controller Forces of Operation and Feedback During Telemanipulation*, Proceedings — NASA Conference on Space Telerobotics, Wisconsin, Madison, pp.99–107.

7. Merrit, H.E., 1967, *Hydraulic Control Systems*, John Wiley & Sons, New York.

8. Hronsky, P. and Martens, H.R., 1973, *Computer Techniques for Stiff Differential Equations*, Proceedings — Summer Conference in Computer Simulation, Montreal, Canada, pp.131–139.

9. Mahmoud, M.S. *et al.*, 1985, *Large-Scale Control Systems; Theories and Techniques*, Marcel Dekker Inc., New-York.

10. Sepehri, N., 1990, *Dynamic Simulation and Control of Teleoperated Heavy-Duty Hydraulic Manipulators*, Ph.D. Thesis, Department of Mechanical Engineering, University of British Columbia.

11. Iyengar, S.K.R. and Fitch, E.C., 1975, A Systematic Approach to the Analysis of Complex Fluid Power Systems, *Proceedings — 4th International Fluid Power Symposium*, Sheffield, England, pp.A219–A234.

12. Sepehri, N., Sassani, F. and Lawrence, P.D., 1990, On Numerical Simplification of Robot Structure Dynamics for Fast Simulation, *Proceedings – 9th IASTED International Conference on Modeling, Identification and Control*, Innsbruck, Austria, pp.311–315.

13. Watton, J., 1987, Dynamic Performance of an Electro-Hydraulic Servovalve/Motor System with Transmission Line Effects, *ASME Journal Dynamic Systems, Measurement and Control*, **190**, 14–18.

14. Bowns, D.E. and Rolfe, A.C., 1975, The Digital Computation of Pressure and Flows in Interconnected Fluid Volumes, using Lumped Parameter Theory, *Proceedings — 4th International Fluid Power Symposium*, Sheffield, England, pp.A1.1–A1.17.

15. Paul, R.P., 1983, *Robot Manipulators, Mathematics, Programming and Control*, MIT Press, Cambridge.

16. Sepehri, N. and Lawrence, P.D., 1992, Model-Based Sensor-Based Velocity Control of Teleoperated Heavy-Duty Hydraulic Machines, *Proceedings — IEEE/RSJ International. Conf. on Intelligent Robots and Systems*, Raleigh, NC, pp.859–862.

17. Viersma, T.J., 1980, *Analysis, Synthesis and Design of Hydraulic Servosystems and Pipelines*, Elsevier Scientific Pub. Co., New-York.

18. Kumbla, K., Moya, J., Baird, R., Rajagopalan, S. and Jamshidi, M., 1992, Fuzzy Control of Three Links of a Robotic Manipulator, Robotics and Manufacturing; *Proceedings — ISRAM'92* (M. Jamshidi *et al.*, eds.), ASME Press, New York, pp.687–694.

19. Ken, C., Jinn-Ya, L. and Xiang, L.Y., 1988, Fuzzy Control of Robot Manipulator, *Proceedings — IEEE International Conf. on Systems, Man, and Cybernetics*, Beijing, China, pp.1210–1212.

20. Zhao, T. and Virvalo, T., 1993, Fuzzy Control of a Hydraulic Position Servo with Unknown Load, *Proceedings — IEEE International Conf. on Fuzzy Systems*, San Francisco, CA, pp.785–788.

21. Chou, C.H. and Lu, H.C., 1993, Design of a Real-Time Fuzzy Controller for Hydraulic Servo Systems, *Computers in Industry*, **15**, 129–142.

22. Lee, C.C., 1990, Fuzzy Logic in Control Systems: Fuzzy Logic Controller — Parts I & II, *IEEE Transactions on Systems, Man and Cybernetics*, **20**, 404–433.

23. Kickert, W.J.M. and Mamdani, E.H., 1978, Analysis of a Fuzzy Logic Controller, *Fuzzy Sets and Systems*, **1**, 29–44.

24. Morse, R., Day, C. and Stoddard, K., 1988, *Positional Control Method and System Utilizing Same*, US Patent 4727303.

25. Khayyat, A.A., Heinrichs, B. and Sepehri, N., A Modified Rate-Varying Integral Controller, *International Journal of Mechatronics*, **6**, 367–376.

26. Whitney, D.E., 1969, Resolved Motion Rate Control of Manipulators and Human Prostheses, *IEEE Transactions on Man-Machine Systems*, **10**, 47–53.

27. O'Hara, J.M., 1987, Telerobotic Control of a Dextrous Manipulator Using Master and Six–dof Hand Controllers for Space Assembly and Servicing Tasks, *Proceedings — Human Factors Society; 31st Annual Meeting, Human Factors Society*, pp.791–795

28. Lawrence, P.D., Sepehri, N. and Sassani, F., 1994, Coordinated Hydraulic Actuation Systems, *Proceedings — 2nd International Conference on Machines Automation (ICMA '94)*, Tampere, Finland, pp.355–367.

29. Wallersteiner, U., Stager, P. and Lawrence, P.D., 1988, A Human Factors Evaluation of Teleoperator Hand Controls, *Proceedings — International Symposium on Teleoperation and Control*, Bristol, England, pp.291–296.

30. Wallersteiner, U., Lawrence, P.D. and Sauder, B., 1993, A Human Factors Evaluation of Two Different Machine Control Systems for Log Loaders, *Ergonomics*, **36**, 927–934.

31. Hecht-Nielsen, R., 1989, *Neurocomputing*, Addison-Wesley, Monlo Park, CA, pp.430–440.

32. Singh, H., Zghal, H., Sepehri, N., Balakrishnan, S. and Lawrence, P.D., 1995, Coordinated-Motion Control of Heavy-Duty Industrial Machines with Redundancy, *Robotica*, **13**, 623–633.

33. Kermanshahi, B.S. and Sepehri, N., 1994, Human-Machine Interface Modelling in Teleorobotics, *Proceedings — JSME International Conference on Advanced Mechatronics*, Tokyo, Japan, pp.766–770.

34. Field, G. and Stepanenko, Y., 1993, Model Reference Impedance Control of Robotic Manipulators, *Proceedings — IEEE Pacific Rim Conference*, pp.614–617.

35. Hogan, N., 1985, Impedance Control: An Approach to Manipulation, Parts I–III, *ASME Journal of Dynamic Systems, Measurement, and Control*, **107**, 1–24.

36. Hogan, N., 1987, Stable Execution of Contact Tasks using Impedance Control, *Proceedings — IEEE Conference on Robotics and Automation*, Raleigh, North Carolina, pp.1047–1054.

37. Raibert, M.H. and Craig, J.J., 1981, Hybrid Position/Force Control of Manipulators, *ASME Journal of Dynamic Systems, Measurement, and Control*, **102**, pp.126–133.

38. Volpe, R. and Khosla, P., 1993, A Theoretical and Experimental Investigation of Impact Control for Manipulators, *International Journal of Robotics Research*, **12**(4), 351–365.

39. Heinrichs, B., Sepehri, N. and Thornton-Trump, A.B., 1996, Position-Based Impedance Control of an Industrial Hydraulic Manipulator, *Proceedings — IEEE Conf. on Robotics and Automation*, Minneapolis, Minnesota, pp.284–290.

40. Nakashima, M. *et al.*, 1995, Application of Semi-Automatic Robot Technology on Hot-Line Maintenance Work, *Proceedings — IEEE Conference on Robotics and Automation*, pp.843–850.

41. Pelletier, M. and Doyon, M., 1994, On the Implementation and Performance of Impedance Control on Position Controlled Robots, *Proceedings — IEEE Conference on Robotics and Automation*, Atlanta, Georgia, pp.1228–1233.

42. McGhee, R.B. and Frank, A.A., 1968, On the Stability Properties of Quadruped Creeping Gait, *Mathematical Biosciences*, **3**, pp.331–351.

43. McGhee, R.B. and Iswandhi, G.I., 1979, Adaptive Locomotion of a Multilegged Robot Over Rough Terrain, *IEEE Transactions on Systems, Man and Cybernetics*, **9**, 176–182.

44. Song, S.M. and Waldron, K.J., 1989, *Machines that Walk*, MIT Press, Cambridge, MA, pp.100–118.

45. Davidson, J.K. and Schweitzer, G., 1990, A Mechanics-Based Computer Algorithm for Displaying the Margin of Static Stability in Four-Legged Vehicles, *ASME Journal of Mechanical Design*, **112**, pp.480–487.

46. Messuri, D. and Klein, C.A., 1985, Automatic Body Regulation for Maintaining Stability of a Legged Vehicle During Rough-Terrain Locomotion, *IEEE Transactions on Robotics and Automation*, **1**, pp.132–141.

47. Ghasempoor, A. and Sepehri, N., 1995, A Measure of Machine stability for Moving Base Manipulators, *Proceedings — IEEE Conference on Robotics and Automation*, Nagoya, Japan, pp.2249–2254.

48. Sepehri, N., Dumont, G.A.M., Lawrence, P.D. and Sassani, F., 1990, Cascade Control of Hydraulically–Actuated Manipulators, *Robotica*, **8**, 207–216.
49. Goldberg, D.E., 1989, *Genetic Algorithms in Search, Optimization and Machine Learning*, Addison-Wesley Pub., New-York.
50. Sepehri, N., Wan, F.L.K., Lawrence, P.D. and Dumont, G.A.M., 1994, Hydraulic Compliance Identification Using A Parallel Genetic Algorithm, *International Journal of Mechatronics*, **4**, 617–633.